AF758071

HYDROPOWER – PRACTICE AND APPLICATION

Edited by **Hossein Samadi-Boroujeni**

INTECHOPEN.COM

Hydropower – Practice and Application
Edited by Hossein Samadi-Boroujeni

Published by InTech

Janeza Trdine 9, 51000 Rijeka, Croatia

Edition 2014

Copyright © 2012 InTech

All chapters are distributed under the Creative Commons Attribution 3.0 license, which allows users to download, copy and build upon published articles even for commercial purposes, as long as the author and publisher are properly credited, which ensures maximum dissemination and a wider impact of our publications. After this work has been published by InTech, authors have the right to republish it, in whole or part, in any publication of which they are the author, and to make other personal use of the work. Any republication, referencing or personal use of the work must explicitly identify the original source.

As for readers, this license allows users to download, copy and build upon published chapters even for commercial purposes, as long as the author and publisher are properly credited, which ensures maximum dissemination and a wider impact of our publications.

Notice

Statements and opinions expressed in the chapters are these of the individual contributors and not necessarily those of the editors or publisher. No responsibility is accepted for the accuracy of information contained in the published chapters. The publisher assumes no responsibility for any damage or injury to persons or property arising out of the use of any materials, instructions, methods or ideas contained in the book.

Publishing Process Manager Ivana Zec

Technical Editor Teodora Smiljanic

Cover Designer InTech Design Team

Additional hard copies can be obtained from orders@intechopen.com

Hydropower – Practice and Application, Edited by Hossein Samadi-Boroujeni
p. cm.
ISBN 978-953-51-0164-2

INTECH

open science | open minds

Contents

Preface

Hydroelectric power or hydroelectricity is called a renewable energy source because the water on Earth is continuously replenished by precipitation. As long as the water cycle continues, we won't run out of this energy source. Hydroelectric energy is the most widely used form of renewable energy, accounting for 16 percent of global electricity consumption, and 3,427 terawatt-hours of electricity production in 2010, which continues the rapid rate of increase experienced between 2003 and 2009. In the recent years, measures against global warming have become an important issue for the world. The use of fossil fuels like oil has raised concerns about carbon dioxide emissions, which cause global warming, and the future depletion of energy sources. Against this background, hydropower generation is drawing renewed attention, with renewable energy sources such as solar power and windpower. Hydro is also a flexible source of electricity since plants can be ramped up and down very quickly to adapt to changing energy demands. However, damming interrupts the flow of rivers and can harm local ecosystems, and building large dams and reservoirs often involves displacing people and wildlife and requires significant amounts of carbon-intensive cement.

This book is primarily based on theoretical and applied results obtained by the authors during a long time of practice devoted to problems in the design and operation of a significant number of hydroelectric power plants in different countries. That is why the authors have endeavored to offer a large variety of carefully chosen case studies covering a wide range of engineering issues on the field of hydropower plants. Therefore, the subjects being mentioned comprise all the main components of a hydro power plant, from the upstream end, with the basin for water intake, to the downstream end of the water flow outlet.

The book is addressed equally to engineers and practitioners in the design, construction and operation of hydropower projects, as well as to research workers and students at universities.

The book comprises 14 chapters - chapters 1 and 2 focus on sustainability of hydropower, chapters 3 to 8 cover basic issues of studying hydropower such as hydrological, hydraulic and geological aspects, chapters 9 to 11 deal with operation and planning, chapter 12 concentrates on the integration of small hydro turbines into existing water infrastructure, chapter 13 covers manufacture management system, and chapter 14 presents the challenges of rapid hydropower development in China and India as a case study.

Hossein Samadi-Boroujeni
Shahrekord University,
Iran

1

Sustainable Hydropower – Issues and Approaches

Helen Locher and Andrew Scanlon
Hydro Tasmania, Australia

1. Introduction

Meeting the growing demands for electricity creates difficult decisions for many countries. The context for decision-making is also changing, particularly in light of climate change imperatives encouraging a move away from greenhouse gas emitting energy sources.

Hydropower is a mature technology, harnessing the energy moving from higher to lower elevations. It comes in various shapes and sizes from large reservoir projects to small run-of-river facilities. Hydropower is renewable, and has low greenhouse gas emissions. It is a premium energy source, providing a range of services. These include baseload and peak load generation, and support for other forms of electricity generation, particularly renewables.

Despite these strengths, hydropower developments over the past decades have been highly controversial due to accompanying social and environmental concerns. A challenge for hydropower developers and operators, as well as government planners and regulators, has been to develop tools that promote good practice and sustainable hydropower projects. Financiers and development partners have similarly developed their own approaches.

Importantly, there has been some convergence in these efforts to assess and guide hydropower sustainability. At this point in time there is a good global understanding of the key sustainability issues that must be addressed by the hydropower sector, and also of the pathways towards continuous good practice for those different issues.

2. Understanding the term "sustainability"

Sustainability is a major challenge facing the world. Do a simple google search on sustainability and 97 million results are presented in an 11 second search. Looking at these results shows that all over the world, countries, regions, institutions, businesses and projects are trying to figure out how this word applies to them and what they should be doing about it. Major global conferences, think-tanks and processes have been in train for decades around this theme, and countries and states are increasingly creating departments and legislation which have sustainability as part of their mandate.

Figure 1 shows some of the leading global events which have shaped our understanding of the term sustainability, and the progression of actors involved.

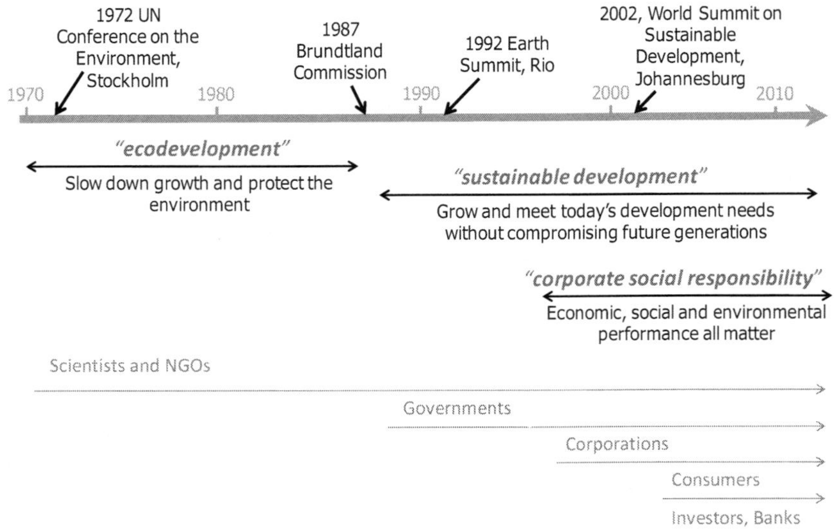

Fig. 1. Evolution of understanding of sustainability and sustainable development.

The term "sustainable development" came from the 1987 Brundtland Commission, which defined it as development that "meets the needs of the present without compromising the ability of future generations to meet their own needs" (United Nations 1987). An important embedded concept is inter- and intra-generational equity, meaning that those within the same generation as well as those who will inherit the world from us get equal opportunities to gain the benefits of natural resources without bearing unfair and inequitably distributed costs.

In the last few decades, corporate social responsibility has received increasing attention and at this point in time is almost business-as-usual for modern corporations. It broadens consideration of corporate performance beyond financial to also encompass social and environmental, often expressed as the "triple bottom line" or "people-planet-profit".

Consumers and investors have increasingly found avenues to promote their interest in sustainability through green choice schemes, sustainability certification schemes, and socially responsible investment indices and offerings. In the finance sector, many commercial banks have signed up to the Equator Principles, which commits them to ensure sustainability expectations are met by any loan recipients.

Looking forward, Rio+20 is to be held in 2012, with a focus on the "green economy". A green economy is defined by the United Nations Environment Programme (UNEP) as one that results in improved human well-being and social equity, while significantly reducing environmental risks and ecological scarcities[1]. In its simplest expression, a green economy can be thought of as one which is **low carbon, resource efficient** and **socially inclusive**.

[1] UNEP (2010) *Green Economy Report: A Preview.* http://www.unep.org/GreenEconomy/LinkClick.aspx?fileticket=JvDFtjopXsA%3d&tabid=1350&language=en-US United Nations Environment Programme. Retrieved 1 September 2011.

Practically speaking, a green economy is one whose growth in income and employment is driven by public and private investments that reduce carbon emissions and pollution, enhance energy and resource efficiency, and prevent the loss of biodiversity and ecosystem services.

Alongside these trends, there is international recognition that climate change presents one of the world's greatest sustainability challenges. If sustainability considerations address how we act now to ensure viability of our societies and their functions in the future, then climate change is inherently part of this discussion. The Kyoto Protocol, the 1997 international agreement linked to the United Nations Framework Convention on Climate Change (UNFCCC), set targets for reducing greenhouse gas (GHG) emissions[2]. The Intergovernmental Panel on Climate Change (IPCC) was set up to provide the world with a clear scientific view on the current state of knowledge in climate change and its potential environmental and socio-economic impacts[3]. UNFCCC conferences subsequent to the Kyoto Convention, based on information arising from IPCC reports, have raised the global imperatives of both mitigating and adapting to climate change. This has influenced thinking on hydropower as part of the climate change solution. This has been thoroughly examined in the hydropower chapter of the IPCC Special Report on Renewable Energy Sources and Climate Change Mitigation (Kumar et al 2011).

In summary, "sustainable hydropower" is considered to have three critical components:

1. The long-term viability of a hydropower project;
2. The contribution of the project to sustainable development; and
3. The integrated consideration of the different sustainability dimensions (social, environmental, financial/economic, technical, governance).

3. Sustainability issues in the international hydropower sector

Sustainable development requires attention to a wide range of social and environmental objectives. These are captured well by the Millennium Development Goals (MDGs)[4]. The MDGs are eight international development goals that all 193 United Nations member states and at least 23 international organizations have agreed to achieve by the year 2015:

1. Eradicate extreme poverty and hunger
2. Achieve universal primary education
3. Promote gender equality and empower women
4. Reduce child mortality rates
5. Improve maternal health
6. Combat HIV/Aids, malaria and other diseases
7. Ensure environmental sustainability
8. Develop a global partnership for development.

Whilst significant progress has been made on these goals, challenges still remain, particularly with respect to addressing disparities between rural and urban areas, supporting the most vulnerable and advancing sustainable development (UN 2011).

[2] http://unfccc.int/kyoto_protocol/items/2830.php, Retrieved 1 September 2011.
[3] http://www.ipcc.ch/index.htm, Retrieved 1 September 2011.
[4] http://www.un.org/millenniumgoals/bkgd.shtml, Retrieved 1 September 2011.

Sustainable hydropower, perhaps well beyond other potential sources of electricity, has significant potential to support progress towards the MDGs. Hydropower is the largest source of renewable energy in the electricity sector, contributing 16% of worldwide electricity supply as of the end of 2008 (Kumar et al 2011). Hydropower is a mature and long-lived technology, with some projects in operation for more than a century. Some parts of the world, such as Quebec, Tasmania and Norway, have built their economies around hydropower, and have a long history of development and management of hydropower operations. Other parts of the world are seeking to utilise their considerable hydropower resources as a major vehicle to advance their economic development (e.g. Lao PDR, Nepal, Sarawak). There are still large opportunities for continued hydropower development worldwide, particularly in Africa, Asia and Latin America. For example, as much as 92% of the technical potential for hydropower remains undeveloped in Africa. There is also considerable potential to upgrade and modernise existing hydropower facilities, or to add hydropower generation to water storages (Kumar et al 2011).

Unlike other forms of electricity, hydropower can provide both energy and water solutions, and consequently can promote economic and social development. Hydropower can be developed at many scales, and to fit many electricity supply needs. Large hydropower projects can have important multiplier effects, and multipurpose hydropower projects may provide the financial means to deliver water services beyond just electricity generation. As a water management measure, hydropower can help address drinking water, irrigation, flood control and navigation services needs. Additionally, hydropower offers significant potential for carbon emissions reductions (Kumar et al 2011).

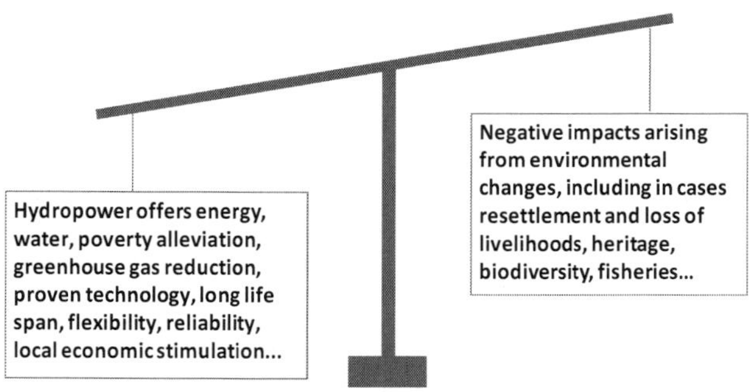

Fig. 2. Sustainable hydropower - ensuring benefits outweigh costs.

The degree to which hydropower can advance sustainable development objectives depends on careful planning, and attention to optimising the positive and minimising the negative in project development and operation. Because hydropower can fundamentally alter landscapes and regions, considerable care must be taken in planning for a development. With careful planning, hydropower projects can help address poverty eradication and regional development needs through provision of electricity and water supply. There is also the potential to leverage additional benefits such as clean water, sanitation, transport, health and educational facilities, other industries, and local capacity building. Without careful assessment and planning, hydropower may undermine sustainable development objectives through negative effects on natural habitats and river flows as well as on project-affected communities, livelihoods and living standards. The extent of both positive and negative impacts can be managed through choices around project siting and design, and attention paid to recognising and addressing social and environmental issues from the outset. The intent is to ensure that benefits are maximised and negative impacts avoided, minimised, mitigated and compensated (see Figure 2).

Sustainability issues relating to hydropower cover all aspects of the triple bottom line. Important environmental issues encompass habitats, biodiversity, invasive species, water quality, erosion, reservoir sedimentation, and downstream flow regimes. Reservoir sedimentation can greatly limit the life of a hydropower project, and can be exacerbated by catchment practices beyond the control of the hydropower facility. Passage of aquatic species past the physical barrier presented by dams has been a challenge for the hydropower industry. Increasingly with climate change, reliability of the water resource and avoidance of greenhouse gas emissions from reservoirs need careful consideration. Important social issues include livelihoods and living standards of project-affected communities, physical and economic displacement, indigenous communities and vulnerable social groups, public health, safety, labour and working conditions, and cultural heritage. Particular areas of concern for both environmental and social issues have related to failure of mitigation measures, lack of adequate compensation or follow-up, and cumulative impacts. Increasingly, the need for community engagement and acceptance, and a "social licence to operate", are recognised as important for successful developments, and attention to human rights. With economic issues, the major concerns have been with delivery of expected benefits, and distribution of costs and benefits; because of these concerns, the concept of "benefit-sharing" has received increasing focus with hydropower developments. Local capacity building is also an important economic issue relating to hydropower sustainability. Alongside these environmental, social and economic issues are issues relating to technical considerations (e.g. infrastructure safety, asset reliability and efficiency) and governance (e.g. institutional capacities, the policy context, and ethical practices).

An important over-arching framework for sustainable hydropower development is Integrated Water Resource Management (IWRM) and basin development planning. Many hydropower projects are evaluated in isolation of an overall basin planning framework, and issues arise due to competing or conflicting needs and uses of the basin resources. IWRM has a focus on understanding and rationalising use of and impacts on basin resources. With respect to hydropower, this may result in measures to ensure maintenance of ecosystem services (e.g. fish passage or sediment through-flow); protection of undeveloped river reaches; more coordinated operation of different hydropower facilities to achieve better water resource efficiencies; delivery of environmental flow regimes; and/or increased multi-purpose hydropower facilities to offer a variety of services such as navigation, irrigation, water supply, aquaculture or recreation.

Transboundary issues can represent a particular challenge for hydropower projects, particularly where the benefits of the project accrue to one country with costs borne by downstream countries in terms of hydrological changes and their environmental and social consequences. Considerable attention has been paid globally to the creation and effective functioning of transboundary river commissions (e.g. the Danube, Zambezi, Nile, and Mekong rivers) with the purpose of enabling information exchange and better transboundary cooperation with water infrastructure development and operation. These issues may be

	ADVANTAGES	DISADVANTAGES
ECONOMIC ASPECTS	Provides low operating and maintenance costs Provides long life span (50 to 100 years and more) Meets load flexibly (i.e hydro with reservoir) Provides reliable service Includes proven technology Can instigate and foster regional development Provides highest energy efficiency rate (payback ratio and conversion process) Can generate revenues to sustain other water uses Creates employment opportunities Saves fuel Can provide energy independence by exploiting national resources Optimizes power supply of other generating options (thermal and intermittent renewables)	High upfront investment Precipitation dependent In some cases, the storage capacity of reservoirs may decrease due to sedimentation Requires long-term planning Requires long-term agreements Requires multidisciplinary involvement Often requires foreign contractors and funding
SOCIAL ASPECTS	Leaves water available for other uses Often provides flood protection May enhance navigation conditions Often enhances recreational facilities Enhances accessibility of the territory and its resources (access roads and ramps, bridges) Provides opportunities for construction and operation with a high percentage of local manpower Improves living conditions Sustains livelihoods (freshwater, food supply)	May involve resettlement May restrict navigation Local land use patterns will be modified Waterborne disease vectors may occur Requires management of competing water uses Effects on impacted peoples' livelihoods need to be addressed, with particular attention to vulnerable social groups Effects on cultural heritage may need to be addressed
ENVIRONMENTAL ASPECTS	Produces no atmospheric pollutants Neither consumes nor pollutes the water it uses for electricity generation purposes Produces no waste Avoids depleting non-renewable fuel resources (i.e., coal, gas, oil) Very few greenhouse gas emissions relative to other large-scale energy options Can create new freshwater ecosystems with increased productivity Enhances knowledge and improves management of valued species due to study results Can result in increased attention to existing environmental issues in the affected area.	Inundation of terrestrial habitat Modification of hydrological regimes Modification of aquatic habitats Water quality needs to be monitored/managed Greenhouse gas emissions can arise under certain conditions in tropical reservoirs Temporary introduction of methylmercury into the food chain needs to be monitored/managed Species activities and populations need to be monitored/managed Barriers for fish migration, fish entrainment Sediment composition and transport may need to be monitored/managed, including measures to limit reservoir sedimentation Introduction of pest species needs to be monitored/managed

Table 1. Economic, social and environmental advantages and disadvantages of hydropower.

between upstream and downstream countries, or those that have the sameriver as a national border. Itaipu Binacional is a classic example of a hydropower development bridging two countries, in this case Paraguay and Brazil. In such cases well-designed cooperative agreements are essential, as well as ongoing processes to anticipate and address any emerging issues and maintain harmony.

After more than a century of experience, hydropower's strengths and weaknesses are well understood. Whilst not all negative impacts of hydropower can be eliminated, much can be done to mitigate them. A summary of economic, social and environmental aspects of hydropower is provided in the Table 1, taken from the Sustainable Hydropower Website[5]:

4. International initiatives relating to sustainable hydropower challenges

4.1 Overview of major international initiatives

Figure 3 provides a timeline of major international initiatives influential in shaping the global understanding of hydropower sustainability. Increasingly through the 1990s, attention at a global level was directed at the sustainability issues relating to dams development broadly, and hydropower specifically. One of the first international level initiatives to better define these issues and mitigation measures to address them was through the International Energy Agency's (IEA) Implementing Agreement on Hydropower Technologies. A major and intensive focus was cast on the dams sector globally through the World Commission on Dams between 1998 and 2000, and its follow-up the UNEP Dams and Development Project. In the last decade, the most far-reaching and influential initiatives addressing sustainability in the hydropower sector have been driven by the International Hydropower Association. These are expanded on in the following sections.

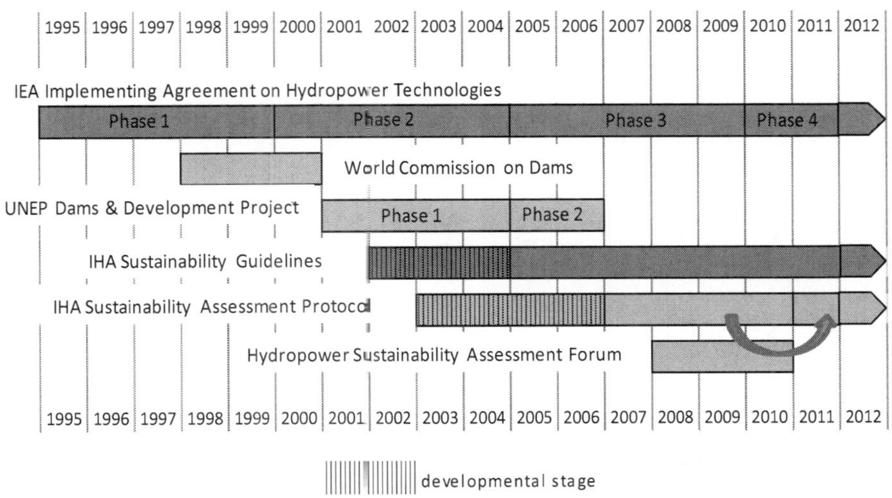

Fig. 3. International initiatives addressing hydropower sustainability issues.

[5] www.sustainablehydropower.org. Retrieved 1 September 2011

4.2 International Energy Agency Implementing Agreement on hydropower technologies

The IEA is an autonomous organisation which works to ensure reliable, affordable and clean energy for its 28 member countries and beyond[6]. The Hydropower Implementing Agreement[7] is a working group of IEA member countries and others that have a common interest in advancing hydropower worldwide. The Implementing Agreement's programme is carried out by task forces called Annexes. Two annexes have been particularly relevant to sustainable hydropower: Annex III - Environmental and Social Impacts of Hydropower, and Annex VIII - Hydropower Good Practices.

Reports from the Task Force on Environmental and Social Impacts[8] examined the positive and negative environmental and social impacts of hydropower (IEA 2000a), the effectiveness of mitigation measures (IEA 2000b), and issued guidelines in relation to the above (IEA 2000c). Annex VIII documented successful mitigation measures for ten key issues in the design and operation of hydropower projects, which included sixty extensively documented Case Histories collected from 20 countries (IEA 2006). These initiatives have provided an important foundation upon which other international inititives have drawn.

4.3 World Commission on Dams

The World Commission on Dams was a highly intensive review of the global dams sector undertaken between 1998 and 2000, with two objectives:

- To review the development effectiveness of large dams and assess alternatives for water resources and energy development; and
- To develop internationally acceptable criteria, guidelines and standards, where appropriate, for the planning, design, appraisal, construction, operation, monitoring and decommissioning of dams.

Key findings documented both the benefits and the costs of dams, and prompted a significant focus on the need to make concerted efforts to address sustainability issues. Important conclusions capturing these concerns were that *"Dams have made an important and significant contribution to human development, and the benefits derived from them have been considerable. In too many cases an unacceptable and often unnecessary price has been paid to secure those benefits, especially in social and environmental terms, by people displaced, by communities downstream, by taxpayers and by the natural environment"*(WCD 2001).

The WCD report set out a series of strategic priorities, policy principles, criteria and guidelines. The seven strategic priorities are:

1. Gaining public acceptance
2. Comprehensive options assessment
3. Addressing existing dams

---

[6] http://www.iea.org/about/index.asp. Retrieved 1 September 2011.

[7] http://www.ieahydro.org/IEA_Hydropower_Agreement.html. Retrieved 1 September 2011.

[8] International Energy Agency (IEA) Implementing Agreement for Hydropower Technologies & Programs: Technical Reports. 1995-2009. http://www.ieahydro.org/agreement.htm. Retrieved 1 September 2011.

4. Sustaining rivers and livelihoods
5. Recognising entitlements and sharing benefits
6. Ensuring compliance
7. Sharing rivers for peace, development and security.

Responses to the WCD report ranged from acceptance to rejection across nations and institutions. Unfortunately, even ten years later, positions on the outcomes remain divisive. One of the challenges was its complexity. With three grounding Global Norms, five Core Values, five key Decision Points, seven Strategic Priorities, 33 associated Policy Principles, and 26 Guidelines, converting these into operational practices has proven to be a difficult. Most institutions and stakeholders broadly accepted the core principles and Strategic Priorities, but have had polarised views on the guidelines (Moore et al 2010).

The WCD has been the focus of considerable review and evaluation (e.g. Dubash et al 2001; Fujikura & Nakayama 2009). The Water Alternatives journal issued a special edition ten years post-WCD to reflect on what had progressed or changed as the result of this process. Critical themes identified from the review of the WCD ten years on (Moore et al 2010) include:

- Diverse perceptions: Perspectives differ on the impact of the WCD Report and process
- Changing drivers:
  - Water and energy demands continue to rise and drive dam development
  - Climate change is now a greater driver of hydropower expansion
  - New financiers are changing the loci and framework for decision-making processes
- Environment and social justice: Negative consequences of dams on the environment and livelihoods of dam-affected communities remain critical issues
- New assessment tools: The quest for new decision-making tools and approaches continues, from assessment protocols to economic analysis
- Advances in participation and accountability: How can participation, compliance, accountability, and performance be ensured?
- Negotiation: Multi-stakeholder platforms continue to show promise for informing and shaping negotiated agreements that result in better sharing of the resources, benefits, and costs associated with dams.

4.4 International Hydropower Association sustainability initiatives

The International Hydropower Association (IHA) addresses the role of hydropower in meeting the world's growing water and energy needs as a clean, renewable and sustainable technology. With members active in more than 80 countries, IHA is a non-governmental, mutual association of organisations and individuals. Its membership is open to all those involved in hydropower. IHA was formed under the auspices of UNESCO in 1995, as a forum to promote and disseminate good practice and further knowledge about hydropower. IHA's mission is to advance sustainable hydropower's role in meeting the world's water and energy needs. It has 85 corporate members spanning six continents.

In 2004 IHA adopted Sustainability Guidelines (IHA 2004), providing a framework for good practice to which IHA committed to work towards in cooperation with government, business, civil society, consumers and individuals. In 2006 IHA adopted its Sustainability

Assessment Protocol (IHA 2006) after having internally trialed a previous five versions. The Protocol's purpose was to evaluate performance of hydropower projects against the IHA Sustainability Guidelines. Also in 2006, IHA launched the Sustainable Hydropower Website (www.sustainablehydropower.org), a joint initiative with the International Energy Agency to demonstrate projects that have successfully implemented sustainability measures on specific sustainability issues.

The IHA initiative which has had the most momentum has been the IHA Sustainability Assessment Protocol (IHA 2006). This was developed as an industry self-assessment tool, and provides a framework for projects to rate their performance on a number of sustainability aspects covering economic, social, environmental issues on a scale of 1 to 5. Scores are for each aspect, not an overall project score, so areas of strength and weakness, and opportunities for improvement, can be clearly identified.

4.5 Hydropower Sustainability Assessment Forum

The Hydropower Sustainability Assessment Forum was a cross-sector collaboration that reviewed the IHA Sustainability Assessment Protocol (IHA 2006) between 2008 and 2010. The Forum involved developing and developed world governments, commercial and development banks, social and environmental NGOs, and the hydropower industry. The mission was to develop a consensus product reflecting common views of the important issues and criteria for a sustainability assessment of a hydropower project at its different life cycle stages. The process involved global consultation (see Arup 2009) and a major trialling program for the draft Protocol (IHA 2009). Reference groups to the Forum members and open global consultation periods were built into the process to obtain views beyond the immediate Forum membership. The Forum's objective was to agree on a measurement tool that is practical, objective, and able to be implemented globally across a range of contexts. The aim was to facilitate objective decision-making on critical hydropower sustainability issues. It was hoped that the thorough and inclusive process would result in commitment by the hydropower sector and endorsement by external organizations.

Identified opportunities in the work of the Forum included:

- Broader endorsement outside of the hydropower sector to see wider promotion and application;
- Greater harmonisation of the Protocol with other standards;
- Improvements on emerging concepts;
- Increased objectivity; and
- Improved support information e.g. technical guidance notes.

After 9 meetings, 10 webinars, 2 global consultation phases, and trialling of Draft Protocol on 6 continents, the Forum recommended the updated Hydropower Sustainability Assessment Protocol to IHA in September 2010. It was formally adopted by IHA in November 2010, and internationally launched, along with an independent and multi-stakeholder governance council, in June 2011.

A fundamental premise of the work of the Forum was that an industry driven and owned initiative has far-reaching potential to influence performance in the hydropower sector. The lack of adoption by industry of the WCD guidelines was a disappointing outcome at the end of such an investment in time, money, stakeholder input and analysis. If an industry owned tool incorporates much of the outcomes embedded in WCD and other standards, disparate approaches start to converge. Locher et al (2010) provides more detail on the relationship of the Forum process with WCD, points of commonality and departure, and the work undertaken by the Forum to embed WCD outcomes into the Protocol. Significantly, the Hydropower Sustainability Assessment Protocol (IHA 2011) benefits from many developments beyond WCD that have been happening in the area of project and corporate sustainability performance. These include but are not limited to the Equator Principles, International Finance Corporation Performance Standards, multi-national development bank safeguards policies (e.g. the World Bank Group, and the Asian Development Bank), the Global Reporting Initiative, Social Responsible Investment assessment tools (e.g. Dow Jones Sustainability Index, FTSE4Good), best practice experiences in the hydropower sector, and corporate experiences with annual sustainability assessment and reporting approaches. The Protocol also incorporates the latest experience in addressing governance issues at the national, sectoral, institutional and project levels.

5. Sustainability assessment tools and approaches for hydropower

5.1 Sustainability assessment

Sustainability assessment tools and approaches are increasingly being recognised and used to ensure more comprehensive consideration of a broad range of criteria (e.g. Gibson et al 2005). Project-level assessments have traditionally been based on Environmental Impact Assessments (EIAs), then increasingly accompanied by Social Impact Assessments (SIAs) either within or alongside the EIA. An integrated impact assessment might be called a Social & Environmental Impact Assessment (SEIA), or some equivalent labelling, but a true sustainability assessment is likely to go beyond this to incorporate economic, financial, governance and technical considerations.

The following sections focus on tools beyond EIA and SIA and government regulatory processes that are applied to hydropower developments to reflect international level standards and expectations. The most recent effort, the Hydropower Sustainability Assessment Protocol (IHA 2011), is explained in some detail since it is the outcome of a collaborative process drawing on other approaches. Development bank safeguards policies such as those of the World Bank, and International Finance Corporation Performance Standards used by the Equator Principles Financial Institutions (the "Equator Banks"), are also briefly described. Also described briefly is a regional initiative assessing the sustainability of river basins in which hydropower is a predominant activity, modelled on the Hydropower Sustainability Assessment Protocol and ultimately aspiring to be a useful interactive tool with the Protocol.

5.2 Hydropower Sustainability Assessment Protocol

The Hydropower Sustainability Assessment Protocol (IHA 2011) is a sustainability assessment framework for hydropower development and operation. It enables the production of a sustainability profile for a project through the assessment of performance within important sustainability topics.

Assessments rely on objective evidence to support a score for each topic that is factual, reproducible, objective and verifiable. The Protocol is designed for, and best used with, repeated application, and is likely to be highly effective if it can be embedded into business systems and processes. If used early in a particular project stage, it can be a gap analysis and can guide further activities. Used at the end of a project stage may help inform decisions to progress to the next stage, or in the case of the Operations tool to guide continuous improvement measures. Assessment results may be used to inform decisions, to prioritize future work and/or to assist in external dialogue.

To reflect the different stages of hydropower development, the Protocol includes four assessment tools - Early Stage, Preparation, Implementation, Operation - which have been designed to be used as stand alone documents.

Through an evaluation of basic and advanced expectations, the Early Stage tool may be used for risk assessment and dialogue prior to advancing into detailed planning. The Early Stage assessment tool includes key topics relating to the strategic environment; first reviewing existing needs, options and policies, then looking at the political situation and institutional capacities, followed by an assessment of the technical, social, environmental and economic risks (Table 2). This tool is not a scoring tool, unlike the other three Protocol assessment documents. It is a guide to consideration of important Early Stage issues, recognising that this project stage may be characterised by limited information and the need for a certain degree of commercial confidentiality.

Assessment of the Project's Strategic Context	Assessment of the Project Issues and Risks
Demonstrated need Options assessment Policies & plans Political risks Institutional capacity	Technical issues and risks Social issues and risks Environmental issues and risks Economic and financial issues and risks

Table 2. Protocol Early Stage Topics.

Integrative perspective	Environmental perspective	Social perspective	Technical perspective	Economic & financial perspective
Demonstrated need & strategic fit Communications & consultation Governance Integrated project management Environmental & social issues management	Downstream flows Erosion & sedimentation Water quality Biodiversity & invasive species Noise, dust & waste management	Project affected communities & livelihoods Resettlement Indigenous peoples Cultural heritage Public health Labour & working conditions	Siting & design Hydrological resource Asset reliability & efficiency Reservoir planning, filling and management Infrastructure safety	Economic viability Financial viability Project benefits Procurement

Table 3. Protocol Preparation, Implementation and Operation Stage Topics and Content.

Communications & Consultation	This topic addresses the identification and engagement with project stakeholders, both within the company as well as between the company and external stakeholders (e.g. affected communities, governments, key institutions, partners, contractors, catchment residents, etc). The intent is that project stakeholders are identified and engaged in the issues of interest to them, and communication and consultation processes establish a foundation for good stakeholder relations throughout the project life.
Governance	This topic addresses corporate and external governance considerations for the project. The intent is that the developer has sound corporate business structures, policies and practices; addresses transparency, integrity and accountability issues; can manage external governance issues (e.g. institutional capacity shortfalls, political risks including transboundary issues, public sector corruption risks); and can ensure compliance.
Demonstrated Need & Strategic Fit	This addresses the contribution of the project in meeting demonstrated needs for water and energy services, as identified through broadly agreed local, national and regional development objectives and in national and regional policies and plans. The intent is that the project can demonstrate its strategic fit with development objectives and relevant policies and plans can be demonstrated, and that the project is a priority option to meet identified needs for water and energy services.
Siting & Design	This addresses the evaluation and determination of project siting and design options, including the dam, power house, reservoir and associated infrastructure. The intent is that siting and design are optimised as a result of an iterative and consultative process that has taken into account technical, economic, financial, environmental and social considerations.
Environmental & Social Impact Assessment & Management	This addresses the assessment and planning processes for environmental and social impacts associated with project implementation and operation throughout the area of impact of the project. The intent is that environmental and social impacts are identified and assessed, and avoidance, minimisation, mitigation, compensation and enhancement measures designed and implemented.
Integrated Project Management	This addresses the developer's capacity to coordinate and manage all project components, taking into account project construction and future operation activities at all project-affected areas. The intent is that the project meets milestones across all components, delays in any component can be managed, and one component does not progress at the expense of another.
Hydrological Resource	This addresses the level of understanding of the hydrological resource availability and reliability to the project, and the planning for generation operations based on these available water inflows. The intent is that the project's planned power generation takes into account a good understanding of the hydrological resource availability and reliability in the short- and long-term, taking into account other needs, issues or requirements for the inflows and outflows as well as likely future trends (including climate change) that could affect the project.
Infrastructure Safety	This addresses planning for dam and other infrastructure safety during project preparation, implementation and operation. The intent is that tife, property and the environment are protected from the consequences of dam failure and other infrastructure safety risks.

Financial Viability	This addresses both access to finance, and the ability of a project to generate the required financial returns to meet project funding requirements, including funding of measures aimed at ensuring project sustainability. Access to carbon finance may be important to this. The intent is that the project proceeds with a sound financial basis that covers all project funding requirements including social and environmental measures, financing for resettlement and livelihood enhancement, delivery of project benefits, and commitments to shareholders/investors.
Project Benefits	This addresses the additional benefits that can arise from a hydropower project, and the sharing of benefits beyond one-time compensation payments or resettlement support for project affected communities. The intent is that opportunities for additional benefits and benefit sharing are evaluated and implemented, in dialogue with affected communities, so that benefits are delivered to communities affected by the project.
Economic Viability	This addresses the net economic viability of the project. The intent is that there is a net benefit from the project once all economic, social and environmental costs and benefits are factored in.
Procurement	This addresses all project-related procurement including works, goods and services. The intent is that procurement processes are equitable, transparent and accountable; support achievement of project timeline, quality and budgetary milestones; support developer and contractor environmental, social and ethical performance; and promote opportunities for local industries.
Project-Affected Communities & Livelihoods	This addresses impacts of the project on project affected communities, including economic displacement, impacts on livelihoods and living standards, and impacts to rights, risks and opportunities of those affected by the project. The intent is that livelihoods and living standards impacted by the project are improved relative to pre-project conditions for project affected communities with the aim of self-sufficiency in the long-term, and that commitments to project affected communities are fully delivered over an appropriate period of time.
Resettlement	This addresses physical displacement arising from the hydropower project development. The intent is that the dignity and human rights of those physically displaced are respected; that these matters are dealt with in a fair and equitable manner; and that livelihoods and standards of living for resettlees and host communities are improved.
Indigenous Peoples	This addresses the rights, risks and opportunities of indigenous peoples with respect to the project, recognising that as social groups with identities distinct from dominant groups in national societies, they are often the most marginalized and vulnerable segments of the population. The intent is that the project respects the dignity, human rights, aspirations, culture, lands, knowledge, practices and natural resource-based livelihoods of indigenous peoples in an ongoing manner throughout the project life.
Labour & Working Conditions	This addresses labour and working conditions, including employee and contractor opportunity, equity, diversity, health and safety. The intent is that workers are treated fairly and protected.

Cultural Heritage	This addresses cultural heritage, with specific reference to physical cultural resources, at risk of damage or loss by the hydropower project and associated infrastructure impacts (e.g. new roads, transmission lines). The intent is that physical cultural resources are identified, their importance is understood, and measures are in place to address those identified to be of high importance.
Public Health	This addresses public health issues associated with the hydropower project. The intent is that the project does not create or exacerbate any public health issues, and that improvements in public health can be achieved through the project in project-affected areas where there are significant pre-existing public health issues.
Biodiversity & Invasive Species	This addresses ecosystem values, habitat and specific issues such as threatened species and fish passage in the catchment, reservoir and downstream areas, as well as potential impacts arising from pest and invasive species associated with the planned project. The intent is that there are healthy, functional and viable aquatic and terrestrial ecosystems in the project-affected area that are sustainable over the long-term, and that biodiversity impacts arising from project activities are managed responsibly.
Erosion & Sedimentation	This addresses the management of erosion and sedimentation issues associated with the project. The intent is that erosion and sedimentation caused by the project is managed responsibly and does not present problems with respect to other social, environmental and economic objectives, and that external erosion or sedimentation occurrences which may have impacts on the project are recognised and managed.
Water Quality	This addresses the management of water quality issues associated with the project. The intent is that water quality caused by the project is managed responsibly and does not present problems with respect to other social, environmental and economic objectives, and that external water quality occurrences which may have impacts on the project are recognised and managed.
Reservoir Planning	This addresses the planning for management of environmental, social and economic issues within the reservoir area during project implementation and operation. The intent is that the reservoir will be well managed taking into account power generation operations, environmental and social management requirements, and multi-purpose uses where relevant.
Downstream Flow Regimes	This addresses the flow regimes downstream of hydropower project infrastructure in relation to environmental, social and economic impacts and benefits. The intent is that flow regimes downstream of hydropower project infrastructure are planned and delivered with an awareness of and measures incorporated to address environmental, social and economic objectives affected by those flows.

Table 4. Sustainability topics in the Protocol's Preparation Assessment Tool.

The remaining three assessment tools, Preparation, Implementation and Operation, set out a graded spectrum of practice calibrated against statements of basic good practice and proven best practice. The graded performance within each sustainability topic provides the opportunity to promote structured and continuous improvement. Each project stage assessment tool has in the order of twenty topics that cover the range shown in Table 3.

Each topic is scored with respect to criteria which may include assessment, management, stakeholder engagement, stakeholder support, conformance and compliance, and outcomes. Within the scoring statements for these topics, a number of important cross-cutting issues are represented, encompassing human rights, climate change, transboundary issues, transparency, gender, and integrated water resources management.

To provide greater insight beyond the labels, Table 4 provides more detail on what the sustainability topics in the Preparation assessment tool address, and what their intent is. This table well-illustrates the comprehensiveness of the Protocol content with respect to critical hydropower sustainability issues.

5.3 Development bank safeguards policies

The World Bank Safeguard Policies apply to all World Bank lending - they establish a range of social and environmental obligations that must be met by recipients of World Bank project finance, and formalise the World Bank's own commitments in relation to related areas such as transparency and consultation. The World Bank Safeguard Policies are designed to help staff promote socially and environmentally sustainable approaches to development as well as to ensure that Bank operations do not harm people and the environment. The World Bank conducts Environmental Assessments (EA) of each proposed investment loan to determine the appropriate extent and type of environmental impact analysis to be undertaken, and whether or not the project may trigger other safeguard policies. Compliance with the safeguards policies is the expected standard, in addition to compliance with applicable local, national, and international laws.[9]

There are eight environmental and social safeguard policies that are used for investment lending:

- OP/BP 4.01 Environmental Assessment
- OP/BP 4.04 Natural Habitats
- OP/BP 4.09 Pest Management
- OP/BP 4.10 Indigenous Peoples
- OP/BP 4.11 Physical Cultural Resources
- OP/BP 4.12 Involuntary Resettlement
- OP/BP 4.36 Forests
- OP/BP 4.37 Dam Safety

The policies include tests relating to the level of social/environmental risk associated with a project, so that whilst the policies are generally applicable the specific requirements vary depending on the assessed level of risk for a particular project. Funding recipients must meet the requirements prior to receiving World Bank finance, and/or must agree to implement the requirements over the project implementation period. There is considerable supporting information provided by the World Bank for many of these policies, for example the World Bank Source Book on Resettlement. Governance of the Safeguard Policies is the responsibility of the World Bank.

[9] http://search.worldbank.org/all?qterm=safeguards%20policies

5.4 International Finance Corporation Performance Standards

The Equator Principles Financial Institutions (EPFIs) are commercial banks who have committed to only provide loans to projects where the borrower complies with the social and environmental policies and procedures that implement the Equator Principles. The Equator Principles were developed by private sector banks - led by Citigroup, ABN AMRO, Barclays and WestLB - and were launched in June 2003. The banks chose to model the Equator Principles on the environmental standards of the World Bank and the social policies of the International Finance Corporation (IFC). Seventy-two financial institutions have adopted the Equator Principles, which have become the de facto standard for banks and investors on how to assess major development projects around the world. In July 2006, the Equator Principles were revised, increasing their scope and strengthening their processes.

The IFC Performance Standards are equivalent to the World Bank Safeguard Policies but are implemented by the IFC, the member of the World Bank Group responsible for private sector lending. They apply to lending to private sector companies rather than government entities. The IFC Performance Standards apply both to IFC's own lending as well as its lending in consortia together with private sector banks.

The eight IFC Performance Standards incorporate essentially the same risk assessment approach as the World Bank:

- PS 1: Social and Environmental Assessment and Management Systems
- PS 2: Labor and Working Conditions
- PS 3: Pollution Prevention and Abatement
- PS 4: Community Health, Safety and Security
- PS 5: Land Acquisition and Involuntary Resettlement
- PS 6: Biodiversity Conservation and Sustainable Natural Resource Management
- PS 7: Indigenous Peoples
- PS 8: Cultural Heritage

These are supported by Guidance Notes. Companies must meet, or agree to meet, the requirements of the Performance Standards in order to receive IFC financing. Governance of the Performance Standards is the responsibility of the IFC.

The Equator Principles are closely linked to the IFC Performance Standards. The Equator Principles are a voluntary framework adopted by a number of private sector banks referred to as the Equator Principles Financial Institutions (EPFI), or the "Equator Banks". These principles specify a common due diligence framework, underpinned for application in emerging economies by the IFC Performance Standards. In signing up to the Equator Principles, the Equator Banks agree to apply such policies to any project lending over US$10 million in value. As for the World Bank Safeguards and the IFC Performance Standards, the application of the policy is based on a risk assessment. Governance of the Equator Principles as a system is the responsibility of the Equator Principles Association, an association of the EPFIs.

Whilst the IFC Performance Standards are accompanied by sector specific guidelines, there are no such guidelines at present for the hydropower sector. This was one factor that led to the participation by the Equator Banks in the Hydropower Sustainability Assessment Forum.

5.5 Rapid Basin-wide Hydropower Sustainability Assessment Tool

The sustainability assessment approaches outlined above are applied at the hydropower project level. A limitation for hydropower sustainability is that issues can arise due to cumulative impacts and poor integrated basin development planning and management.

In the Mekong River basin where there are ambitious hydropower development agendas, work has been undertaken during the past few years on a basin sustainability assessment tool that is designed to address the types of sustainability issues that arise with an accumulation of hydropower developments. The Rapid Basin-wide Hydropower Sustainability Assessment Tool (RSAT)[10] is a basin / sub-basin assessment and dialogue tool that brings together the major actors in a river basin to undertake a structured analysis of sustainability issues. The RSAT was designed to replicate the structured and comprehensive approach of the Protocol, within a graded spectrum promoting continuous improvement, and hence can serve as a complementary tool to the Protocol. RSAT application in a basin can be enhanced if hydropower projects within that basin have applied the Protocol, and these could potentially be applied in an iterative manner to progressively improve basin outcomes.

The RSAT is a product of several years of conceptualization, preparation, and stakeholder engagement in the Mekong region under the partnership initiative called the Environment Criteria for Sustainable Hydropower (ECSHD). ECSHD partners are the Asian Development Bank (ADB), Mekong River Commission (MRC) and the World Wide Fund for Nature (WWF). The ECSHD was formalised in 2006 as a platform to develop tools that will assist decision-making for sustainable hydropower development in the Mekong River Basin.

The RSAT is currently being trialed in the four lower Mekong River basin countries - Cambodia, Lao PDR, Thailand and Vietnam. It is planned to have an updated tool at the end of 2011.

6. Moving forward

Each of the major initiatives outlined in Section 4 and the diversity of tools and approaches outlined in Section 5 have played a role in building an understanding of and addressing hydropower sustainability. The work of the IEA established a foundation of knowledge that was hydropower specific. The World Commission on Dams (WCD) process brought together a diversity of sectors and captured their knowledge and views on dam development and operation. The initiatives of IHA have been important steps for the hydropower industry to digest the findings and outcomes of earlier processes, and interpret this and other information into tools that industry members can understand and apply. The Hydropower Sustainability Assessment Forum brought the IHA initiatives, and a broader set of tools and approaches used by governments and banks, together to build a cross-sectoral consensus on a structured view of hydropower sustainability.

The review of the WCD process (Moore et al 2010) concluded that controversy around dams has not gone away. It is apparent, however, that the framework surrounding dam controversies has shifted due to changing regional development pressures, due to evolution

[10] http://www.mrcmekong.org/ISH/sustainability-assessemt-tool.htm. Retrieved 1 September 2011.

of international norms and processes relating to sustainability in general, and due to particular sustainability issues such as climate change. Of interest has been a swing in international NGOs towards cautious support of hydropower provided sustainability principles (such as those underlying the Protocol) are strongly embedded in its development and operation.

An example of this swing can be seen in the work of M-Power (the Mekong Programme for Water, Environment & Resilience) which is actively supporting more sustainable hydropower governance through its project activities. M-Power is introducing and building capacity for the utilisation of new governance tools, such as the Protocol[11] in response to rapid advancement of hydropower agendas in the Mekong Region, and growing concerns of a range of stakeholders in the potential for adverse social and environmental impacts. M-Power's work seeks to influence the way hydropower is designed, developed and managed, through facilitating constructive engagement with stakeholders (civil society, government, developers and banks) about hydropower development issues in the Mekong Region. M-Power also aims to work with major hydropower companies operating in the region to make demonstrable gains in corporate social and environmental responsibility.

The Protocol is an evolutionary tool, and not the final answer on sustainable hydropower. It is an important consensus product of a rigorous 2½ year process that establishes a platform from which further steps can be taken. The scoring system, the structure, the method of application, the content and detail on the sustainability topics and criteria will all benefit from testing. Understanding of issues such as environmental flows, benefit sharing, and climate change is continuously improving. Learning from experience will be equally important for some long-term and highly sensitive issues such as stakeholder engagement, indigenous peoples, human rights, transboundary issues, corruption and legacy issues. All of these are challenges for many sectors beyond hydropower. The Protocol provides a framework for sharing of good practice amongst hydropower sector practitioners. This is already demonstrated by the IHA's formation of a Sustainability Network with sub-groups on key issues such as Indigenous Peoples.

It is necessary to be clear on what the Protocol is not. The Protocol is not an international standard specifying essential performance requirements or the basis for gaining a sustainability label or stamp of certified sustainability for a hydropower project; these may be developed in the future building on the platform provided by the Protocol. The Protocol is not a replacement for regulatory processes, although it is able to guide governments' understanding of sustainability issues and how these can be addressed. The Protocol does not make decisions on hydropower projects; decisions are always made by those authorities mandated with this responsibility and will be made based on their criteria, but may be better informed by application of the Protocol. The Protocol is not the "solution" to achieving hydropower sustainability; it is a voluntary industry support tool, and sustainability will best be achieved where all actors work in support of this objective. Sustainability outcomes can be well achieved if Protocol application to individual projects is alongside government attention to the context for sustainable hydropower development, e.g. through energy master planning, water development planning, basin development

[11] http://www.mpowernetwork.org/Major_Projects/Hydropower_Sustainability/index.html, Retrieved 1 September 2011.

planning and integrated water resource management, needs and options assessments incorporating sustainability considerations.

An important legacy of the Forum experience has been a recognised need to demonstrate the further evolution and application of the Protocol is not an industry-controlled process. An independent Hydropower Sustainability Assessment Council (HSAC), chaired by WWF, governs the use, application, quality control and any future reviews of the Protocol, and its official use must comply with Terms & Conditions of Use (see www.hydrosustainability.org). Crediblity in application is also a very important objective. Work is underway on the development of Protocol assessor accreditation courses that will be overseen by the HSAC. IHA, as the Protocol's management entity under the Council Charter, has formed IHA Sustainability Partnerships[12] with a diversity of global institutions to offer formalised training, capacity building and supported application of the Protocol so that it can be used in a consistent manner.

Evolution of the Protocol into future versions is alongside evolution of the other sustainability assessment tools reviewed in this paper. The World Bank has reengaged with hydropower as an improtant vehicle for sustainable development (World Bank 2009); it was an observer and in-kind contributor to the Forum process, has expressed its support for the Protocol, and continues to act as observer to the HSAC. The World Bank announced in March 2011 that it has embarked on a two year process of updating and consolidating its environmental and social safeguard policies into an integrated environmental and social policy framework[13]. IFC is presently in the midst of a two year review process for the standards, with revised standards expected to become operational in 2012[14]. Future work is being planned to take the Rapid Basin-wide Hydropower Sustainability Assessment Tool through an international testing process to provide a globally applicable tool to support river basin planning and management.

7. Conclusion

This paper captures numerous important and complex issues that must be taken into account, and the potential sustainable develoment benefits that can be realised, if sustainable hydropower projects are implemented. Increased scrutiny of the dam sector and sustainability issues with hydropower projects in the 1990s caused some organisations to turn away from support for this industry. The last decade has seen a renewed interest, particularly with respect to the role hydropower can play in helping to address climate change. This renewed interest is now accompanied by the ability to access tools to guide sustainable hydropower development. There is also a high degree of convergence across sectors in terms of expectations for sustainable hydropower projects.

Basic good practice for hydropower requires good environmental and social impact assessments, and effective mitigation and self-sustaining compensation measures. Attention needs to focus on values-based downstream flow regimes, equitable distribution of project

[12] http://www.hydrosustainability.org/Sustainability-Partners.aspx. Retrieved 1 September 2011.
[13]http://web.worldbank.org/WBSITE/EXTERNAL/PROJECTS/EXTPOLICIES/EXTSAFEPOL/0,,contentMDK:22849125~pagePK:64168445~piPK:64168309~theSitePK:584435,00.html. Retrieved 1 September 2011.
[14] http://www.ifc.org/ifcext/policyreview.nsf/Content/Process. Retrieved 1 September 2011.

costs and benefits, adaptive management, genuine stakeholder engagement processes, and local capacity building.

Sustainable development of hydropower is not simple. Tools and approaches are increasingly available to guide and advance sustainable hydropower. These need to be accompanied by awareness-raising, education, incentives, open dialogue, and willingness. In the end, sustainable hydropower will only be successful where there is genuine commitment.

8. References

Arup (2009) *HSAF Phase I Consultation: Consultation Outcomes Report*. Report to the Hydropower Sustainability Assessment Forum on the outcomes of the HSAF Phase 1 Consultation, 27 February 2009. Leeds: Ove Arup & Partners Ltd.

Dubash N.K., Dupar M., Kothari S. & Lissu T. (2001) *A Watershed in Global Governance? An Independent Assessment of the World Commission on Dams*. World Resources Institute.

Fujikura R. & Nakayama M. (2009) *Lessons Learned from the World Commission on Dams*. International Environmental Agreements, Vol. 9, pp. 173-190.

Gibson R.B., S. Hassan, S. Holtz, J. Tansey and G. Whitelaw (2005) *Sustainability assessment: criteria, processes and applications*. London: Earthscan Publications Ltd.

IEA (2000a) *Hydropower and the Environment: Survey of the Environmental and Social Impacts and Effectiveness of Mitigation Measures in Hydropower Development*. The International Energy Agency - Implementing Agreement for Hydropower Technologies and Programmes, Annex III Hydropower and the Environment.

IEA (2000b) *Hydropower and the Environment: Effectiveness of Mitigation Measures*. The International Energy Agency - Implementing Agreement for Hydropower Technologies and Programmes, Annex III Hydropower and the Environment.

IEA (2000c) *Hydropower and the Environment: Present Context and Guidelines for Future Action*. The International Energy Agency - Implementing Agreement for Hydropower Technologies and Programmes, Annex III Hydropower and the Environment.

IEA (2006) *Hydropower Good Practices. Environmental Mitigation Measures and Benefits*. The International Energy Agency - Implementing Agreement for Hydropower Technologies and Programmes, Annex VIII Hydropower Good Practices.

IHA 2004. *International Hydropower Association Sustainability Guidelines*. International Hydropower Association, London, UK.

IHA 2006. *International Hydropower Association Sustainability Assessment Protocol*. International Hydropower Association, London, UK.

IHA 2009. *Draft Hydropower Sustainability Assessment Protocol – August 2009*. International Hydropower Association, London, UK.

Kumar, A., T. Schei, A. Ahenkorah, R. Caceres Rodriguez, J.-M. Devernay, M. Freitas, D. Hall, Å. Killingtveit, Z. Liu (2011) Hydropower. In IPCC *Special Report on Renewable Energy Sources and Climate Change Mitigation* [O. Edenhofer, R. Pichs-Madruga, Y. Sokona, K. Seyboth, P. Matschoss, S. Kadner, T. Zwickel, P. Eickemeier, G. Hansen, S. Schlömer, C. von Stechow (eds)], Cambridge University Press, Cambridge, United Kingdom and New York, NY, USA.

Locher, H.; Hermansen, G.Y.; Johannesson, G.A.; Xuezhong, Y.; Phiri, I.; Harrison, D.; Hartmann, J.; Simon, M.; O'Leary, D.; Lowrance, C.; Fields, D.; Abadie, A.;

Abdel-Malek, R.; Scanlon, A.; Nyman, K. (2010) Initiatives in the hydro sector post-World Commission on Dams - The Hydropower Sustainability Assessment Forum. Water Alternatives 3(2): 43-57

Moore, D.; Dore, J. and Gyawali, D. (2010) *The World Commission on Dams + 10: Revisiting the large dam controversy*. Water Alternatives 3(2): 3-13.

UN (1987) *Report of the World Commission on Environment and Development*. General Assembly Resolution 42/187, United Nations, 11 December 1987.

UN (2011) Millenium Development Goals Progress Report 2011. United Nations, New York.

UNEP (2007) *Dams and Development: Relevant Practices for Improved Decision-Making*. United Nations Environment Program.

World Bank (2009) *Directions in Hydropower*. Produced by the Sustainable Development Vice Presidency, International Bank for Reconstruction and Development / World Bank.

WCD (2000) *Dams and Development: A New Framework for Decision-Making*. The Report of the World Commission on Dams. London: Earthscan Publications Ltd.

2

Hydropower – The Sustainability Dilemma

Wilson Cabral de Sousa Junior[1] and Célio Bermann[2]
[1]*Instituto Tecnológico de Aeronáutica,*
[2]*Universidade de São Paulo, São Paulo,*
Brazil

1. Introduction

Since the development of the agriculture, at least, hydropower has been used for irrigation and other engines such as watermills and domestic lifts (provision of water). It was only in the late 19th century, after the discovery of the electrical generator, that hydropower could be converted into electricity. The early 20th century was the turning point of the hydroelectricity, as we know. Since then, the hydroelectricity sector has technologically advanced and the current engineering arrangements are considerably improved compared with the pioneers.

Most of this improvement is related to the scale of the hydropower plants, which have increased from some kilowatts to gigawatts. Nowadays, the produced hydroelectricity is transmitted for considerable distance between where it is created to where it is consumed and the complete arrangement includes reservoirs, turbines, generators, power houses and long distance power lines. It involves more complex, local, regional and global impacts, which have to be considered in a sustainability analysis.

As a renewable resource, the hydropower was historically treated as a clean source of energy. However, the scientific researches launched in the last decades have produced arguments that brought another balance to this discussion. In this context, this chapter aims to explore the impacts of the hydropower plants under the sustainability's viewpoint. Despite the use of the "hydropower" term, our focus is on the hydroelectricity and its projects, which represent the major parcel of the sector.

2. The economic growth and its energy demands

Energy is an important factor of production, which is one of the main objects of the economy. The other economic piece is consumption. Based on the binomial "production and consumption", the economics mainstream has established the agenda of the global capitalist system that has been running almost in the entire world. Therefore, the energy has an essential role in this context. The world energy consumption is shown in Figure 1.

A country's energy system has complex impacts on its economy. In general, a contraction of energy supply restrains the economic activity, which can provoke impacts like the reallocation and even rationing of energy, as well as changes in technology to emphasize energy efficiency.

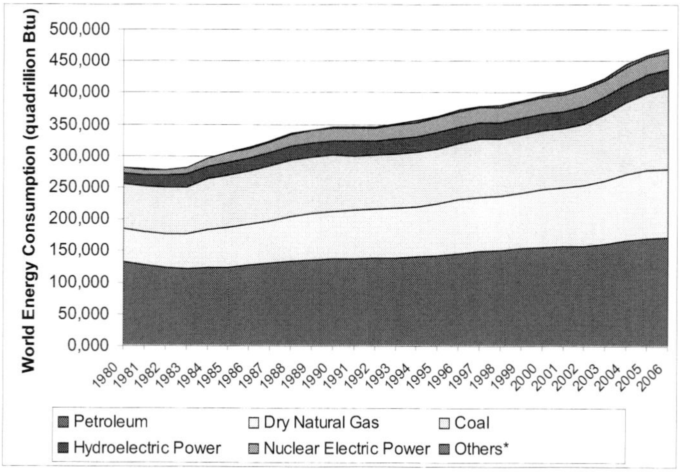

Source: IEA (2011).

Fig. 1. The World energy consumption, 1980 - 2006.

Given the relationship between energy and the economy, many development models place strong emphasis in the energy-economic production correlation. Various studies associate energy availability with gross domestic product (GDP) (Nilsson, 1993; Schipper, 2000). But, according to Cohen (2005), the energy-economic development relationship merits closer analysis because GDP hides a series of economic problems, including inequity among regions and social classes, not to mention uncounted environmental costs. All these characteristics is the key to energy planning geared to the true economic goals of a country or region.

Princen (1999) argues that excessive energy consumption in northern hemisphere nations and among southern hemisphere elites needs to be brought into closer balance with energy use in southern countries and less-privileged classes. This notion has gained increasing acceptance within environmentalist circles. Still public policies in developed countries have tended to focus almost exclusively on energy efficiency without addressing the overall consumption, which will continue to drive high energy use, even with efficiency improvements. This same pattern is being emulated by developing countries, which strive to increase energy supply (sometimes from cleaner sources) more than managing and reducing demand, as signaled by Sunkel (1979), and corroborated by many researchers over the last few decades.

The possibility of electricity savings illustrates that consumption do not need exactly to track economic growth and that, indeed, countries have the potential to reduce energy consumption per unit GDP (Totten et al., 2010).

A comparison between the world GDP and energy demand growth is presented in the Figure 2.

Note that the rise of the GDP is followed by the energy demand, although the level of the growth rate is different in each case. In average, for the period of 1980 until 2006, each $n\%$ of raising the world GDP was followed by the energy demand raise in $n\text{-}1\%$.

Even though the world energy demand is increasing, as shown in Figure 1, the Hydropower is decreasing its share.

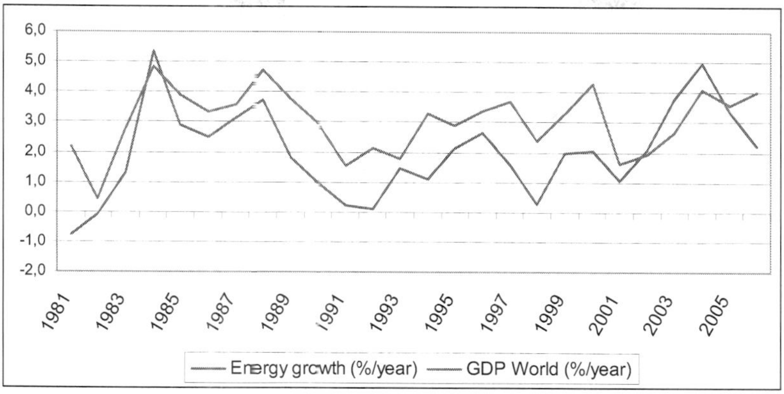

Source: adapted from World Bank (2011) and USEIA (2011).

Fig. 2. The world GDP and energy demand growth.

3. The economics context of "hydropowering"

There is a significant participation of the OECD countries, covering mainly the developed countries, concentrating about 42% of hydropower worldwide. Asia, in turn, owns 26% of hydropower, and hydropower production in China is the most significant one. Latin America also stands out with 20% of hydropower worldwide.

Producers	TWh*	% of world total	Installed capacity (GW)**
People's Rep. of China	616	18.5	168
Brazil	391	11.7	78
Canada	364	10.9	75
United States	298	9.0	100
Russian Federation	176	5.3	47
Norway	127	3.8	30
India	107	3.2	37
Venezuela	90	2.7	15
Japan	82	2.5	47
Sweden	66	2.0	16
Rest of the world	1,012	30.4	324
World	3,329	100.0	952

Notes: *2009 data; ** end-2008 data. Source: IEA (2011) and WEC (2011).

Table 1. Top-ten hydroelectricity producers on the world.

Table 1 presents the annual energy produced, total share in world production and the installed capacity of 10 countries which produce energy from hydroelectric sources in the world. It is observed that 67.1% of the world's electricity is produced from the burning of fossil fuels, while 13.4% is generated from nuclear power plants. Only 19.5% of the electricity generated worldwide is produced from renewable energy.

The distribution of hydroelectric production, estimated at around 3,329 TWh (2009), is shown in Figure 3, according to the major regions in the world (IEA, 2011).

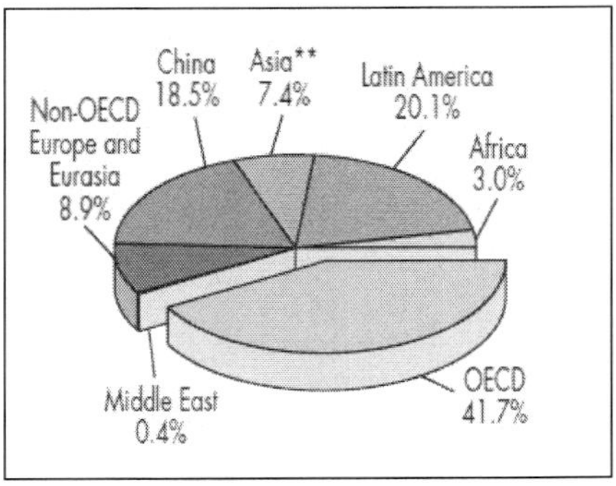

Notes: Includes pumped storage. **Asia excludes China.
Source: IEA (2011).

Fig. 3. Regional shares hydro production, 2009.

The use of water resources for a significant production of electricity is spread by several countries with very different levels of development. In addition to Canada, United States, Norway, Japan and Sweden, which are among the 10 largest producers of hydroelectricity, other European countries like France (60 TWh / 21 GW) and Italy (42 TWh / 18 GW) are also prominently featured in the global context.

China, Brazil, Russia, Canada and USA lead the countries which have the biggest hydroelectricity plants in the world. The areas of these countries and the existence of large drainage basins could explain this concentration. However, in terms of the electricity production related to the total power demand, countries like Norway, Brazil, Venezuela, Canada and Sweden appear between the countries to which hydroelectricity represents the main source.

4. The hydropower´s role in this context, as a renewable source of electricity

Hydropower remains the largest source of renewable energy in the electricity sector. On a global basis, the technical potential for hydropower is unlikely to constrain further deployment in short and medium terms. Hydropower is technically mature, it is often economically competitive with current market energy prices and it is already being deployed at a rapid pace. Situated at the crossroads of two major issues for development, water and energy, hydro reservoirs can often deliver services beyond electricity supply. The significant increase in hydropower capacity over the last 10 years is anticipated in many scenarios to continue in short terms (2020) and medium terms (2030), with various environmental and social concerns representing perhaps, the largest challenges to a continuous deployment if not carefully managed.

Hydropower production represented about 16% (1,010 GW) of global electricity production in 2010 and accounted for about 76% of electricity from renewable sources. An estimated 30 GW of capacity was added during the year, with the existing global capacity reaching an estimated 1,010 GW. Asia (led by China) and Latin America (led by Brazil) are the most active regions for new hydro development (REN21, 2011).

5. The agenda of hydropower for next decades

In the case of hydroelectric projects planned or forecasted in the medium term, the available information is not accurate. However, it is possible to be shorter in the initiative of some regions. In this context, it stands out in African countries like the Democratic Republic of Congo (Congo Democratic Republic) with 43 GW planned, and Nigeria (Niger) with 12 GW planned. In America, Brazil, with 68 GW planned and Canada, with 15 GW, are the ones that shall expand their hydroelectric capacity. In Asia, China, with 65 GW planned, Turkey, and Vietnam with 23 GW, 14 GW are where the main initiatives are focused

Anyway, in the long run, the forecast growth of hydroelectricity in the world energy supply is not expressive. According to WEC (2011), the share of hydroelectricity in Total Primary Energy Supply (Total Primary Energy Supply) might grow from 2.3% to 2.4% in the Business-as-Usual scenario, and 3.5% of the scenario based on the premise of stabilizing the concentration of greenhouse gases at 450 ppm equivalent CO_2.

The expansion of hydropower is limited due to the reduction of areas with greatest potential, most of which are already exploited, leaving regions of lower potential, where the social and environmental impacts are of greater magnitude. On the other hand, the supply of new sources of electricity generation such as wind and solar ones, has contributed to the falling costs of these modes, making it an increasingly competitive scenario for the new investments.

6. The best alternative: non generation?

The most efficient investment in energy supply is one that concentrates on reducing the consumption at the maximum efficient point. There are several ways to apply this issue. Regarding this, Leite and Balay (2007) estimate that the energy consumption could be reduced by 20% solely through energy-efficiency measures. That research focused on the main consuming sectors, industry, other commercial users, residential consumers and agriculture, in a non developed country scenario. Totten et al. (opus cit) present evidence that a value-adding water planning process can be achieved by shifting from the focus on supply expansion to one that concentrates on efficient delivering services at and near the point of use.

Some studies have already demonstrated the potential of investments in existing dams, upgrading their power production. A range from 20% to 40% of new energy could be provided by investments in existing dams (Bermann et al., 2004). However, it is a proposal that must be considered as a part of a whole energy plan, since the refurbishment or upgrading could not fulfill, in isolation, the growing energy demand.

7. Discussing the main environmental impacts of hydropower

The modern economy's critical dependency on energy underscores the need for a more rational and effective use by society as a whole. Large projects in the energy sector come up against financial, environmental and social restrictions. Regarding hydroelectric plants, these issues are more critical and involve conflicts with various actors: landowners (livestock ranchers), farm workers, traders, the urban and rural population that has to be moved, as well as loggers, indigenous communities, social movements and non-governmental organizations (NGOs). This web of interests makes analysis of these projects complex.

The impact on ecosystems and biodiversity shall also be highlighted. The direct and indirect effects include the alteration of the natural habitat (in this case, largely a change in the freshwater ecosystem), consequently impacting biotic interaction; saturation of adjacent soils; micro-climate alterations; and compartmentalization of habitats (formation of islands in the reservoirs and the segregation of tracts along the transmission lines). Such effects have unpredictable results on biodiversity, which in turn is hard to measure, contributing to the underestimation of environmental impacts in environmental assessments (Sousa Junior & Reid, 2010).

In terms of hydrology, the formation of a reservoir increases the hydrostatic pressure on springs situated along river banks and on rivers that are dammed. Such situation leads to alterations in the natural feeding and draining of aquifers. Consequently alterations to aquifers lead to ecological and economic impacts, as they modify the land use patterns. This has occurred at some hydroelectric plants, requiring the projects to compensate for land that had not been included for expropriation. For instance, according to Muller (1996), in Samuel dam (Brazil), groundwater elevation also resulted in the hydromorphization of about 8000 ha.

Goodland et al. (1993) analyzed various hydroelectric plants in tropical forest regions and identified situations in which such projects should be avoided. These situations include projects in pristine forest regions, places where the local population would have to be removed, areas of species endemism, and areas where there would be a possibility of biodiversity lost, among others.

Other important environmental impacts of the hydropower projects are presented and briefly discussed below:

- **Productive flooding areas**

The flooding area of a hydropower dam can cover productive sites determining losses on agricultural and livestock activities. Cultivated areas can be flooded, with lost net income currently derived from farming and ranching in those areas. Also sites with great potential for tourism can be covered. In these cases, a survey of the immediate added value plus the potential over the period of the reservoir operation would represent the opportunity cost of the activities.

- **Greenhouse gases emissions**

There are also impacts from inundated forest biomass. Not cutting down the forest, in addition to making it difficult to use the reservoir for other purposes, alters the water quality and favors the proliferation of insects, both of which affect public health and human migration patterns. Historically, there have been few cases of pre-flooding forest clearing.

Significant amounts of methane are produced by hydroelectric dams. According to Fearnside (1995), these amounts in some cases can be higher than power plants running on fossil fuels (in terms of carbon equivalent). This carbon is released when the reservoir is initially flooded. After the first decay, organic matter settling on the reservoir bottom decomposes in anoxic conditions, resulting in a build-up of dissolved methane. This is released into the atmosphere mainly by degassing after water flows through the reservoir turbines. A continuous supply of organic matter is provided by the seasonal changes in reservoirs levels, what means that there is a regular flow of methane from them, especially those located on tropical regions.

The precise contribution of hydroelectric reservoirs to GHG emissions is still a matter of discussion. There is controversy, even in the scientific world, as can be seen in the debate that has lasted for over ten years on the methodologies and results of GHG emission estimates for tropical reservoirs in Brazil (Fearnside, 1995; Rosa et al., 1996; Fearnside, 2004; and Rosa et al., 2004). The main point of contention is the accounting of gases, mainly methane, emitted by the hydroelectric plants' spillways and turbines. Methane, concentrated at depths of around 30 meters, is said to be quickly moved at lower pressures and higher temperatures, becoming volatile in contact to the atmosphere (Fearnside, 2004; Kemenes et al., 2007).

Furthermore, greenhouse gases are also released during the production of materials and fossil fuels used on the dams building process. A set of average greenhouse gases emissions from several energy technologies is shown in Figure 4, considering the full operational life cycle of each one.

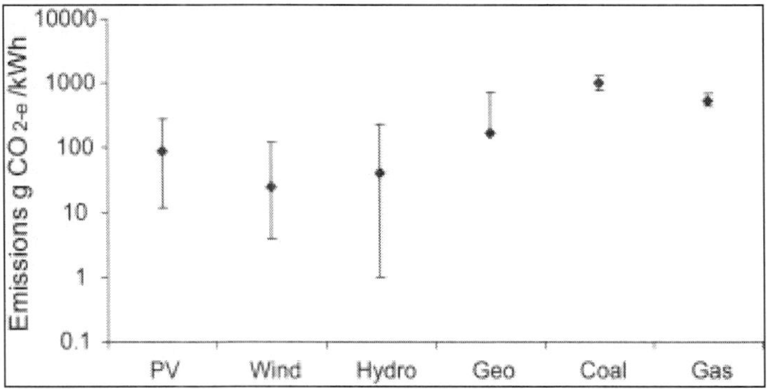

Source: Evans et al. (2008).

Fig. 4. GHG emissions from energy technologies.

Note that, although the hydropower is well positioned, in terms of GHG emissions average, its upper limit reaches values around the lower limit of the Gas technology emissions. Indeed, the range of hydropower GHG emissions is the largest one: it varies from the small run-of-river plants to the big dams emissions.

- **People´s resettlement**

As long as the social aspects are concerned, specifically in relation to the riverside people affected by the undertakings, they are generally disregarded before the perspective of the irreversible loss of their production and social reproduction conditions, established by the formation of the reservoir. The undertakings cause the compulsory displacement of those people and the resettling process, when there is any, doesn´t ensure the maintenance of life conditions that existed before.

The construction of a hydropower plant has often represented the destruction of life projects for those people. It imposes their discharge from the land without offering compensations that could at least ensure the maintenance of their reproduction conditions in the same level as before the implantation of the enterprise. The wearing away of the reservoirs, due to the lack of control of the territorial occupation pattern in its headwaters, is sometimes subject to processes of deforestation and removal of the riparian forest.

Regarding indigenous areas, the main direct and indirect impacts of the construction of large hydroelectric reservoirs are resettlement of communities (affecting lifestyle), flooding of areas (including places of spiritual value), loss of hunting and farming plots, and an increase in infectious disease (Santos and Andrade, 1990).

- **Breaking the fish fauna mobility**

The impact on ecosystems and biodiversity must also be highlighted. The direct and indirect effects include the alteration of the natural habitat (in this case, largely a change in the freshwater ecosystem), consequently impacting biotic interaction; saturation of adjacent soils; micro-climate alterations; and compartmentalization of habitats (formation of islands in the reservoir and the segregation of tracts along the transmission lines). Such effects have unpredictable results on biodiversity, which in turn is hard to measure, contributing to the underestimation of environmental impacts in environmental assessments.

- **Impacting the multipurpose use of the water**

The regulation bodies sometimes face some difficulties to ensure the multiple usage of waters, due to the historical character of prioritizing the electric generation instead of the other possible uses such as irrigation, leisure, fishing, among others. This issue has a political approach in the sense of which group or sector sets the water agenda. In the financial point of view, the electricity generation is generally the most profitable activity among other water uses and when no equitable rule is established, this sector prevails over the others. The constraints could be related to the water level control, the transportation barrier on the rivers or reservoirs, the limitation of the water withdrawals and the imposition of limits to the leisure uses.

- **Impacts from non-dam infrastructure**

Non-dam infrastructure required for the construction of the plant, which includes transmission lines, sub-stations, maintenance areas, and roads, are in fact part of the plant's activities and hence should be considered when analyzing its feasibility. For instance, a 500 kV transmission line in general takes up a space of 65 meters in width. These works very often affect archaeological sites, indigenous villages, forest parks or ecological reserves as much as the plant itself.

Box 1 brings a case of energy expansion running in Brazil, under environmental criticism.

To increase supply, Brazil began damming Amazonian rivers on a large scale in the mid-1980s. The country turned to the North because nearly all the hydropower potential in the densely populated Southeast had been exhausted by that point. Nevertheless, Amazon dam projects face divisions in public opinion. Industrial projects, particularly energy projects, now face higher standards and scrutiny since the promulgation of new environmental regulations and the advent of stricter environmental licensing procedures. At the same time, the multiple economic and political interests in large projects have limited the impact and efficacy of these environmental procedures. The new planning approach in the country's electricity sector points to the need for socio-environmental evaluation at the stage when potential projects are being compiled in inventories, long before specific projects are in advanced planning stages. We would add that to enable this sort of pro-active planning, old inventories of priority projects in the Amazon need to be discarded in favor of up-to-date lists that reflect a more comprehensive and holistic vision of energy development, particularly in undeveloped watersheds like the lower Xingu, where is building up the Belo Monte Hydropower Complex.

Source: Sousa Junior & Reid (2010).

Box 1. The Brazilian's electric expansion plan over the Amazon region.

8. Hydropower and social issues

Historically and coincidentally, many hydropower plants are installed in social spaces originally conceived by riverside people for them to produce their subsistence through fishing and agriculture. The projects for the construction of hydropower plants end up occupying the spaces for social/cultural reproduction of land owners and non-owners alike (sharecroppers, tenants, holders, wage earners, etc.) and determining the beginning of conflicts. The essence of which is the seizure of the geographical space as a form of specific commodity for hydroelectric power generation; and the social and socio-cultural reproduction use as a way of life. On the one hand, the entrepreneurs try to hide or muffle conflicts, trying to go on with their projects and using essentially economic criteria. On the other hand, the affected people, along with religious and environmental authorities, try to make the conflicts evident, showing that certain rights are not being considered. They use essentially environmental, social and humanitarian criteria.

It is a logic that invades regions that are not totally included in the market economy and that supposedly need incentives for their inclusion. The hydropower plants are directed towards the development of extensive territorial areas that have not been included in the market economy yet. Besides, the same logic will only be conceived when the invaded space offers conditions for capital reproduction and exploitation of the natural space as a commodity: "*The projects identify entire regions, very extensive basins, rich meadows transformed in energy mines*" (Vainer & Araújo, 1992, p.71). As a rule, the regional development programs assume that the region has some kind of ability for hydroelectric installation to become feasible.

There is a great range of issues that involves hydroelectric projects. Another aggravating factor is the difficulty of participation of those interested in the decision making process about the installation of the undertaking or not. The involvement of society in the issues related to hydroelectric installation is limited, when it exists at all.

The predominance of a reductionist and hegemonic conception determines that the ways of life and the forms to use natural resources act according to the market logic and that they prevent the communities affected by dams from being acknowledged as *"subjects that are active and have discussion and deliberation margin"* (Zhouri et al., 2005, p.98-9).

The non-identification of the subjects and their interests, their history and culture by the investing agent, actually constitutes a previously defined element to conceive the invisibility phenomenon for riverside people. According to Leroy (2002, p. 9), *"for the government, the multilateral Banks, the construction companies and the consultants that elaborate Environmental Impact Studies, they don't exist"*, and, since they don't exist, they are not considered in the decision making process, and their interests and proposals are not taken into account. Using the invisibility strategy means denying the rights and duties of the investing agent himself in relation to families and riverside communities and cities. Operating the invisibility resource means not to observe the existence of subjects, cultures, developed social organization, building and being rebuilt in the identified area while fit to receive the hydroelectric plant. It ends up favoring the involuntary displacements of people and the withdrawal of families from agricultural work to try to encourage the regional development.

Using invisibility as a tool to control the costs of investments and non-acknowledgement of the social groups historically constituted in a certain region, reduces the range of politics as a field for negotiations and possibilities, although it doesn't mean the non-existence of both social and environmental problems.

9. How risky can the hydropower investments be?

Several variables can affect the economic feasibility of a hydropower project. Nevertheless, these variables are not taking into account when projects are launched. These include the construction time, real costs and the future prices of energy. Since the analysis does not include all categories of social costs (due to data or information constraints) and risk factors, it has almost certainly underestimated their possible total value. Therefore, we cannot conclude decisively whether the project passes the basic test of economic feasibility - a net present value greater than zero. Generally these factors are more important, and relevant, in large dams projects, although they can affect projects at any scale.

Cost overruns and delays are clear factors that impact the feasibility of large infrastructural projects. The WCD report (WCD, 2000) cites an average cost overrun of 56% in a review of 81 large dams worldwide. Though the variation is wide - the worst cases are from India - the numbers show that planning and technical difficulties are endemic to large dams. Another important and typical problem is corruption in public contracts, particularly when these processes lack transparency, which has been the case of many Brazilian projects, notably Itaipú, the world's second largest hydroelectric plant. McCully (2001) presents various cases in which dams were delayed and fraught with corruption. The author cites the Itaipú dam as the worst case of cost overruns. Schilling and Canese (1991) estimated that an amount of around $20 billion was spent on the project, while the original budget was $3.4 billion.

Technical difficulties often delay dam construction and decrease economic returns by delaying the onset of revenues. The WCD report (cited before) showed that of 99 projects studied, 50% were completed on time, 30% with delays of one to two years, and 15% with delays of three to six years. Four other projects had delays longer than 10 years. The Tucuruí dam, another Brazilian case, was slowed for over nine years due to financing and other difficulties.

Another critical issue when analyzing the financial risk of these projects is related to the use of the discount rate in the benefit-cost analysis (see Box 2).

The value of time in benefit-costs analysis can make a great difference in terms of weighting short- and long-run values. Some authors advocate the use of minimum values when analyzing sustainability. Fearnside (2009) proposes evaluations for a period of 100 years with discount rates at 1%/y, in cost-benefit analysis of tropical forest carbon storage projects. Row et al. (1981) propose 4% as the discount rate for long-term forestry projects. The financial assessment generally uses 10%/y as proposed by the World Bank (Belli et al., 2001).

High discount rates tend to overestimate the short-term values. Environmental conservation projects, when analyzed with high discount rates become less attractive to traditional investors. Similarly, infrastructural projects, when analyzed at discounted rates around 12%/y can appear unattractive unless costs are spread over the first 10 years. When costs are concentrated in the first years of construction, these projects are difficult to justify economically.

Source: Sousa Junior & Reid (2010).

Box 2. Discount rates and their influence on an environmental economics analysis.

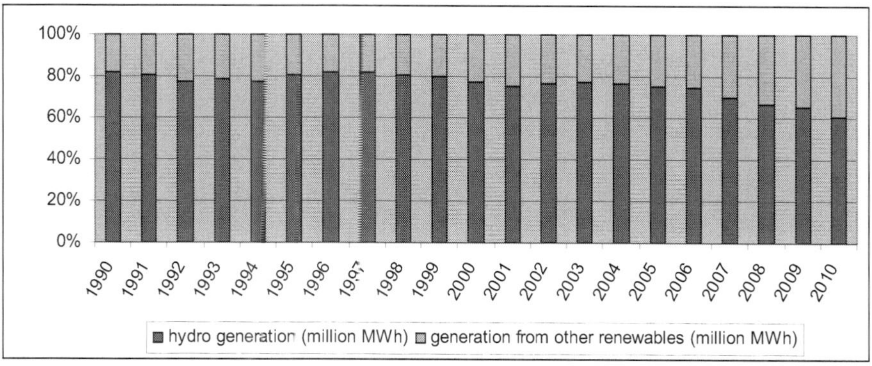

Source: IEA (2011).

Fig. 5. Hydropower generation vs other renewable sources, USA.

In general, investors avoid projects with such an elevated degree of risk. In this sense, private projects are usually more feasible. In the case of public projects, such risks can be assumed for political reasons and spread across the entire base of taxpayers (or utility ratepayers) in the form of subsidies from the public treasury or approval of electricity rates high enough to pay back the construction costs. In most cases governments assume part of the financial risk and transfer it to the general public through tax exemptions and subsidized credit granted to the firms selected to build the dams.

In another direction, as pointed out by Totten et al. (opus cit), many hydropower schemes are at risk from irregular flow regimes resulting from drought and climate change, while increased land-use intensity leads to sedimentation rates that diminish the reservoir storage capacity.

According to IEA (2011), in the USA, despite being the hydroelectric power the main renewable energy source, it is tightly to decline while other sources are going up. The main reason is the hydropower susceptibility to climate oscillations and possible changes (Figure 5).

10. When is hydropower an object of strong or weak sustainability?

One of the most effective indicators to assess the sustainability of projects for power generation is emergy. The term "emergy" can be understood as a combination of the words "embodied" and "energy". Originated from the systems ecology, the term defines the amount of embodied energy over the life cycle of the product or process, taking the solar thermal energy as a primary source of energy to the Earth.

According to Odum (1983), the creator of the concept, emergy *"is the available energy of one kind that has to be used up directly and indirectly to make the product or service."* From the solar transformity (amount of solar energy needed to generate one calorie of a particular good or service), it is possible to establish comparison tables between energy modals and other goods or services in order to support the analysis on the use of one or other modalities. From the solar energy, whose transformity is 1, by definition, the range reaches its maximum point for human services, information and formation of species.

From the standpoint of energy efficiency, a high transformity service or good should not be used for the production of goods or services of lower transformity. As an example, it would be inefficient to use electricity to heat water.

A comparison in terms of emergy, between energy sources is shown in Table 2.

Electric power source	Energy yield ratio
Solar voltaic cell electricity	0.41
Ocean-thermal power plant	1.5
Wind electro-power, strong steady wind regime	2-?
Coal-fired power plant	2.5
Rainforest wood power plant	3.6
Nuclear electricity	4.5
Hydroelectricity, mountain watershed	10.0
Geothermal electric plant, volcanic area	13.0
Tidal electric, 25 ft tidal range	15.0

Source: Odum & Odum (2001)

Table 2. Emergy yield ratios among electric power sources.

As can be seen in Table 2, hydroelectricity is a prime source of high transformation, portability, versatility and flexibility of use. According to Odum (1996), its use should be restricted to activities of greater complexity, ensuring efficient use of natural resources as a principle of sustainability

The terms "weak sustainability" and "strong sustainability" are derived from the concept of sustainable development, which was arisen from the discussions that followed the World Environment Conference in Stockholm, 1972. Assuming the most widespread concept of sustainable development as one which acting on a foundation of efficient usage, meets the current demands without compromising the demands for future generations, it is clear that various actions and initiatives which were taken from this concept, interpreted in a context or localized scale, lead to conceptual distortions when viewed holistically. There is therefore, a considerable difference in human activities in terms of sustainability, which allows the inferences that some of them contribute in a more solid way, "strong" while others contribute in a less cohesive one. "weak" for sustainable development.

According to Neumayer (1999), the fundamental difference between the concepts of weak and strong sustainability is the concept of replaceability of natural capital in both concepts: the first considers this possibility and rejects the second. Nowadays this difference of concepts is one of the pillars of the rift between traditional economics and ecological economics.

The analysis of sources of electricity generation is used to meet certain demands, this distinction may consider sustainability, especially when performed in a broad context, which involves since the offer of resources, the end use of electricity and even the products resulting from this usage, when it comes to production systems. In this sense, it is possible to identify situations where the inclusion of sources of electrical generation meets the assumptions of strong or weak sustainability, in a preliminary analysis. Thus, an electricity generation project is the object of **weak sustainability** if:

- it disregards, or just considers in a little emphatic way, the possibility of generating it by a more efficient power, viewing the use of natural resources and/ or generation or socio-environmental impacts, or even of less value in terms of emergy;
- it generates little or no local benefits and is directly or indirectly attached to a context of widening inequality;
- it is primarily for the productive sector demands, market-based government grants;
- it meets the demands of sectors whose energy usage is wasteful, to which efficient programs could provide the biggest part of the new energy requirements.

On the other hand, an electricity generation project is the object of **strong sustainability** when it:

- is the most efficient source, from the viewpoint of using natural resources and generating social and environmental impacts;
- meets activities which demand energy whose power transformations are smaller than the projected source;
- contributes to reducing inequalities and caters primarily to local needs;
- meets the demands of sectors whose energy usage is efficient, and therefore, the added power will have high marginal productivity

Regarding to hydroelectric generation projects, they can be established on a basis of weak or strong sustainability, depending on how the issues presented are dealt with. In general, the greater the scale of the hydroelectric project is, the more it approaches the weak sustainability criteria presented here. However, this does not mean that small-scale projects such as small hydro power plants are always established based on strong sustainability.

11. Some important assets: the WCD reports and hydropower

The former World Comission on Dams (WCD) has developed an important role on bringing light to the hydropower conflicts.

The main recommendations of the WCD (Dubash et al., 2001) are summarized on Table 3.

Principle	Explanation
Gaining public acceptance	Decision-making processes and mechanisms are used that enable informed participation by all groups of people, and result in the demonstrable acceptance of key decisions. Where projects affect indigenous and tribal people, such processes are guided by their free, prior and informed consent.
Comprehensive options assessment	Alternatives to dams often do exist. To explore these alternatives, needs for water, food and energy are assessed and objectives are clearly defined. The appropriate development response is identified from a range of possible options. In the assessment process, social and environmental aspects have the same significance as economic and financial factors.
Addressing existing dams	Opportunities exist to optimize benefits from many existing dams, address outstanding social issues and strengthen environmental mitigation and restoration measures. Benefits and impacts may be transformed by changes in water use priorities, physical and land use changes in the river basin, technological developments, and changes in public policy expressed in environmental, safety, economical and technical regulations.
Sustaining rivers and livelihoods	Understanding, protecting and restoring ecosystems at river basin level is essential to foster equitable human development and the welfare of all species. Options assessment and decision-making around river development prioritize the avoidance of impacts, followed by the minimization and mitigation of harm to the health and integrity of the river system.
Recognizing entitlements and sharing benefits	Joint negotiations with adversely affected people result in mutually agreed and legally enforceable mitigation and development provisions. Successful mitigation, resettlement and development are fundamental commitments and responsibilities of the State and the developer. Accountability of responsible parties to agree mitigation, resettlement and development provisions is ensured through legal means, such as contracts, and through accessible legal recourse at national and international levels.
Ensuring compliance	Ensuring public trust and confidence requires that governments, developers, regulators and operators meet all commitments made for the planning, implementation and operation of dams. A set of mutually reinforcing incentives and mechanisms is required for social, environmental and technical measures.

Sharing rivers for peace, development and security	Storage and diversion of water on transboundary rivers have been a source of considerable tension between countries and within countries. As specific interventions for diverting water, dams require constructive cooperation. Consequently, the use and management of resources increasingly becomes the subject of agreement between States to promote mutual self-interest for regional cooperation and peaceful collaboration. External financing agencies support the principles of good-faith negotiations between riparian States.

Table 3. WCD's recommendations for hydropower projects.

In addition to the principles established by the WCD, the Commission also presented guidelines for reservoir implementation (WCD, 2000), which are elaborations of the principles. Guideline 11 presents criteria for good economic risk analysis (WCD, 2000). According to Fujikura and Nakayama (2002), this is one of the easiest of the guidelines for governments and investors to implement, given that the analytical tools are readily available and in compliance with the guideline does not entail a final decision on whether or not to build a given dam. Nevertheless, in most cases, especially in developing countries, no risk analysis of any kind is presented to society. Further risk studies could be developed incorporating factors for which data have so far been unavailable. Such contributions would enrich the debate and clarify aspects of the projects which would remain obscure without them.

12. Sustainability and hydropower: is there a real dilemma?

The hydropower is one of the most efficient technologies to generate electricity and it has an important role in the world energy matrix. Even when considering just the renewable sources, the most advanced hydropower techniques are at the top in terms of sustainability. However, the implementation of a hydropower plant has to be analyzed under a comprehensive framework which needs to incorporate the local, regional and global context.

To evaluate the sustainability profile of a hydropower project it is important to consider: i) the macro context in which a hydropower project is conceived, e. g. the energy matrix, the environmental opportunities costs for more efficient alternatives, etc.; ii) the participatory framework of the decision making process; iii) the social context and how the project could address this issue; iv) the physical-chemical-biological problems resulting from its implantation and operation, and its interaction with the environmental characteristics of the place where it is built.

Another important discussion under the sustainability coverage is about the virtual use of water and the ecological dumping. A good part of the hydroelectric power in the world (in absolute numbers) is addressed to the commodities production (mineral and agricultural) in subsidies contexts. This follows the classic pattern of privatizing the benefits and socializing the costs of development and infrastructural projects, especially, but not only, in non developed countries.

Unfortunately, the decision-making process on planning and building hydropower plants has not been as open and participatory as demanded by society. The pervasive costs and problems mentioned above speak for themselves with regard to the necessity of the countries to adopt better standards in the way hydroelectric plants are planned and built. However, the electricity sector generally took shape around a technical bureaucracy that centralizes decision-making to the exclusion of institutions with related interests. The inner circle of institutions linked directly to the electrical sector in the hydropower intensive countries - government bodies, generation companies, and electricity research bodies, as well as regulatory bodies, share decision-making among themselves.

Furthermore, as reported by The Economist (2003), if the World Bank and other international agencies were to far away from financing big dams, many bigger countries would go ahead on their own. When it happens, "it is a racing certainty that their dams will involve more kickbacks and corruption - and that they will ignore the WCD guidelines altogether" (The Economist, 2003). Nowadays, it is factual.

To meet sustainability, the planning process, which is based on management of supply to meet the constant and unmanaged expansion in demand, has to be changed to reach overall efficiency. This includes investments on the demand management and new production arrangements with renewable energy sources. In this context, micro and small hydropower schemes could be interesting whether they are linked to local demands and irrelevant environmental impacts.

13. References

Bermann, C. 2004. Repowering hydroelectric utility plants as an environmentally sustainable alternative to increasing energy supply in Brazil. In: Becker, M. (Ed). Research Report, Vol. X. Brasília: WWF-Brazil.

Cohen, C. 2005. Padrões de consumo, energia and meio ambiente. Discussion Paper No. 185. Rio de Janeiro, Brazil: Universidade Federal Fluminense.

Dubash, N.K.; Dupar, M.; Kothari, S. and Lissu, T. 2001. A watershed in global governance? An independent assessment of the World Commission on Dams. Executive Summary. Washington, DC: World Resources Institute.

Evans, A.; Strezov, V.; Evans, T. J. 2008. Assessment of sustainability indicators for renewable energy technologies. Renewable and Sustainable Energy Reviews, 13(5): 1082-1088.

Fearnside, P.M. 1995. Hydroelectric dams in the Brazilian Amazon as sources of greenhouse gases. Environmental Conservation 22(1): 7-19.

Fearnside, P.M. 2004. Greenhouse gas emissions from hydroelectric dams: Controversies provide a springboard for rethinking a supposedly 'clean' energy source: An editorial comment. Climatic Change 66(1-2): 1-8.

Fujikura, R. and Nakayama, M. 2002. Study on feasibility of the WCD guidelines as an operational instrument. Water Resources Development 18(2): 301-314.

Goodland, R.J.A.; Juras, A. and Pachauri, R. 1993. Can hydro-reservoirs in tropical moist forests be environmentally sustainable? Environmental Conservation 20(2): 122-130.

International Commission on Large Dams, ICOLD. World Register of Dams. Available at http://www.icold-cigb.org/. Accessed on 18 Sep 2011..

International Energy Agency, IEA. Key World Energy Statistics. Available at www.iea.org. Accessed on 10 Jul 2011.

Kemenes, A.; Forsberg, B.R. and Melack, J.M. 2007. Methane release below a tropical hydroelectric dam. Geophysical Research Letters 34(12-809): 1-5.

Leite, A.A.F. and Bajay, S.V. 2007. Impactos de possíveis novos programas de eficiência energética nas projeções da demanda energética nacional. Revista Brasileira de Energia 13(2): 21-33.

Leroy, J. P. 2002. Prefácio. In: Bermann, C. Energia no Brasil: para quê? Para quem? - Crise e alternativas para um país sustentável. São Paulo: Livraria da Física, Fase, p.7-9.

McCully, P. 2001. Silenced rivers: The ecology and politics of large dams. London: Zed Books.

Muller, A.C. 1996. Hidrelétricas, meio ambiente e desenvolvimento. São Paulo: Makron Books.

Neumayer, E. 1999. Weak versus strong sustainability: Exploring the limits of two opposing paradigms. Cheltenham: Edward Elgar Publishing Ltd.

Nilsson L.J. 1993. Energy intensity trends in 31 industrial and developing countries 1950-1988. Energy 18(4): 309-22.

Odum, H. T. 1983. Systems ecology. New York: John Wiley. 644 p.

Odum, H. T. 1996. Environmental accounting, emergy and decision making. New York: John Wiley. 370 p.

Odum, H. T.; Odum, E. C. 2001. A prosperous way down: principles and policies. Boulder, Colorado: University Press of Colorado.

Princen, T. 1999. Consumption and the environment: Some conceptual issues. Ecological Economics 31(3): 347-363.

Renewable Energy Policy Network for the 21st Century, REN21. Renewables 2011 - Global status report. Available at www.ren21.net. Access on 11 Nov 2011.

Rosa, L.P.; Santos, M.A.; Matvienko, B.; Santos, E.O. and Sikar, E. 2004. Greenhouse gas emissions from hydroelectric reservoirs in tropical regions. Climatic Change 66(1-2): 9-21.

Rosa, L.P.; Schaeffer, R. and Santos, M.A. 1996. Are hydroelectric dams in the Brazilian Amazon significant sources of greenhouse gases? Environmental Conservation 23(2): 2-6.

Santos, O.A.L. and Andrade, M.M.L. 1990. Hydroelectric dams on Brazil's Xingu river and indigenous peoples. Cambridge, MA: Cultural Survival.

Schilling, P. and Canese, R. 1991. Itaipú: Geopolítica e corrupção. São Paulo: Centro Ecumênico de Documentação e Informação.

Schipper, L. 2000. On the rebound: The interaction of energy efficiency, energy use and economic activity. An introduction. Energy Policy 28(6-7): 351-355.

Sousa Junior, W.C. and Reid, J. 2010. Uncertainties in Amazon hydropower development: Risk scenarios and environmental issues around the Belo Monte dam. Water Alternatives 3(2): 249-268.

Sunkel, O. 1979. Estilos de desarrollo y medio ambiente en America Latina: Una interpretacion global. E/CEPAL/PROY.2/R.50. Santiago de Chile: Economic Commission for Latin America and The Caribbean.

The Economist. 2003. Damming evidence. 17 Jul 2003.

The World Bank. 2011. GDP growth. Available at: http://data.worldbank.org/indicator. Accessed at 11 Nov 2011.

Totten, M.P.; Killeen, T.J. and Farrell, T.A. 2010. Non-dam alternatives for delivering water services at least cost and risk. Water Alternatives 3(2): 207-230.

United States Energy Information Administration, USEIA. 2011. International energy annual 2006. Available at: http://38.96.246.204/iea/. Accessed at 10 Nov 2011.

Vainer, C. B.; Araújo, F. 1992. Grandes projetos hidrelétricos e desenvolvimento regional. Rio de Janeiro: Cedi.

World Commission on Dams, WCD. 2000. Dams and development: A new framework for decision-making. London: Earthscan.

World Energy Council, WEC. Survey of Energy Resources 2010. Available at www.worldenergy.org. Accessed on 10 Jul 2011.

Zhouri, A. et al. 2005. Desenvolvimento, sustentabilidade e conflitos socioambientais. In: Zhouri, A. et al. (Org.) A insustentável leveza da política ambiental. Belo Horizonte: Autêntica, p.11-24.

3

Hydrological Statistics for Regulating Hydropower

Anders Wörman
The Royal Institute of Technology, Sweden

1. Introduction

This chapter describes the basic statistical tools needed to analyse water availability in the context of hydropower production. The main topics concern the estimation of water availability from autocorrelated and limited records of river discharge and time-series analysis for the purposes of optimising hydropower output and understanding the physical-statistical effects in regulated river systems.

Conflicting water needs exist in various parts of the world, e.g., the use of water in irrigation, hydropower production and municipal consumption or in sustaining natural biological habitats. Freshwater availability has been identified as one of the most severe near-term problems (Niijssen et al., 2001; Alley, 2004; Barnet, et al., 2005; National Science and Technology Council, 2004). Furthermore, to counteract global warming, there is an urgent need to adapt global energy production to use more sustainable energy sources (Frossard et al., 2009; Mo et al., 2006). Globally, hydropower potential is only a limited part of total energy consumption (International Energy Agency, 2010), but it can still play a major role as a regulator of energy production. Renewable energy sources such as wave and wind power or bioenergy are not easily regulated. The intermittent nature of wave and wind power may require special regulation strategies if these energy sources are to make substantial contributions to powering the electrical grid.

A range of factors may limit hydropower potential; these include river discharge and its variation, landscape topography and environmental considerations. Landscape topography controls both the head available, which determines hydroelectric energy yield (see Section 2.1), and the conditions for constructing reservoirs. Additional limiting factors for hydropower potential include technical capacity, e.g., turbine design, limitations of the electrical transmission system, technical flaws (failures) and the functionality of the energy market (demand fluctuations). The implication is that water availability alone does not determine the potential for hydropower nor uncertainties in its estimated potential. This chapter, however, is dedicated to evaluating the statistical basis for analysing hydropower potential by considering river discharge and reservoir regulation, and neglecting these other possible limitations.

The annual mean discharge of rivers limits the overall energy output from hydropower plants. Often, however, discharge records are comparatively short and subject to fluctuations over different periods that may persist for several years. Time series that exhibit such systematic variations in addition to completely random fluctuations are said to be autocorrelated (see Section 3.2). For example, if the data record of an autocorrelated discharge time series is too limited, there is a risk that the record may represent a comparatively dry or wet period that is not representative of the long-term average. Fluctuations in discharge records are generally also characterised by different typical periods. This study is concerned with both annual and short-term river discharge statistics. The uncertainties in their estimation depend on the annual variation, decadal discharge trends and the length of the time series available. Time-series analysis offers methods by which overall trends and cyclic patterns can be analysed. The information obtained from time-series analysis can be used to forecast the future evolution and describe the overall statistical behaviour of river discharge.

2. River regulation for hydropower use

2.1 Physical characteristics of water regulation for hydropower use

The availability of water is primarily controlled by the level of precipitation. However, due to evaporation and deep infiltration, not all precipitation becomes runoff and river discharge. The percentage variation in runoff—the basin discharge coefficient —typically varies widely, ranging from less than 20% to close to 100% (Chow, 1988; Dunne and Leopold, 1978). Furthermore, temporary storage of water in the river basin introduces significant distortion of the precipitation time series when translated into river discharge. In rural watersheds, the water balance can be expressed by the following relationship (Lascano and van Bavel, 2007; Oki and Kanae, 2006):

$$Q_R = A_W(P - ET - D) - dS/dt \tag{1}$$

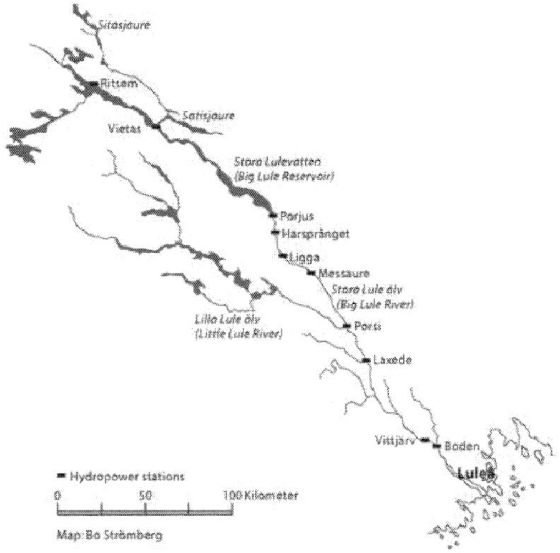

Fig. 1. A map showing the series of hydropower stations along the Luleälven River and the major upstream regulation reservoirs.

Here, Q_R is the river runoff from the watershed (m^3/s), P is the level of precipitation (m/s), ET is the level of evapotranspiration (m/s), A_W is the area of the watershed (m^2), D is the net groundwater discharge (m/s) and dS/dt is the rate of change of water storage in the watershed with respect to time (m^3/s), where S is the amount of storage (m^3). The river runoff Q is the total flow of water that can be used for irrigation, hydropower generation and sustaining biota, especially aquatic ecosystems. The actual discharge available for hydropower production is, therefore, limited by these competing needs and the temporal availability of water.

In addition to discharge, the potential power P (W) of a hydropower plant depends on the average gross head of water assessed for the specific site (i.e., the head of water that can be utilised at the site, m) and the efficiency of the plant η (-) according to

$$P = \eta \, Q_P \, \rho \, g \, h \quad (2)$$

where ρ is the density of water (kg/m^3), and g is the acceleration due to gravity (m/s^2). The gross head, h, is a landscape-specific factor but can be altered by constructing a dam that localises the loss of head at the power plant. A reservoir has the dual purposes of creating gross head and storing water (see Section 2.3). Because hydroelectric turbines have limited capacities, an upper limit to the discharge generally exists, which is much lower than the peak discharge of the river. This implies that a portion of the annual river discharge is a spill discharge, Q_S; the remaining portion is utilised for the generation of hydropower, Q_P. The potential power production is that given by the total discharge, i.e. assuming that $Q_S = 0$.

2.2 Water regulation strategy

The potential for hydropower is usually assessed in terms of overall energy output (in units of joules), the maximum suitable installed power generation capacity (in units of watts) and typical variations in power production. Several factors affect this potential, e.g., fluctuations in demand, limitations in the power transfer grid, river discharge fluctuations and the feasibility of water storage and creating a head fall for energy production. Fluctuation in the demand for hydroelectric production depends on behavioural patterns and the extent to which it is possible to operate a hydropower plant in concert with other production sources such as hydropower plants in the same river basin or other power plants.

Figure 2 illustrates the discharge variations over one year at the Vietas hydropower station on the Luleälven River (see Fig. 1) in northern Sweden. The red shaded area in Fig. 2 indicates the volume of water stored in the reservoir during the warm season, when there is less demand for electricity for home heating, and the blue shaded area indicates the volume of water released during the cold season, which is used for electricity production. The volume of water stored during the period of water surplus varies between years in the interval 3-6 × 10^9 m^3, where the upper limit corresponds to the difference in reservoir volume between the lowest and highest permissible reservoir levels.

From Eqn. (2), the stored energy E can be expressed as $E = \eta \, V \rho g h$, where V is the volume of water. Using this relationship, we calculated the annual energy storage in this single reservoir to be somewhat less than 1 TWh or approximately 1.5% of the total annual hydropower production of Sweden. Without the storage volume of the reservoir, large quantities of water would have to be spilled when they are not directly required for generation, causing a corresponding energy deficiency during the periods when energy is needed.

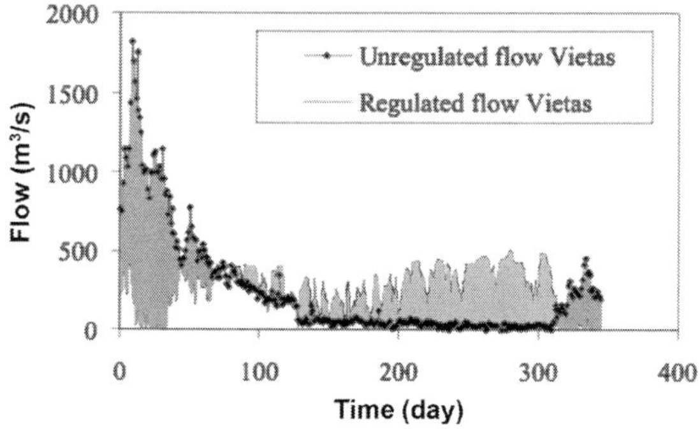

Fig. 2. Example of the flow pattern after regulation of the Luleälven River. The figure shows the observed regulated discharge and the estimated unregulated discharge at Vietas. The period illustrated is from June 20, 2000 to May 31, 2001 (adapted from Jonsson and Wörman, 2005).

This example shows that water availability for hydropower production depends significantly on the timing of 1) energy needs, 2) runoff statistics and 3) water storage. The interactions of such factors are commonly analysed using simulation models for the economic optimisation of power production (see Section 4). Furthermore, water storage in a reservoir affects all hydropower plants located downstream of the reservoir. Such storage generally has only a minor impact on the annual mean discharge (due to its effect on evaporation) but can have a major impact on the temporal fluctuations in water availability and, therefore, on the economic potential of hydropower. Each hydropower plant is associated with a specific time-lag distribution between the release and arrival of discharged water. When water is released from the Suorva Reservoir, there is practically no time lag for the Vietas hydropower station, whereas all 8 units located downstream are subject to various degrees of time lag, with times ranging up to weeks for the entire river (see Fig. 1). The availability of water for power production is, therefore, dependent on the coordination of the power plants and reservoirs within the river basin and, to some extent, on operations occurring between river basins and other energy sources on the grid.

2.3 Energy output

The rate of change of the reservoir volume with time can be expressed as a balance between the river inflow and outflows through power production and spillways as follows:

$$\frac{dV}{dt} = Q_R - Q_P - Q_S \tag{3}$$

Because the reservoir volume is a function of the head, of the form V = A h, the power as expressed by Eqn. (2) depends on both the storage head (i.e., h) and the water discharge used for production (Q_P), where A is the reservoir area. By substituting Eqn. (2) into Eqn. (3) and assuming zero spill (Q_S = 0), we obtain a relationship that can be used to optimise energy output:

$$Q_R(t) = \frac{P(t)}{\eta\rho g}\frac{1}{h(t)} + A\frac{dh(t)}{dt} \quad (4)$$

This is a first-order differential equation with variable coefficients that has no general solution, except for specific variation of the coefficients $Q_R(t)$ and P(t). However, we can derive some insight into the physical-statistical nature of the problem by analysing the special case of a stepwise constant power function, i.e., where P(t) is a constant. For this analysis, we assume no river inflow (Q_R = 0) and, as a compensation, introduce an initial increment in the water level on the form $h(t = t_i+\delta) = h(t = t_i) + \Delta R$, where i is a time step, δ is an infinitesimally small change in time and $\Delta R(t_i)$ is the elevation change due to river inflow; $\Delta R(t_i) = Q_R(t_i)\,\Delta t/A$ and $\Delta t = t_{i+1} - t_i$ (>> δ). Hence, the solution becomes

$$h(t_i < t < t_{i+1}) = \sqrt{2}\left(\frac{(\Delta R_i + h_i)^2}{2} - \frac{P_i}{A\eta\rho g}\Delta t\right)^{1/2} \quad (5)$$

Because the energy output for each time step $E_i = P_i\,\Delta t$, we can express the total energy output as the sum over all time steps N. By evaluating the limit when Δt tends to zero, hence neglecting higher-order terms (containing $(\Delta t)^2$), we obtain

$$E = \sum_{i=1}^{N} E_i = \eta\rho g\left[A\frac{h_0^2 - h_N^2}{2} + \int_0^T Q_R(t)h(t)dt\right] \quad (6)$$

where $T = t_N - t_1$. The second term on the right-hand side of Eqn. (6) follows directly from Eqn. (2), but here we can see that this result is consistent with a fluctuating production, regardless of the h(t) curve described by the regulation.

Because the long-term change of the reservoir level should normally be zero (i.e., $h_0^2 - h_N^2 =$ 0), we can identify the second term as the main source of energy. The integrand represents a product of the arithmetic averages of the discharge Q_R and the reservoir head h plus the cross-correlation between the river inflow and reservoir head. In other words,

$$E = \eta\rho g T\left(E[Q_R]E[h] + Cov[Q_R,h]\right) \quad (7)$$

where E[...] is the expectation operator, Cov[...] is the covariance operator (defined as $Cov[Q_R,h] = E[(Q_R - \mu_{QR})(h - \mu_h)]$), and μ denotes an arithmetic mean value (of Q_R and h, respectively). Consequently, maximising the energy output requires maximisation of both the product of the mean values and the cross-covariance of the discharge and reservoir level. Maximising the cross-covariance means that the frequencies in the river discharge should be followed by corresponding variations in the reservoir elevation. Consequently, relevant discharge statistics in hydropower contexts involve both correlation properties and Fourier spectra.

3. Analysis of discharge time-series and the characteristics of regulated rivers

3.1 Frequency and return period

Frequency and return period are common tools used in the analysis of river discharge (Raghunath, 2006). The probability that a river discharge is lower than the discharge at a certain percentile p is given by

$$P(Q_R < Q_{R,p}) = \int_{Q_R=0}^{Q_{R,p}} f(Q_R)dQ_R \tag{8}$$

in which f is the frequency function. Eqn. (8) can be conceived of as a definition of both the frequency function and the percentile (which is a number between 0 and 1). The frequency function has the inverse units of discharge and, hence, can be defined for discharge data representing different time intervals such as daily or annual maximum discharge (Fig. 3). The frequency function can thus be calculated for data covering a specific time interval, as shown in Fig. 4 for the Dalälven River, Sweden. Typically, the regulation of river discharge implies an alteration of its discharge statistics. Because the Dalälven River has been regulated for the purposes of hydropower production since 1922, when Lake Siljan was first used as a reservoir, a shift in the frequency distributions after regulation commenced is clearly evident.

Given a specific time interval of the data series, the corresponding return period T_R is given by (Chow, 1988):

$$T_R = \left[1 - P\left(Q < Q_p\right)\right]^{-1} \tag{9}$$

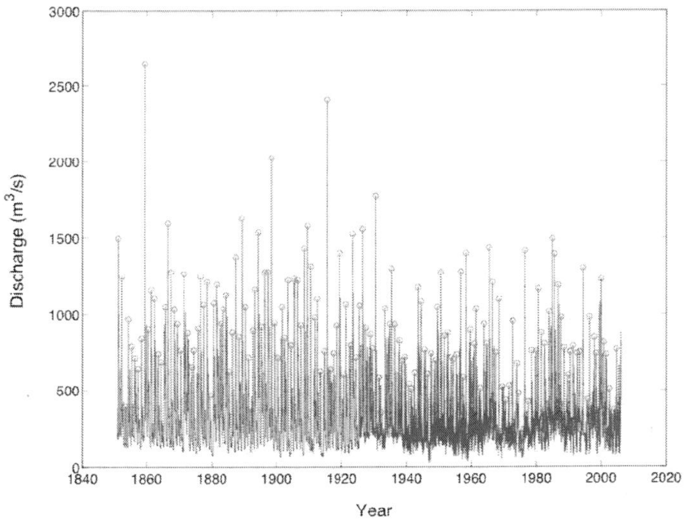

Fig. 3. Discharge time series from the Dalälven River at Fäggeby, Sweden. The blue curve represents daily data, and the red dots represent annual maxima.

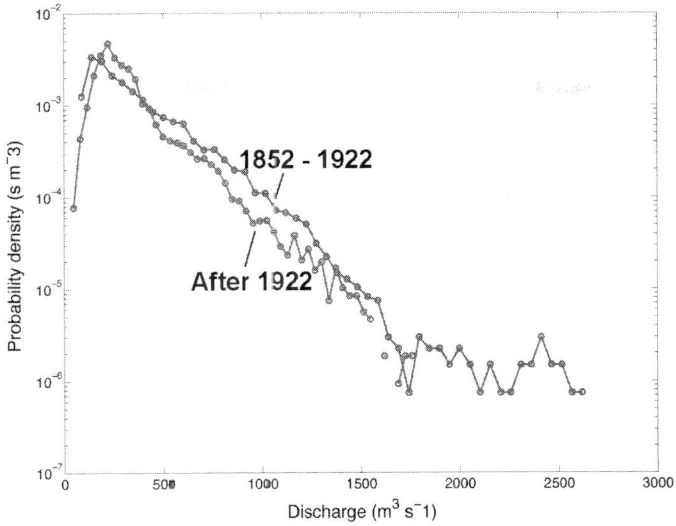

Fig. 4. Frequency distribution for the daily discharge in the Dalälven River at Fäggeby before and after regulation commenced in the early 1920s.

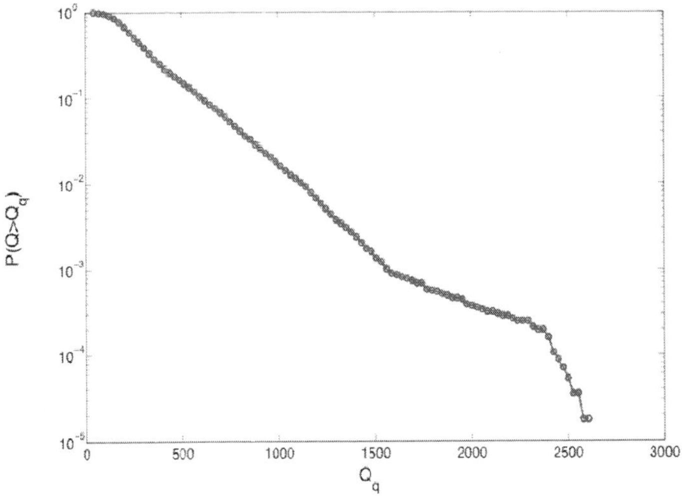

Fig. 5. The probability that the daily discharge exceeds the pth percentile Q_p.

It is also possible to use the probability for the complementary event that the discharge at the pth percentile is exceeded, i.e., $P(Q>Q_p) = [1 - P(Q<Q_p)]$. The maximum Q_p obtained during this long record, starting in 1852, is somewhat more than 2,500 m^3/s. At Q_p = 2,500 m^3/s, the corresponding $P(Q>Q_p) = 5.5 \times 10^{-5}$, and the return period is T = 18,182 days (or 50 years). Consequently, even very long records of discharge data seldom empirically cover the range of return periods considered to ensure the structural safety of dams. Because typical design considerations may assume a return period of 1,000 years or even longer, distributions must generally be extrapolated, and predictions using runoff models (Nash and Sutcliff, 1970; Saleh et al., 2000; Lindström et al., 1997) must also be extrapolated outside of the calibration intervals (Seibert, 2003). The time series may be sufficiently long, however, for the consideration of extreme events when planning hydroelectric production.

3.2 Correlations and the uncertainty of estimates

As discussed in Section 2.2, the cross-correlation between the river discharge and the reservoir level is a key statistical property for the regulation of river discharge. More generally, autocorrelation represents the covariation of properties over time. Not only is this autocorrelation essential for the estimation of statistical measures for time series, but river regulation also specifically aims at altering the auto-covariance in discharge. The auto-covariance in discharge is defined as

$$C_{QRQR}(t,s) = E\left[\left(Q_R(t) - \mu_{QR}(t)\right)\left(Q_R(t+s) - \mu_{QR}(t+s)\right)\right] \tag{10}$$

where s is a time lag. The expectation operator can be obtained either as an integration over a known frequency function $f(Q_R)$ or as a summation over discrete samples (Chatfield, 2004; Chow, 1988).

If the lag s = 0, Eqn. (10) defines the variance, which means that the covariance always adopts the variance V_{QR} at s = 0 and decays with increasing s. As illustrated in Fig. 6, this behaviour is demonstrated for the discharge of the Dalälven River, Sweden in terms of the following autocorrelation:

$$r(Q_R) = \frac{C_{Q_R}(t,s)}{\sigma_{Q_R}(t)\sigma_{Q_R}(s)} \tag{11}$$

where σ is the standard deviation.

Fig. 6 demonstrates that the autocorrelation of the discharge in the Dalälven River is cyclic. The periodicity of the discharge is specific for the geographic area and is of significant importance for the estimation of expected values such as the mean or covariance. If the expectation operation is performed for data obtained during a limited period that is not representative of a long time series, the estimation of the mean values becomes uncertain. In the left panel of Fig. 6, we observe that the autocorrelation function for the daily discharge decreases continuously over one year, reflecting seasonal correlations, whereas for longer time lags, the autocorrelation displays a cyclic behaviour, for which the envelope slowly decays with time over several decades. In the annual average data (the right panel of Fig. 6), there is also a continuous decrease of the autocorrelation function for an initial period that is succeeded by a period with more erratic, cyclic behaviour. The uncertainty of the mean of an autocorrelated discharge time series can be expressed as follows (Ballesta, 2004; Zhang, 2005):

$$u_{QR}^2 = Var[\overline{Q}_R] = \left(1 + \sum_{i=1}^{n-1}\left(1 - s_i/T\right)C_{QR}(s_i)\right)\frac{\sigma_{QR}^2}{n} \tag{12}$$

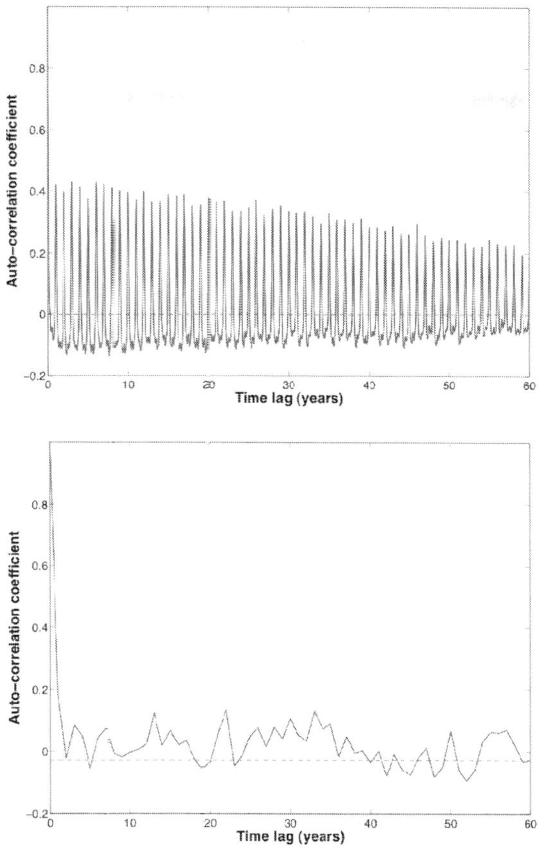

Fig. 6. Autocorrelation coefficients for the daily discharge (left) and the annual average discharge (right) in the Dalälven River at Fäggeby, Sweden.

where $\overline{Q}_R$ is the estimated arithmetic mean, s is the lag in the data from the first data point, and T is the total interval of the data (the duration) used to estimate the mean value. Thus, the total number of samples affects the uncertainty, as does the sampling duration. For a non-autocorrelated time series, the uncertainty of the estimated mean becomes Var[$\overline{Q}_R$] = σ^2_{QR}/n, which is therefore the least possible uncertainty. In a practical management plan for hydropower regulation, the available time series is limited; sometimes, in a statistical sense, the series is too limited for reliable estimation of any of the properties on the right-hand side of Eqn. (12). Such incomplete sampling makes estimation of the uncertainty itself uncertain, as discussed by, e.g., Ballesta (2005).

Due to the inter-annual autocorrelation of discharge, there is a significant fluctuation of the standard deviation of the discharge when it is assessed over a limited time window T. This behaviour can be demonstrated by dividing a multidecadal time series into limited time windows and assessing the statistical properties in terms of the mean and variance for each window (i.e., examining the intra-window properties). The solid curve in Figure 7 represents the coefficient of variation of the annual mean discharge, which increases slightly with the length of the assessment time window. In addition, the bounds of the maximum and minimum values assessed for the intra-window coefficient of variation tend to converge, which reflects an increased accuracy of the assessed coefficient of variation with a longer assessment window (or a greater number of samples). As shown in Fig. 7, the time window would have to be greater than 20 years for the minimum and maximum estimates to converge with a relative deviation from the mean of less than approximately 10% in the daily data series (Syrstad and Fäggeby). For the monthly data obtained at Gilgel Abay, the number of samples in the time window is significantly smaller than in the daily time series. This causes a larger uncertainty in the result. Consequently, the data series can be smoothed (averaged) over a sub-window, e.g., one day or one month, with significant consequences for any statistical property. This type of smoothing is discussed in Section 3.2.

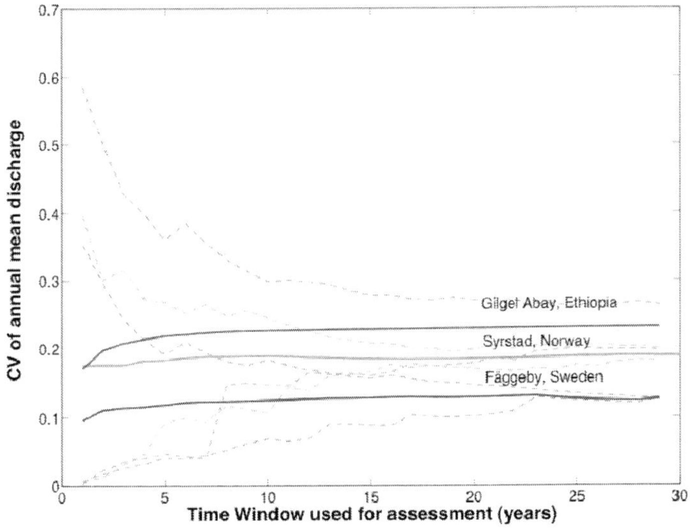

Fig. 7. The coefficient of variation (CV) of annual discharge as a function of the assessment window length at three locations. The dashed curves indicate the lower and upper bounds of the CV assessed for each time window, ranging over the entire available time series. Solid curves indicate the mean of all assessed CVs. Station Fäggeby is located on the Dalälven River, Sweden, and the daily discharge record used here extends from 1852–2006. Station Syrstad is located on the Orkla River, Norway, and the daily data used here extend from 1974–2008. Station Gilgel Abay is located on a tributary of the Blue Nile in Ethiopia, and the monthly discharge record used here extends from 1973–2002.

In addition to smoothing, the uncertainty and auto-covariance in the discharge time series is affected by variations in the underlying meteorological and hydrological conditions. Because the flow network structure in watersheds tends to smooth the highly intermittent (erratic) nature of rainfall, the autocorrelation of discharge increases with the size of the river basin (Wörman et al., 2010; Rinaldo et al., 1991). Figure 8 compares the autocorrelation of a sub-basin for the River Dalälven at Fäggeby with that of an even smaller sub-basin at Ersbo. Typically, the time rate of change of natural variations tend to be smaller with increasing size of the watershed, which is reflected in the increase in autocorrelation with size. Furthermore, the main aim of regulating rivers is to alter natural flow variations; this alteration is manifested both in the autocorrelation function (Wörman et al., 2010) and the frequency function, as discussed in Section 3.1.

3.3 Auto-regressive representation of discharge time-series

Discharge time series can be represented by what are termed auto-regressive, moving average (ARMA) models (Chatfield, 2004). Such models can be used to predict the evolution of time series, but can also be conceived of as a "smoothed" representation of the time series. An auto-regressive (AR) model of order p for a discharge time series can be written as

$$Q_{R,t} = Q_{R,0} + \sum_{i=1}^{p} \alpha_i Q_{R,t-i} + \varepsilon_t \tag{13}$$

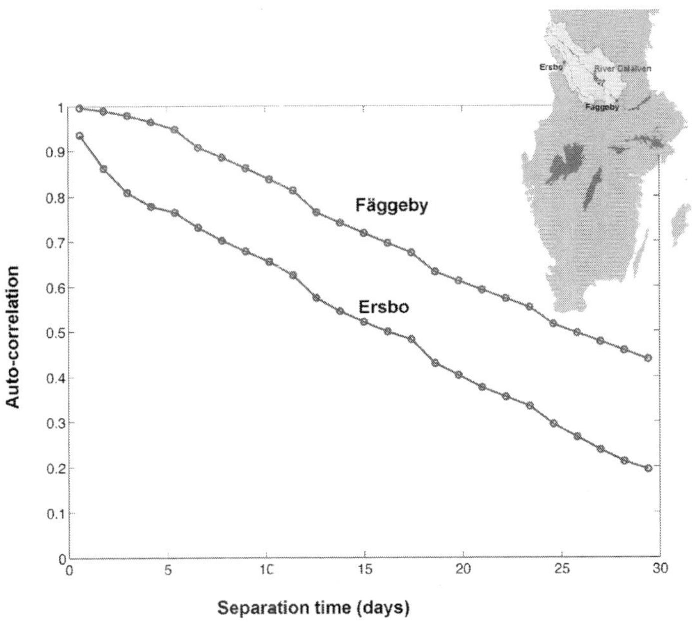

Fig. 8. The auto-correlation function for the discharge at two sub-basins of the Dalälven River.

in which $Q_{R,0}$ is an initial constant, t is a time index (an integer value), the α terms are coefficients and ε is the deviation from the trend curve, i.e., an error. Consequently, Eqn. (13) represents a functional variation of the discharge that carries a memory (i.e., is smoothed) within the time window associated with the order p. In particular, the second two terms on the right-hand side of Eqn. (11) represent the functional trend or the smoothed discharge:

$$\tilde{Q}_{R,t} = Q_{R,0} + \sum_{i=1}^{p} \alpha_i Q_{R,t-i} \tag{14}$$

One example is obtained using constant α-coefficients, $\alpha_i = 1/p$, by which the smoothed discharge is given as an arithmetic average discharge in a moving window defined by the pth order. The individual data points can be represented in relation to the smoothed curve by adding the appropriate error terms to yield the auto-regressive, moving average (ARMA) model

$$Q_{R,t} = \tilde{Q}_{R,t} + \sum_{i=0}^{q} \beta_i \varepsilon_{t-i} \tag{15}$$

This is useful in the stochastic predictive mode of a time series, whereas the error term is numerically generated. For example, a purely random-walk type of discharge representation is obtained for p = q = 0, implying that $Q_{R,t} = Q_{R,0}+\varepsilon_t$. In the representation and interpretation of a discharge time series, we are mostly interested in the smoothed discharge. If we assume a time series with zero autocorrelation, the variance of the smoothed discharge is obtained by inserting $\alpha_i = 1/p$ in Eqn. (14) and it takes the following form:

$$Var[\tilde{Q}_{R,t}] = \sigma^2_{QR,t} \sum_{i=1}^{p} \alpha_i^2 = \frac{\sigma^2_{QR,t}}{p} \tag{16}$$

Consequently, if the auto-regressive representation is based on a single value, the variance of the time series is the same as that for the original data. However, the variance decreases linearly as additional data points are added to the smoothing, as exemplified in Fig. 9 for the discharge of the Dalälven River, which compares the variance of the discharge for a moving one-year average, the annual average and the original daily discharge data. Because of the smoothing, it is easier to discern possible trends in the data compared with the highly erratic daily discharge time series. Fig. 10 illustrates the coefficient of variation in the discharge using two time windows for the averaging: three days and one year. The two smoothing techniques reveal the behaviour of the autocorrelated discharge records and the predominance of different frequencies in the time series. Using the one-year window, we observe a decrease in the discharge variance after regulation commenced in the 1920s. However, the short-term variation in discharge simultaneously increased, and this is more clearly seen when using the shorter, three-day time smoothing window. The need to use different time windows (possibly a large number of windows) to reveal the typical behaviour of discharge time series calls for the use of methods other than ARMA models. In particular, if the variance in the data depends on there being different periods of change in the time series, then the standard deviation on the right-hand side of (16) will depend on the order p. Spectral techniques are specifically designed to study periodic behaviours in time series, as discussed in Section 3.4.

3.4 Decomposing hydropower time-series in spectra

The regulation of rivers for hydropower production implies the alteration of discharge periods rather than the mean or variance of the discharge. The variance can be either positive or negative depending on the time window used for averaging (Fig. 10), which calls for a technique that can specifically differentiate among a distribution of periods in the discharge time series. Spectral techniques offer possibilities for relating typical periods of variation in river discharge, reservoir level (regulation) and electricity production in response to changes in demand. All these time series exhibit typical daily variations due to, e.g., consumption behaviour, and yearly variations due to the seasonality of river discharge. Using a Fourier spectral analysis, we can identify the dominant periods, both short and long, that prevail over a sufficiently long time interval and use this information to statistically evaluate and plan appropriate long-term regulation. However, spectral analysis is not the perfect tool for optimising regulation at a specific point in time in a regulation scheme. Wavelet transforms are also commonly used in hydrological problems (Nakken, 1999) and can be used to identify the variation in spectra over time.

The Fourier transform of a time series f(t) is defined as follows:

$$\bar{f}[k] = \int_{-\infty}^{\infty} f(t)\left(\cos(kt) + i\sin(kt)\right)dt \qquad (17)$$

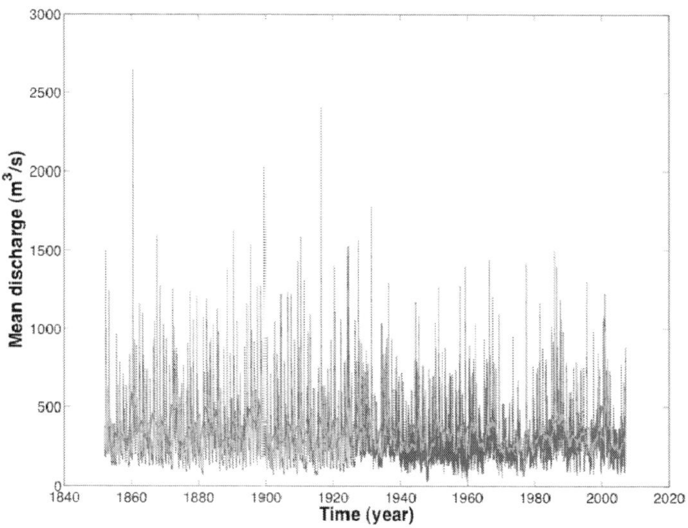

Fig. 9. Mean discharge for the Dalälven River at Fäggeby assessed in three ways. The red curve illustrates the annual average, the green curve illustrates the moving one-year average and the blue curve illustrates the raw daily discharge data.

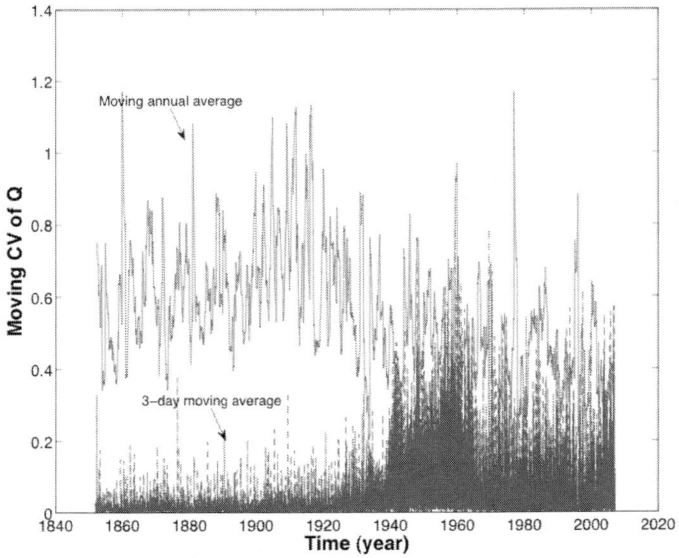

Fig. 10. Coefficient of variation of discharge for the Dalälven River at Fäggeby, assessed as a moving one-year window over daily data and as a three-day moving window of daily discharge data. The coefficient of variation markedly changed after hydropower regulation commenced in the 1920s.

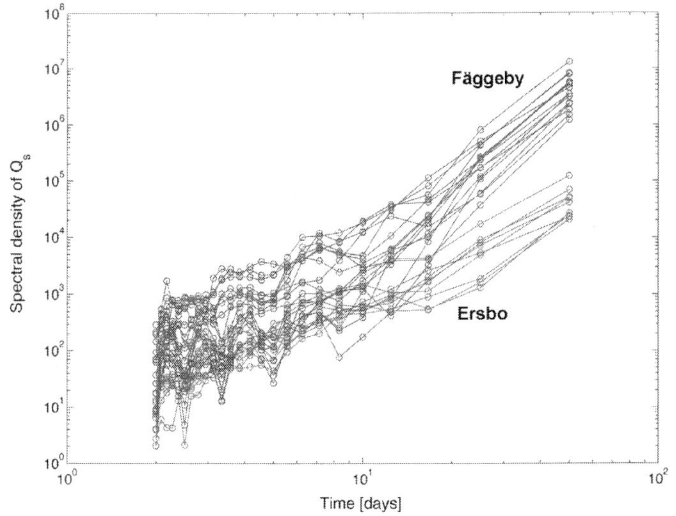

Fig. 11. Spectra derived for various years at two hydrological stations along the Dalälven River. Each curve represents the spectrum from a one-year period distributed along the time series before regulation commenced in the 1920s.

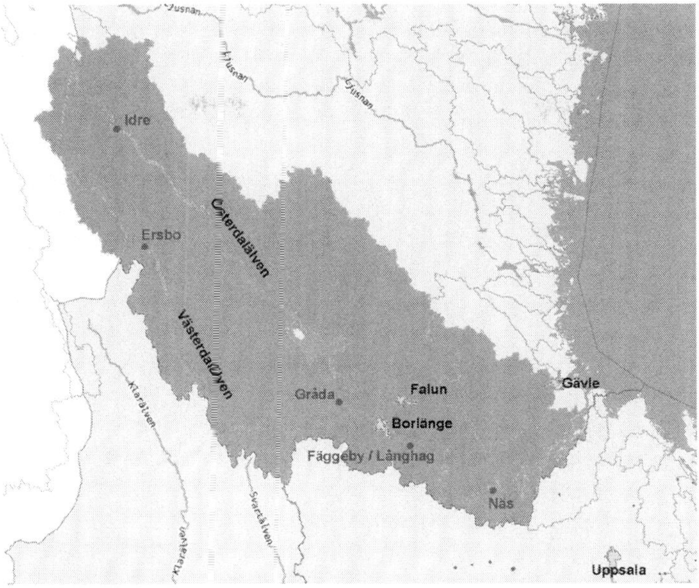

Fig. 12. The Dalälven River Basin and selected hydrological stations. Hydropower plants are located at Gråda, Långhag and Näs.

Here, the wavenumber (frequency) $k = 2\pi/\lambda$, λ is the wavelength (time period), and i is the imaginary part of the complex number. This representation makes it possible to study the variation of the time series as a function of the typical wavelengths of periodic functions rather than as the sum of all periodic functions. Specifically, the power spectrum is given by

$$P(k)=\bar{f}(k)\cdot\bar{f}(k)^{*} \tag{18}$$

where the superscripted star denotes the complex conjugate and is introduced to obtain a real-valued spectrum. The power spectral density is defined as $P(k)/2\Delta k$. Both the power spectrum and the spectral density represent the relative contributions of various wavelengths to the original (real) time series. An advantage, but also a limitation, of the Fourier spectrum is its periodic nature, which makes it possible to directly observe repeated patterns such as self-similarity or fractal distribution in the periodic functions. The power spectrum reflects the relative importance of short and long return periods for the flow and, hence, indicates the degree of periodicity and the randomness of the water-flow time series. A steeper spectrum implies a more organised return-period pattern, whereas a completely flat spectrum reflects white noise (fully random). It is known that river basins have fractal network properties and that this geometric structure consistently organises runoff time series (Rodriguez-Iturbe and Rinaldo, 1997; Snell and Sivapalan, 1994).

Fig. 11 presents the power spectrum of river discharge evaluated at the two stations shown in Fig. 12, Ersbo and Fäggeby. The discharge was evaluated for various years to demonstrate the consistency of the results and the systematic difference in spectral slopes. The slope of the discharge spectrum is significantly steeper for the larger river basin defined by the station at Fäggeby than for the discharge spectrum from the sub-basin defined by the station at Ersbo. This result is consistent with the difference between the autocorrelation spectra shown in Fig. 8. The spectra in Fig. 11 are almost fractal because they follow the power law $P(k) \sim k^a$, where a is a constant exponent. These fractal properties change with systematic alterations of the landscape runoff characteristics and river regulation (Wörman et al., 2010). Previous studies have indicated that climate change may have significant effects on the mean runoff in many parts of the world (e.g., Bergström et al., 2001; Vörösmarty, et al., 2000; Lindström and Alexandersson, 2004). Specifically, physical simulation studies have indicated a long-term relationship between climate change and hydrological time series, which manifests as an alteration of the power spectrum (Blender and Fraedrich, 2006).

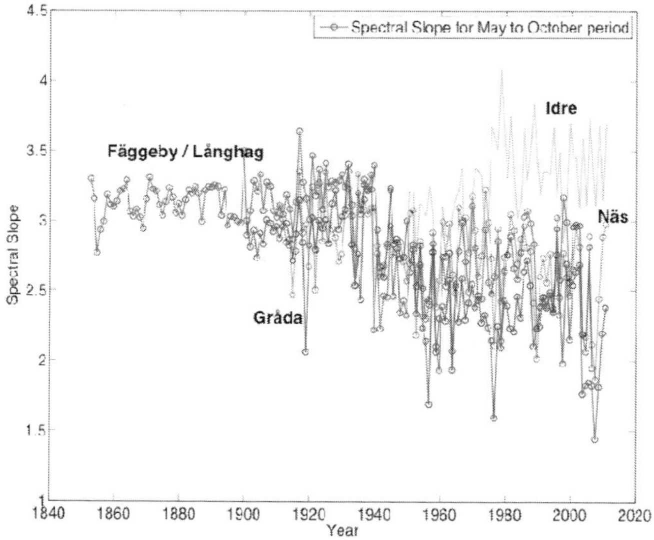

Fig. 13. Evolution of the discharge spectral slopes at selected hydrological stations along the Dalälven River during the 20th century. The spectral slope is evaluated based on one-year data intervals (January to December).

The change in the discharge spectral slopes over time at several hydrological stations along the Dalälven River is illustrated in Fig. 13. Similar to the results obtained during the variance analysis (Fig. 10), the discharge spectra illustrated here demonstrate sudden, significant changes (generally a reduction of the slope) at the time when hydropower regulation was introduced. This behaviour is evident in the results obtained for the hydropower stations at Gråda, Långhag and Näs but is not seen in the results obtained for the station at Idre, which was not affected by regulation. In contrast, at the Idre station, there is a successive increase in the spectral slope, which is generally associated with a change in the drainage capacity of the watershed (Wörman et al., 2010). This led to an inverse relation with time for the discharge spectral slope at Idre compared to those for the stations further downstream.

4. Optimisation of energy output

One aim of hydropower production is to maximise energy output. The estimation of the theoretical potential for hydropower or the optimisation of future production is therefore of interest. The relationships governing the energy output of a river basin hydropower system include the following:

- Power generation as a function of discharge and head, Equation (2)
- The reservoir water balance, Equation (3)
- Information regarding reservoir characteristics and the hydraulic behaviour among hydropower plants
- Meteorological and hydrological statistics
- Power demand statistics (energy demand fluctuations)

As stated in the introduction of this chapter, we are only concerned here with energy output and the forcing due to fluctuations in power use and hydrological conditions. In a market economy, electricity is bought and sold, and power production is generally optimised with respect to the profit generated by industrial activity. This market effect can include price fluctuations based on the availability of power and other factors.

Eqns. (2) and (3) can be combined into Eqn. (4) to represent the impact of the forcing power function and river discharge on optimal reservoir management. On the river-basin level, the power and water mass balance equations are defined for each reservoir using the appropriate time lags in water flow. In a daily optimisation scheme for hydropower production, the future behaviour of the forcing functions due to hydrology and power demand are uncertain, which is why such an optimisation should be based on a stochastic description of the forcing time series. When estimating the (maximum) potential for hydropower, it may suffice to consider only the statistical nature of the time series. For the single-reservoir problem, we can demonstrate that these relationships combine to yield Eqn. (7), which includes the mean value of the discharge and reservoir head, and the covariance of these two time series. Eqn. (10) gives the uncertainty in the estimation of the mean of an autocorrelated discharge time series. The importance of the covariance also follows from the spectral analysis, in which we also see the importance of optimising individual periods of the reservoir elevation with the forcing functions associated with variations in river discharge and power demand.

By applying the Fourier transform to Eqn. (4) and multiplying the transform by its complex conjugate, we obtain the following relationship between time-series spectra:

$$S_{Q_R} = \frac{S_P * S_{1/h}}{(\eta \rho g)^2} + \omega^2 A^2 S_h \tag{19}$$

The convolution between the demand power spectrum SP and reservoir elevation spectrum Sh is linearly proportional to the convolution of the cross-spectrum S_{Ph}, i.e. as $S_{Ph}*S_{1/Ph}$.

5. Conclusion

Statistical methods of hydrology can be used to predict the future evolution of a discharge time series or obtain the typical statistical properties that are important for hydropower planning such as a smoothed representation of the discharge time series, its correlation structure and spectral properties. Typical effects of hydropower regulation include a smoothing of the variance in discharge, which leads to a narrower frequency function, longer periods of significant autocorrelation (Wörman et al., 2010) and a lower slope for the power spectrum. The frequency function (Fig. 4) shows an increase in the peak discharge and peak frequency because the flow regulations tend to prefer discharges close to the mean. The higher curvature of the declining limb implies a smaller spreading of the frequency distribution in the middle range of the discharges but can also indicate a flatter and longer tail. Simultaneously, the power spectrum becomes flatter due to regulation (Fig. 13), which means that the discharge is statistically more random than before regulation. The randomness of the discharge time-series increases as the size of the sub-catchment decreases.

The covariance between discharge and reservoir elevation can be identified as a specific contribution to the energy output from a hydropower plant (Eqn. (7)). Thus, the optimisation of hydropower regulation requires consideration of the cross-correlation between river discharge and reservoir elevation. This consideration can be addressed by relating the spectra for reservoir elevation and river discharge to the spectra for power production or demand (Eqn. (19)).

Global awareness is increasing that the predominant use of fossil fuel is a main driver of global warming and that an increase in the use of renewable energy sources is required (International Energy Agency, 2010). However, renewable energy sources such as wind, waves and tides are intermittent and require the balancing of production sources to meet demand during periods of low production. Hydropower is primarily renewable and can be regulated, i.e., energy can be stored in regulation reservoirs. Therefore, it is likely that not only will hydropower play an important role as a renewable energy source in the future but also as an important regulator of the energy system.

6. Acknowledgements

The research presented in this thesis was conducted at the Swedish Hydropower Centre (Svenskt Vattenkraftcentrum, SVC). The SVC was established by the Swedish Energy Agency, Elforsk and Svenska Kraftnät, together with Luleå University of Technology, The Royal Institute of Technology, Chalmers University of Technology and Uppsala University.

7. References

Alley WM, Leake SA (2004) The journey from safe yield to sustainability. Ground Water 42:12-16.

Ballesta, P.P., 2005. "The uncertainty of averaging a time series of measurements and its use in environmental legislation", Atmospheric Environment 39 (2005) 2003–2009.

Barnet TP, Adam JC, Lettenmaier DP (2005) Potential impacts of a warming climate on water availability in snow-dominated regions. Nature 438:307-309.

Bergström, S., Carlsson,B., Gardelin, M.G., Lindström, G., Pettersson, A., Rummukainen, M. 2001. Climate change impacts on runoff in Sweden- assessments by global climate models. Climate Research 16: 101 - 112.

Bleander R, Fraedrich K. 2006. Lond-term memory of the hydrological cycle and river runoffs in China in a high-resolution climate model. International Journal of Climatology 26:1547 - 1565., DOI: 10.1002/joc.1325.

Chatfield, C., 2004. "The analysis of Time Series", Chapman & Hall/CRC, London.

Chow, Ven Te, 1988. Applied Hydrology. Mc Graw-Hill Book Company, New York.

Dunne and Leopold. 1978. Water in Environmental Planning. W.H. Freeman and Company, New York.

Frossard, A., de Lucena, P., Szklom, A.S., Schaeffer, R., 2009. "Renewable Energy in an Unpredictable and Changing Climate", Modern Energy Review, 1(22-25).

Jonsson, K. and Wörman, A.. 2005. "Influence of Hyporheic Exchange on Solute Transport in a Highly Hydropower-Regulated River", Chapter in text book "Water Quality Hazards and Dispersion of Pollutants", Eds. Czernusznko W., Rowinski P.M., Springer, USA, ISBN 0-387-23321-0, pp. 185-213.

Lascano, R.J., van Bavel, C.H.M., 2007. "Explicit and Recursive Calculation of Potential and Actual Evapotranspiration", Agron. J. 99:585–590 (2007)

Lindström, G., Johansson, B., Persson, M., Gardelin, M., Bergström, S., 1997. Development and test of the distributed HBV-96 hydrological model, J. Hydrology, 201:272-288.

Lindström G, Alexandersson H. 2004. Recent mild and wet years in relation to long observation records and climate change in Sweden. Ambio XXXIII(4–5): 183–186.

Mo, B., Doorman, G., Grinden, B., 2006. "Climate Change - Consequences for the electric system", Report no. CE-5, Hydrological Service - National Energy Authority, Iceland, ISBN 9979-68-194-2.

National Science and Technology Council (NSTC) (2004) Science and Technology to Support Fresh Water Availability in the United States. (NSTC, Washington D.C.).

Nakken M. 1999. Wavelet analysis of rainfall-runoff variability isolating climatic from anthropogenic patters. Environmental Modelling and Software 14: 283–295.

Nash, J. E. & Sutcliffe, J.V., (1970) River flow forecasting through conceptual models part I-A discussion of principles. Journal of Hydrology, 10 (3), 282–290.

Nijssen , B., O'Donnell, G.O., Hamlet, A.F., Lettenmaier, D.P, 2001. "Hydrologic Sensitivity of Global Rivers to Climate Change", Climate Change, 50(1-2), 10.1023/A:1010616428763.

OECD, 2008. "OECD Key Environmental Indicators 2008", OECD, Paris, France.

Oki T, Kanae S (2006) Global hydrological cycles and world water resources Science, 313:1068-1072.

Raghunath, H.M., 2006. Hydrology - Principles, Analysis, Design. New Age Int. Lim. Pub., New Dehli.

Saleh, A., Arnold, J.G., Gassman, P.W., Hauck, L.M., Rosenthal, W.D., Williams, J.R. & McFar-land, A.M.S. (2000) Application of SWAT for Upper North Bosque River Watershed. Trans. ASAE, 45(3), 1077-87.

Rinaldo A, Marani A, Rigon R. 1991. Geomorphological dispersion. Water Resources Research 27(4): 513–525.

Rodriguez-Iturbe I, Rinaldo A. 1997. Fractal River Basin. Cambridge. University Press: United States.

Seibert, J., 2003. Reliability of model predictions outside calibration conditions, Nordic Hydrology, 34: 477–492.

Snell JD, Sivapalan M. 1994. On the geomorphological dispersion in natural catchments and the geomorphological unit hydrograph. Water Resources Research 30(7): 2311–2323.

Wörman, A., Lindström, G., Riml., J., Åkesson, A., 2010. "Drifting runoff periodicity during the 20th century due to changing surface water volume", Hydrological Processes 2010, 24(26), 3772 - 3784, DOI: 10.1002/hyp.7810

Vörösmarty CJ, Green P, Salisbury J, Lammers RB (2000) Global water resources: vulnerability from climate change and population growth. Science 289:284-288.

Zhang, N.F., 2005. "Calculation of the uncertainty of the mean of autocorrelated measurements", Metrologia 43 (2006) S276–S281.

4

Assessment of Impact of Hydropower Dams Reservoir Outflow on the Downstream River Flood Regime – Nigeria's Experience

David O. Olukanni[1] and Adebayo W. Salami[2]
[1]*Department of Civil Engineering, Covenant University, Ota, Ogun State,*
[2]*Department of Civil Engineering, University of Ilorin, Ilorin,*
Nigeria

1. Introduction

Over more than five decades, the energy sector in Nigeria, particularly the rural energy sector, is characterized by lack of access, low purchasing power and over-dependence on traditional fuels for meeting basic energy needs. In an attempt by the government to solving this challenge, the hydropower scheme came on stream as the forerunner in 1968, 1986 and 1990 at Kainji, Jebba and Shiroro respectively. The objective was to improve access to reliable, secure, affordable, climate friendly and sustainable energy services and to boost investment in energy in Nigeria. However, this solution seems to be characterized with some challenges at the downstream sector of the hydropower dams. The communities in the flood plains experience annual flooding when the authorities of Power Holding Company of Nigeria (PHCN) open the gates of the dams to let off water at the peak of the rains. The floods have caused damages and untold hardships to lives and properties. The occurrence of flood has great effect on communities and farming activities downstream of Jebba and Shiroro dams.

Hydro Electric Power (HEP) is one of the few sources of energy that has assumed great significance since the beginning of the twentieth century. Electric power supply in Nigeria is government controlled and operated by the Power Holding Company of Nigeria (PHCN). PHCN has five thermal stations located at Afam, Delta, Egbin, Ijora and Sapele power stations and three hydropower plants located at Kainji, Jebba, and Shiroro hydropower power stations. They have installed capacities of 760 MW, 560 MW and 600 MW respectively and a total output of 1900 MW. The choice of hydro systems to generate peaking power carries a higher economic value of the water resource used and resulting in a substantial increase in the benefits realized. HEP project requires high initial investment cost, but are easy to run and generally have low maintenance cost compared to other sources of energy (Aribisala and Sule, 1998). One major reason that makes HEP attractive is that water, like wind and sun, is a renewable resource and is sustainable through the hydrologic cycle. This chapter presents an assessment of the impact of hydro electric power dams' reservoir outflow of Kainji, Jebba and Shiroro dam on the environment and the mitigation measures. Figure 1.1 displays the World Map showing the location of Nigeria while Figure 1.2 shows location of the three Hydro electric power dams with flood plain hatched.

1.1 Effect of reservoir outflow of dams

1.1.1 Effect of reservoir outflow of dams to the environment

The operation of HEP dams often leads to environmental and ecological problems. When inflows are low, energy output from HEP sources is limited. Water may not be released in adequate quantities from the reservoir, a situation that can affect ecological balance of the river below the HEP dam. On the other hand discharge from HEP dams can entail large water outflow which can cause flooding to adjoining lands downstream of the dam, where the flood plains are regions of economic, social and agricultural activities extensive damages will be incurred in the process. In Nigeria this is particularly so, as the riverbanks are used for farming and are inhabited by farming communities. The operation of hydropower dams in Nigeria has been based on conventional water release rule instead of using scientific analysis to determine the reservoir regulation policies (Sule, 2003). This has led to lack of proper water release plan.

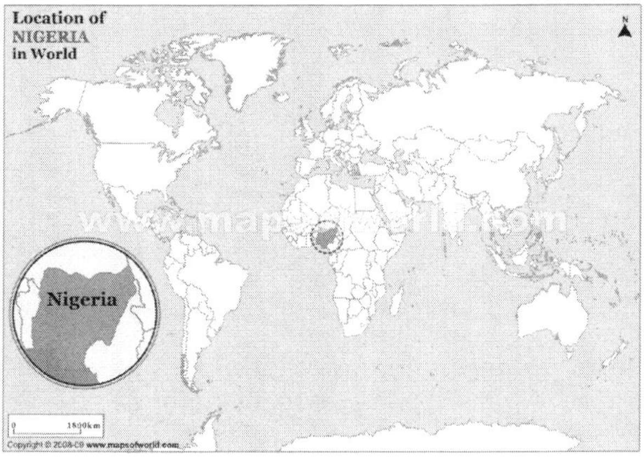

Fig. 1.1 Map of the world showing the location of Nigeria.

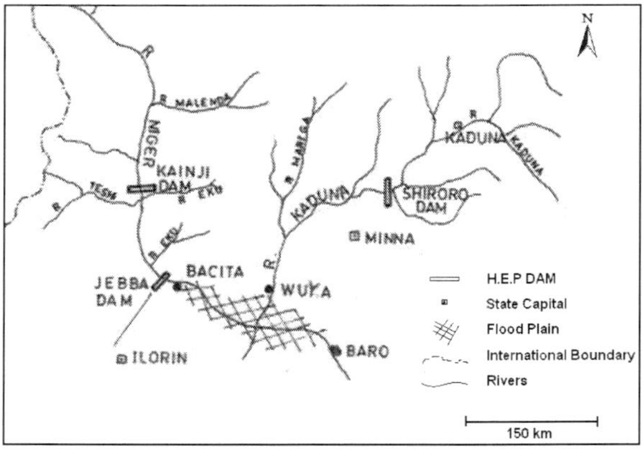

Fig. 1.2 Location of the three Hydro Electric Power Dams and flood plain hatched.

1.1.2 Effect of outflow of Kainji H.P dam on the downstream of Jebba H.P dam

The flow regime of the River Niger downstream of Jebba dam is governed by the operations of the Kanji and Jebba hydroelectric power schemes and runoff from the catchments. Releases from Kainji HEP dam constitutes the major inflow into Jebba HEP dam since it lies directly under it, this mean that the more the release from upper reservoir the faster the downstream reservoir fill up and excess will be discharged thereby leading to flooding. The annual 'white floods' event which usually sets in July and peaks in September does not maintain the same frequency as almost every four years the flood sets in with greater velocity. This has lead to the dam overflowing its banks. When the authority of PHCN opens the spillway gates of the dams to let out torrents of water, it creates flooding and other social impact on the downstream communities and projects. As reported by Lawal and Nagya (1999), the occurrence of flood at Mokwa and its environment in Niger State in 1997 and 1998 destroyed so many properties worth over five hundred million naira and submerged several houses, farmland and crops. The havoc caused by the flooding of the lower Niger in 1998 and 1999 also has its effect on social services to the people of the area (Bolaji, 1999). It has affected the sugar cane plantations, other farming activities and inundated the communities located within the flood plain of the river. In addition, the flooding experienced at the sugar cane plantation in Bacita, Niger State was due to excessive runoff from the catchment areas of the river when reservoirs of both Kainji and Jebba were filled. Over 2,260 ha of sugar cane farm were flooded and remained swamped for over six weeks (Sule, 2003). The flood damaged water conveyance structures, washed away the existing flood protection embankments, impaired roads and caused displacement of settlements and communities along River Niger. A total cost effect of $3.1 million was estimated during the 1994 flood damage to the sugar cane company which increase to $3.7 and $3.3 million in 1998 and 1999 respectively due to the re-occurrence of the same flood effect (Bolaji, 1999). The only alternative left to protect the dams from collapse as at that time was to discharge more water to the downstream areas.

1.1.3 Flood damage at downstream of Shiroro dam

Shiroro HEP dam was built on river Kaduna which forms one of the tributaries to River Niger. After the construction of the dam there were two serious floods in 1985 and 1988. Losses during these periods due to floods were in millions of naira and a large hectarage of arable land with crops submerged. The badly affected area was the Lavun local government area where properties worth millions of naira were destroyed. After 1985 and 1988 flooding, the reoccurrence is more frequent and the damages are higher. As reported by Lawal and Nagya (1999), properties worth over five million naira were destroyed due to the occurrence of flood at Mokwa, Rabba and its environs in 1997 and 1998.

1.2 Highlights on flood control measures

1.2.1 Non-structural measures

This represents an administrative measure of flood plain regulation and management. It involves flood forecasting and flood warning, based on observation of rainfall and river gauge reading in the upstream catchments areas. It is possible to forecast the magnitude and time of occurrence of flood at any downstream point in a river. With modern communication system like the telephone, radio, microwave, radar and artificial satellites, it is then possible to instantly transmit the data observed in the upstream of the catchments area to centrally located flood forecasting stations. The adoption of all flood mitigation measures except flood insurance creates economic benefits by reducing both the expected value of flood losses and the cost of risk taking, and the adoption of flood insurance creates economic benefits by reducing only the cost of risk taking.

1.2.2 Structural measures

The various structural measures prevent inundation of the flood plain in different ways. For example, reservoirs reduce peak flows; levees and flood walls confine the flow within pre-determined channels; improvements to channels reduce peak stages; the flood-ways help divert excess flow. Structural measures alter the stream-flow of rivers and channels resulting in reduction of frequency and severity of floods.

2. Analysis of reservoir inflow data and design of a structural control measure

The existing structural control measures at the stations have been impaired and there is need for redesign. The design of new structural measures would be based on the predicted flood level having 100 yrs return period. The reservoir inflow, turbine discharge and reservoir elevation data were collected from the hydrological unit of the three hydropower stations in Nigeria namely; the Kainji, Shiroro and Jebba hydropower stations. A total of 40 years (1970 - 2010), 27 years (1984 - 2010) and 20 years (1990 - 2010) of inflow data were collected from Kainji, Shiroro and Jebba hydropower stations respectively. The variations of the minimum and maximum reservoir inflow data are presented. The maximum reservoir inflow, turbine discharge and reservoir elevation data were fitted with the Gumbel probability distribution function in accordance to Olukanni and Salami (2008) and Olofintoye et al. (2009) for the prediction of flood of 100 year return period required for the design of flood control structures. This fits the peak and low values of the variables under consideration while the observed and predicted values were plotted. The relationships between the observed and predicted values of the peak and low reservoir inflow are presented. The peak reservoir inflow data were selected for the three hydropower stations and ranked according to Weilbull plotting position. The corresponding return period are determined and plotted against the maximum reservoir inflow data in order to fit the best probability distribution for the prediction of future occurrence of the flood.

3. Hydrological assessment

3.1 Statistical analysis

This section involves the statistical, time series, flow duration curve and probability distribution analyses of the hydrological variables collected at the three hydropower stations. The statistical analysis carried out covered descriptive statistics (i.e. the estimation of the mean, standard deviation, skewness coefficient, minimum and maximum values of the variables). The statistics of the reservoir inflow, turbine discharge and reservoir elevation for Kainji, Jebba and Shiroro are presented in Table 3.1, 3.2 and 3.3 respectively.

3.2 Time series analysis

A time series is plotted for maximum and minimum values to depict the variations of the hydrological variables such as reservoir inflow, turbine discharge and reservoir level. The monthly and annual trends of the maximum and minimum variables were determined.

3.2.1 Kainji hydropower dam

3.2.1.1 Reservoir inflow

The summary of statistics of reservoir inflow at Kainji dam is presented in Table 3.1a. During the 40 years of operation (1970 - 2010), the peak reservoir inflow was 3065.0 m^3/s, while the lowest reservoir inflow was 9.4 m^3/s. The peak value occurs during the month of September in 2000, while the low flow occurred during the month of June 1972. The monthly and annual variation of the reservoir inflow is presented in Figures 3.1a and 3.1b respectively.

Month	Jan	Feb	Mar	Apr	May	Jun	Jul	Aug	Sep	Oct	Nov	Dec
Mean	1319.52	1119.23	506.87	202.26	91.48	130.65	344.23	1020.23	1749.11	1227.54	1350.90	1400.37
S.D	457.68	599.56	457.88	266.50	85.75	107.94	159.72	373.99	584.28	803.42	349.24	386.40
Skew	-0.26	0.22	2.00	3.20	2.39	1.77	1.27	1.34	0.51	-0.25	-1.31	-1.13
Max	2157.00	2340.00	2251.00	1405.00	455.00	522.00	866.00	2267.00	3065.00	2617.00	1801.00	1962.00
Min	468.64	280.55	81.55	13.00	13.46	9.40	97.00	455.00	807.03	29.12	510.74	527.77

Source: Kainji Hydroelectric Power Station (2010)

Table 3.1 a) Statistics of the reservoir inflow at Kainji Hydropower dam (m^3/s) (1970-2010).

where:
Mean= Average value; S.D=Standard deviation; Skew=Skewness coefficient;
Max= Maximum; Min=Minimum

The trend in Figure 3.1b is that reservoir inflow reached a peak in 2000 and has been reducing slightly since then. This may be due to control releases from the upstream reservoir from neighboring country like Niger Republic. Figure 3.1a indicated two peak seasons, occurring in the months of February and September. The first peak inflow is due to black flood resulting from excess releases from upstream reservoirs from neighboring countries which get to Nigeria during dry season, while the second peak flow is due to white flood resulting from excess rainfall within River Niger catchment within Nigeria.

Month	Jan	Feb	Mar	Apr	May	Jun	Jul	Aug	Sep	Oct	Nov	Dec
Mean	784.16	813.77	747.86	793.90	720.41	666.18	632.00	709.35	746.30	752.98	740.50	766.69
S.D	232.94	269.72	234.07	232.14	195.55	171.33	197.99	257.17	259.44	252.30	208.79	232.56
Skew	0.14	0.45	0.34	0.18	0.30	-0.10	0.00	0.45	0.69	0.52	-0.10	0.53
Max	1234.43	1431.96	1203.41	1345.23	1176.13	1026.62	1060.00	1445.34	1396.60	1289.02	1248.93	1401.94
Min	377.06	405.22	404.27	416.17	404.79	337.31	206.42	203.32	300.60	380.23	198.27	382.77

Source: Kainji Hydroelectric Power Station (2010)

Table 3.1 b) Statistics of the turbine discharge at Kainji Hydropower dam (m^3/s) (1970-2010).

Month	Jan	Feb	Mar	Apr	May	Jun	Jul	Aug	Sep	Oct	Nov	Dec
Mean	140.88	140.85	139.94	138.27	136.37	134.52	133.79	134.50	137.39	139.00	140.00	140.57
S.D	0.94	1.00	1.41	1.79	1.95	2.12	1.63	1.43	1.92	1.87	1.47	1.12
Skew	-1.28	-1.17	-0.66	-0.39	-0.11	0.04	-0.26	0.49	0.25	-0.93	-1.00	-1.11
Max	141.89	141.89	141.90	141.10	139.58	138.27	136.89	137.80	141.23	141.61	141.70	141.72
Min	138.49	137.96	136.30	134.22	132.95	130.28	130.33	131.76	133.96	134.16	136.49	137.82

Source: Kainji Hydroelectric Power Station (2010)

Table 3.1 c) Statistics of the reservoir elevation at Kainji Hydropower dam (m) (1970-2010).

Month	Jan	Feb	Mar	Apr	May	Jun	Jul	Aug	Sep	Oct	Nov	Dec
Mean	1064.00	988.20	903.00	913.80	823.52	834.60	757.72	1055.44	1637.00	1642.20	1002.16	1065.76
S.D	320.32	340.85	304.11	276.06	241.37	259.96	288.46	453.32	705.46	942.13	338.30	265.55
Skew	0.15	0.09	0.06	0.30	0.19	0.02	1.17	0.92	0.79	1.08	0.55	0.11
Max	1575.00	1637.00	1422.00	1566.00	1282.00	1332.00	1567.00	2379.00	3182.00	3636.00	1688.00	1565.00
Min	518.00	378.00	417.00	436.00	428.00	359.00	378.00	445.00	750.00	666.00	516.00	610.00

Source: Jebba Hydroelectric Power Station (2010)

Table 3.2 a) Statistics of the reservoir inflow at Jebba Hydropower (m^3/s) (1984-2010).

Month	Jan	Feb	Mar	Apr	May	Jun	Jul	Aug	Sep	Oct	Nov	Dec
Mean	1092.40	1019.65	930.00	922.54	860.04	824.15	767.27	997.15	1319.23	1314.31	1037.65	1014.57
S.D	308.34	351.84	305.81	312.02	237.47	250.00	288.14	407.05	391.22	433.39	336.48	278.64
Skew	0.02	-0.04	0.08	0.04	0.32	0.03	0.90	0.34	0.05	0.30	0.21	0.22
Max	1575.00	1643.00	1466.00	1672.00	1383.00	1340.00	1556.00	1927.00	2079.00	2143.00	1655.00	1606.00
Min	585.00	376.00	425.00	232.00	451.00	362.00	328.00	366.00	633.00	685.00	479.00	514.00

Source: Jebba Hydroelectric Power Station (2010)

Table 3.2 b) Statistics of the turbine discharge at Jebba Hydropower (m^3/s) (1984-2010).

Month	Jan	Feb	Mar	Apr	May	Jun	Jul	Aug	Sep	Oct	Nov	Dec
Mean	102.20	102.13	102.11	101.90	101.95	101.87	101.53	101.71	102.16	102.40	102.14	102.08
S.D	0.45	0.52	0.52	0.65	0.62	0.65	0.57	0.56	0.48	0.55	0.73	0.52
Skew	0.13	-0.93	-0.18	0.01	-0.99	-0.76	0.32	0.25	-0.25	-1.33	-1.04	-0.14
Max	102.87	102.87	102.98	102.98	102.90	102.91	102.65	102.74	103.02	103.05	103.00	102.78
Min	101.52	100.65	100.86	100.60	100.19	100.27	100.76	100.61	101.21	100.82	100.27	101.16

Source: Jebba Hydroelectric Power Station (2010)

Table 3.2 c) Statistics of the reservoir elevation at Jebba Hydropower (m) (1984-2010).

Month	Jan	Feb	Mar	Apr	May	Jun	Jul	Aug	Sep	Oct	Nov	Dec
Mean	43.05	53.04	45.87	29.19	76.67	190.45	434.09	924.50	1043.24	476.63	94.32	68.05
S.D	14.68	79.24	70.00	17.46	46.95	58.77	164.65	187.98	243.34	196.12	35.48	77.04
Skew	-0.11	4.30	4.13	1.67	2.47	0.26	-0.32	1.44	1.12	0.80	0.63	4.11
CV	0.34	1.49	1.53	0.60	0.61	0.31	0.38	0.20	0.23	0.41	0.38	1.13
Max	73.40	385.66	336.02	78.94	245.78	318.79	684.18	1431.29	1752.51	878.17	178.24	387.10
Min	18.97	12.52	10.84	9.87	21.95	97.72	56.49	680.48	627.08	239.51	46.49	26.36

Source: Shiroro Hydroelectric Power Station (2010)

Table 3.3 a) Statistics of the reservoir inflow at Shiroro Hydropower dam (m^3/s) (1990-2010).

3.2.1.2 Turbine discharge

The summary of statistics of turbine discharge at Kainji dam is presented in Table 3.1b. During the 40 years of operation (1970 – 2010), the peak turbine discharge was 1445.34 m^3/s, while the lowest turbine discharge was 198.27 m^3/s. The peak value occurs during the month of August in 1979, while the low turbine discharge occurred during the month of November 2000. The monthly and annual variation of the reservoir inflow is presented in Figures 3.2a and 3.2b respectively. The trend in Figure 3.2b indicated highest discharge value in 1979 and has been decreasing steadily until 1990. The discharge starts to increase again until 1994 after which it start decreasing. The trend exhibited by releases from Kainji has direct influence on reservoir

Month	Jan	Feb	Mar	Apr	May	Jun	Jul	Aug	Sep	Oct	Nov	Dec
Mean	236.99	260.65	255.32	238.34	213.58	231.07	338.20	359.43	339.27	345.93	245.03	225.88
S.D	81.36	71.07	72.44	85.20	79.13	84.18	84.69	140.14	149.38	119.09	121.24	86.59
Skew	0.11	-0.55	-0.30	0.68	0.12	2.41	0.33	1.30	-0.04	-0.26	-0.04	0.04
CV	0.34	0.27	0.28	0.35	0.37	0.36	0.25	0.39	0.44	0.34	0.49	0.38
Max	416.63	390.13	381.98	444.95	407.48	525.85	494.03	792.45	604.63	575.27	504.75	435.86
Min	75.60	99.21	118.17	87.31	35.34	141.63	172.84	127.50	94.33	86.36	22.15	20.80

Source: Shiroro Hydroelectric Power Station (2010)

Table 3.3 b) Statistics of the turbine discharge at Shiroro Hydropower dam (m^3/s) (1990-2010).

Month	Jan	Feb	Mar	Apr	May	Jun	Jul	Aug	Sep	Oct	Nov	Dec
Mean	375.85	373.18	369.98	366.19	362.92	361.28	361.65	410.97	377.23	381.05	380.17	378.07
S.D	2.67	3.00	3.29	3.32	3.03	2.64	2.56	19.32	2.27	1.73	1.79	2.13
Skew	-0.40	0.25	0.65	1.22	1.22	0.20	0.05	-2.08	-0.52	-2.13	-1.68	-1.05
CV	0.01	0.01	0.01	0.01	0.01	0.01	0.01	0.05	0.01	0.00	0.00	0.01
Max	379.99	378.94	377.52	375.29	371.87	366.36	367.80	423.80	380.75	382.20	381.97	381.00
Min	370.43	367.97	365.77	362.28	358.25	355.33	355.44	364.39	372.53	375.56	374.87	372.91

Source: Shiroro Hydroelectric Power Station (2010)

Table 3.3 c) Statistics of the reservoir elevation at Shiroro Hydropower dam (m) (1990-2010).

inflow at Jebba dam. When the releases from Kainji is high, the reservoir inflow at Jebba dam also high, while low releases from Kainji implies low reservoir inflow at Jebba dam. The operation of Kainji dam dictate operation pattern in Jebba dam, excess releases at Kainji will force the reservoir manager at Jebba dam to release so as to accommodate releases from Kainji and thereby causing flooding at the downstream area.

3.2.1.3 Reservoir elevation

The summary of statistics of reservoir elevation at Kainji dam is presented in Table 3.1c. During the 40 years of operation (1970 - 2010), the peak reservoir elevation was 141.90 m, while the lowest reservoir elevation was 130.28 m. The peak value occurs during the month of March in 1973, while the low reservoir elevation occurred during the month of June-July in 2007. The monthly and annual variation of the reservoir inflow is presented in Figures 3.3a and 3.3b respectively. The trend in Figure 3.3b is that the maximum reservoir elevation has remained relative constant since 1970, but lowered in 1978, 1984, 2004, and 2007. The minimum elevation was raised to 136.37 m in 1974 and subsequently lowered to 130.28 m in 2007. This is to allow for more water to be stored in the dam when water is released from the upstream reservoir in the neighboring country or as a result of black flooding. The trend in Figure 3.3a indicated lowest values during the months of June-July and highest value in the month of March.

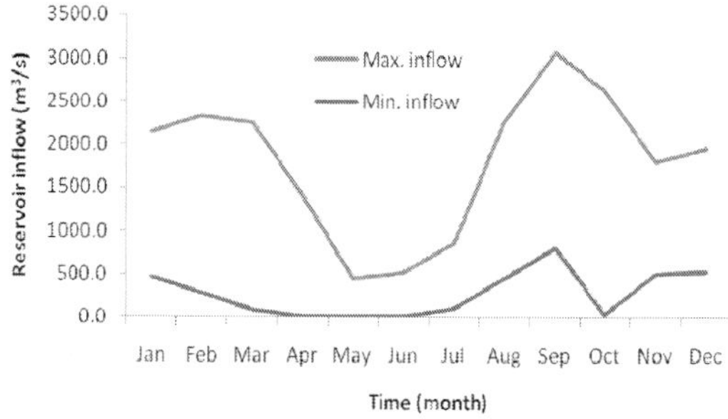

Fig. 3.1 a) Montly variation maximum and minimum reservoir inflow at Kainji HEP dam (1970 - 2009).

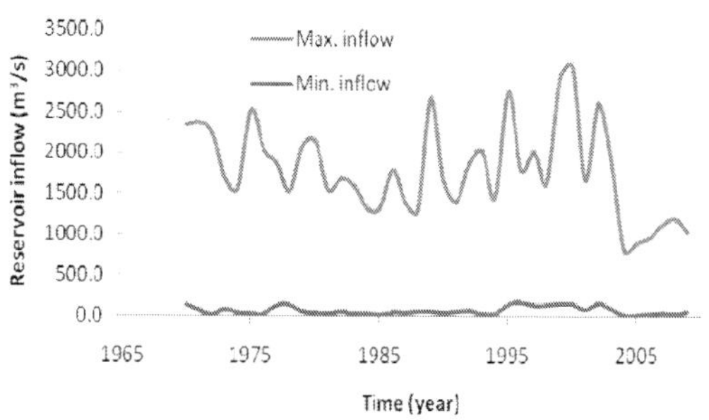

Fig. 3.1 b) Annual variation of maximum and minimum reservoir infloe at Kanji HEP dem (1970 - 2009).

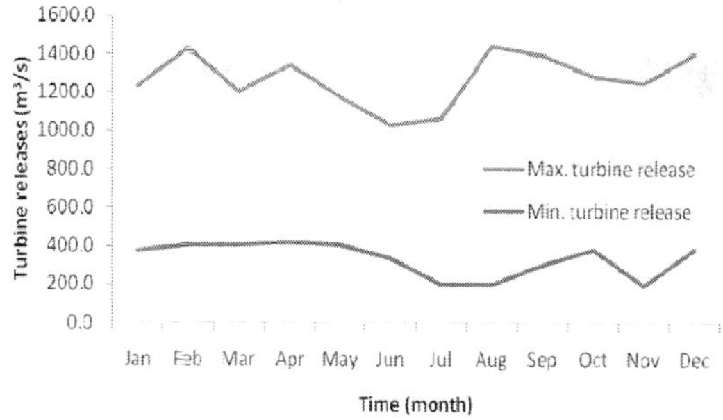

Fig. 3.2 a) Monthly variation of maximum and minimum turbine releses at Kanji HEP dem (1970 - 2009).

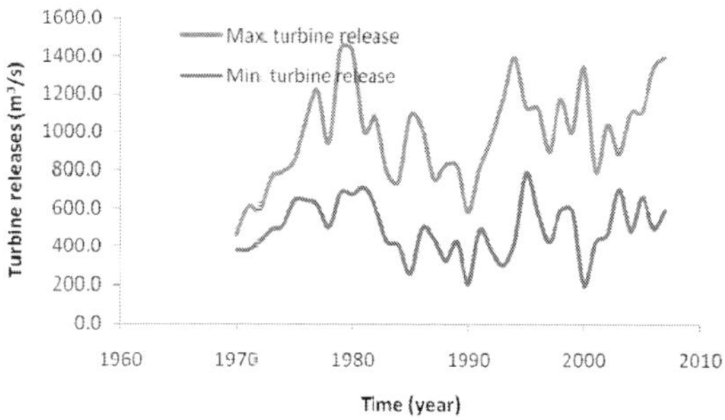

Fig. 3.2 b) Annual variation of maximum and minimum turbine releses at Kanji HEP dam (1970 - 2009).

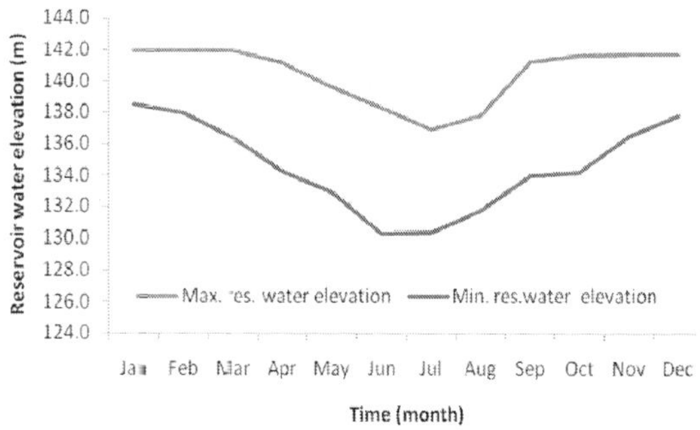

Fig. 3.3 a) Monthly variation of maximum and minimum reservoir water elevation at Kanji HEP dam (1970 -2009).

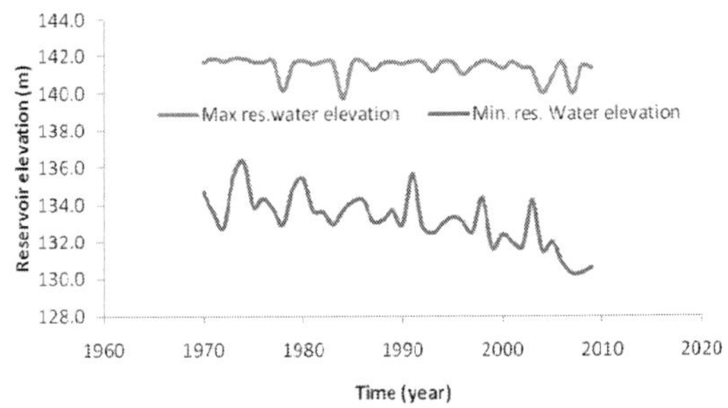

Fig. 3.3 b) Annual variation of maximum and minimum reservoir water elevation at Kanji HEP dam (1970 - 2009).

3.3.1 Jebba hydropower dam

3.3.1.1 Reservoir inflow

The summary of statistics of reservoir inflow at Jebba dam is presented in Table 3.2a. During the 27 years of operation (1984 - 2010), the peak reservoir inflow was 3636.0 m^3/s, while the lowest reservoir inflow was 378.0 m^3/s. The peak value occurs during the month of October in 1998, while the low flow occurred during the month of February 1984. The monthly and annual variation of the reservoir inflow is presented in Figures 3.4a and 3.4b respectively. Reservoir inflow reached a peak in 1998 and has been reducing slightly since then. This is not unconnected with the discharge from Kainji which has also reduced since 1997 but started increasing since 1995.

3.3.1.2 Turbine discharge

The summary of statistics of turbine discharge at Jebba dam is presented in Table 3.2b. During the 27 years of operation (1984 - 2010), the peak turbine discharge was 2143.0 m^3/s, while the lowest reservoir inflow was 232.0 m^3/s. The peak value occurs during the month of October in 2008, while the low flow occurred during the month of April in 1987. The monthly and annual variation of the turbine discharge is presented in Figures 3.5a and 3.5b respectively. The trend in Figure 3.5b is that discharge has been increasing steadily since 1990. Even the lowest release in 2010 was as high as the highest release at the early stages of the dam operation. This means higher likelihood of flooding in recent years compared to pre-1990.

3.3.1.3 Reservoir elevation

The summary of statistics of reservoir elevation at Jebba dam is presented in Table 3.2c. During the 27 years of operation (1984 - 2010), the peak reservoir elevation was 103.05 m, while the lowest reservoir elevation was 100.19 m^3/s. The peak value occurs during the month of October in 1994, while the lowest value occurred during the month of May in 1985. The monthly and annual variation of the reservoir elevation is presented in Figures 3.6a and 3.6b respectively.

3.3.2 Shiroro hydropower dam

3.3.2.1 Reservoir inflow

The summary of statistics of reservoir inflow at Shiroro dam is presented in Table 3.3a. During the 20 years of operation (1990 - 2010), the peak reservoir inflow was 1752.51 m^3/s, while the lowest reservoir inflow was 9.87 m^3/s. The peak value occurs during the month of September in 2003, while the low flow occurred during the month of April 2008. The monthly and annual variation of the reservoir inflow is presented in Figures 3.7a and 3.7b respectively. The trend in Figure 3.7b is that the reservoir inflow reached a peak in 2003 and has been reducing until 2005 when it starts to increase. The first peak value was experienced in 1992 and start decreasing until 2002. The highest peak in 2003 might be due to high rainfall within the catchment of River Kaduna, which is the main source to the Shiroro reservoir. The trends in Figure 3.7a indicate peak inflow during September and low inflow during April.

3.3.2.2 Turbine discharge

The summary of statistics of turbine discharge at Shiroro dam is presented in Table 3.3b. During the 20 years of operation (1990 - 2010), the peak turbine discharge was 792.45 m^3/s, while the lowest turbine discharge was 20.80 m^3/s. The peak value occurs during the month of August in 2004, while the low turbine discharge occurred during the month of December in 2002. The monthly and annual variation of the turbine discharge is presented in Figures 3.8a and 3.8b respectively. The trend in Figure 3.8 is that discharge has been increasing steadily since 1990 until 1998 when it start to decrease. From 2000 it starts to increase again until it get to a peak in 2004 and start to decrease. The fluctuation in the pattern of releases might be connected to reservoir flow pattern. The highest value experienced in some years might lead to flooding at the downstream reaches of the reservoir. The trend in Figure 3.8a indicated that the occurrence of the peak discharge is in the month of August. Hence flooding may be experienced in August annually.

3.3.2.3 Reservoir elevation

The summary of statistics of reservoir elevation at Shiroro dam is presented in Table 3.3c. During the 20 years of operation (1990 - 2010), the peak reservoir elevation was 423.80 m, while the lowest reservoir elevation was 355.44 m. The peak value occurs during the month of August in 1991, while the low reservoir elevation occurred during the month of July in 2009. The monthly and annual variation of the reservoir elevation is presented in Figures 3.9a and 3.9b respectively.

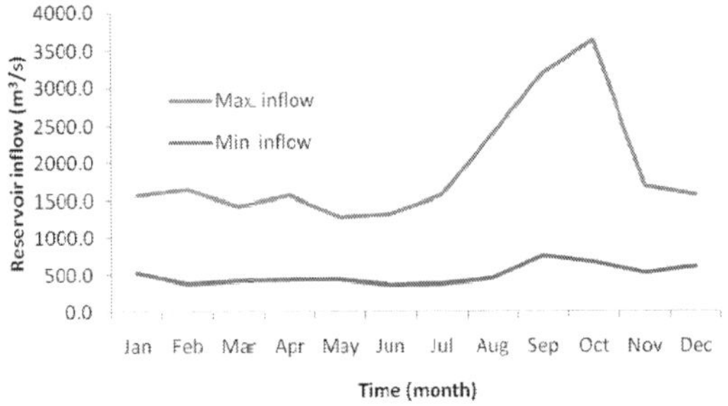

Fig. 3.4 a) Monthly maximum and minimum reservoir inflow at Jabba HEP dam (1984 - 2008).

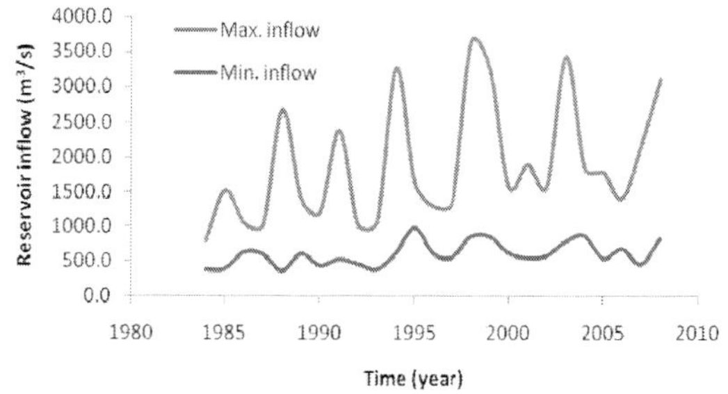

Fig. 3.4 b) Annual variation of maximum and minimum reservoir inflow at Jabba HEP dam (1984 - 2009).

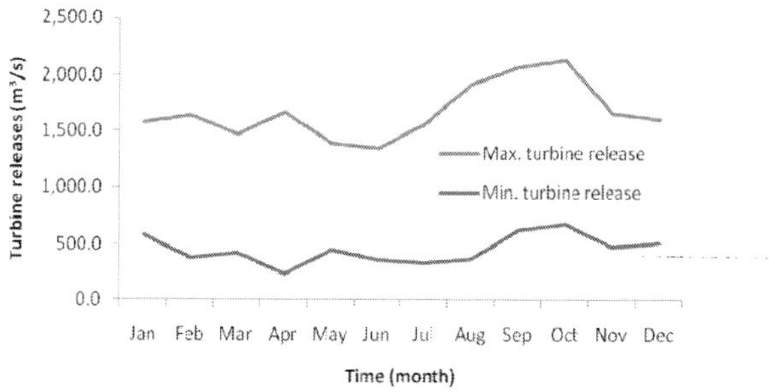

Fig. 3.5 a) Monthly variation of maximum and minimum turbine releases at Jabba HEP dam (1984 - 2010).

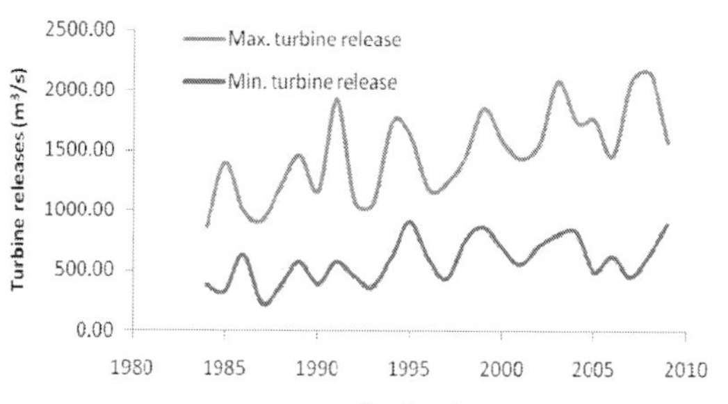

Fig. 3.5 b) Annual variation of maximum and minimum turbine releases at Jabba HEP dam (1984 - 2010).

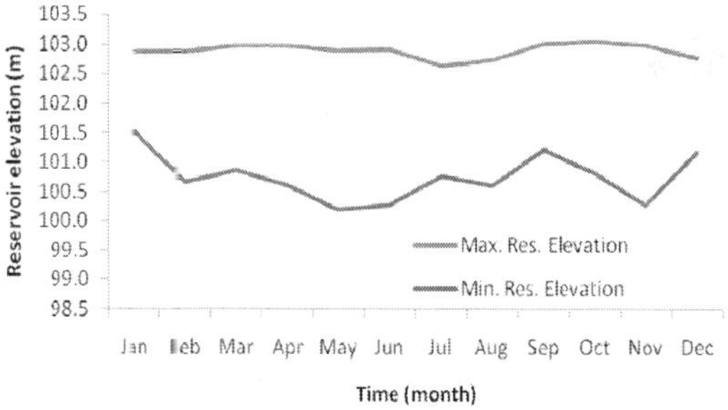

Fig. 3.6 a) Monthly variation of maximum and minimum reservoir elevation at Jabba HEP dam (1984 - 2003).

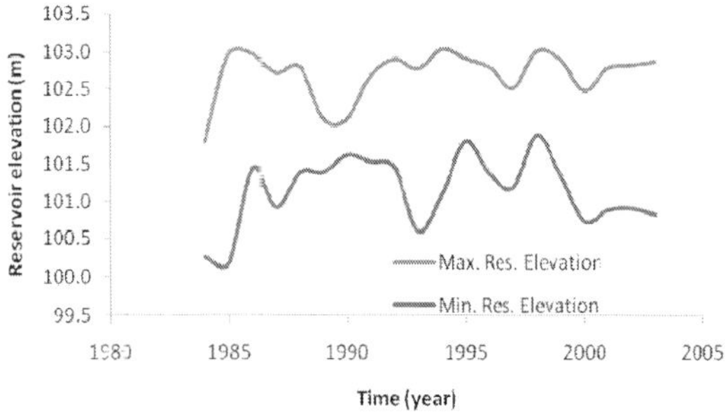

Fig. 3.6 b) Annual variation of maximum and minimum reservoir elevation at Jabba HEP dam (1984 - 2003).

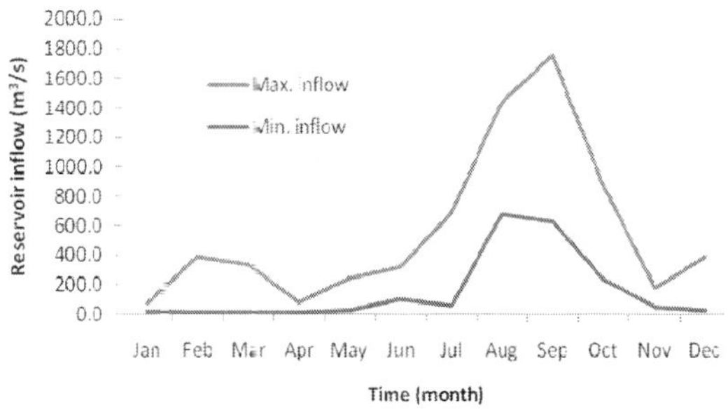

Fig. 3.7 a) Monhly variation maximum and minimum reservoir inflow at Shiroro HEP dam (1990 - 2009).

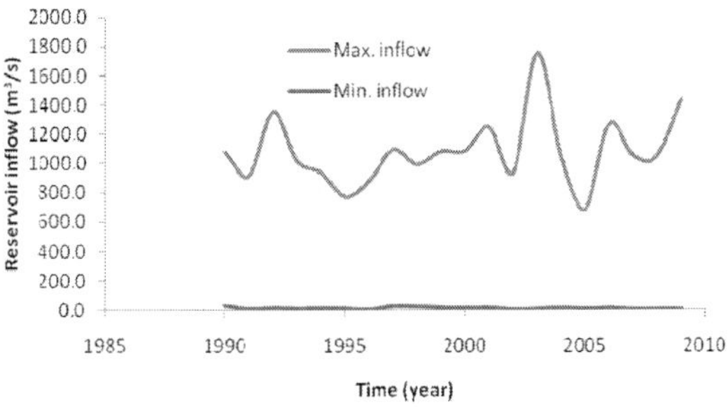

Fig. 3.7 b) Annual variation of maximum and minimum reservoir inflow at Shororo HEP dam (1990 - 2009).

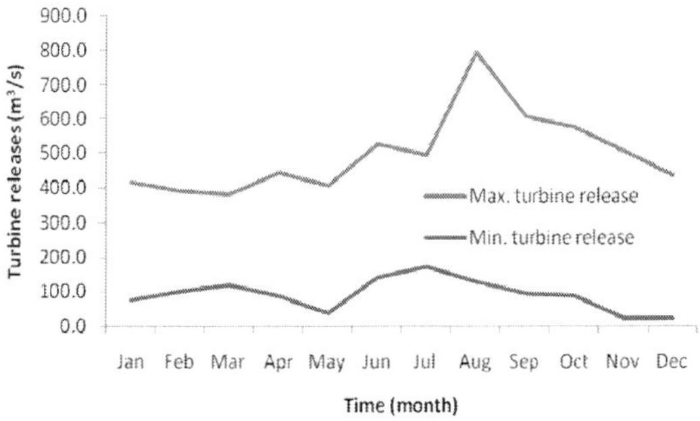

Fig. 3.8 a) Monthly variation of maximum and minimum turbine releases at Shiroro HEP dam (1990 - 2009).

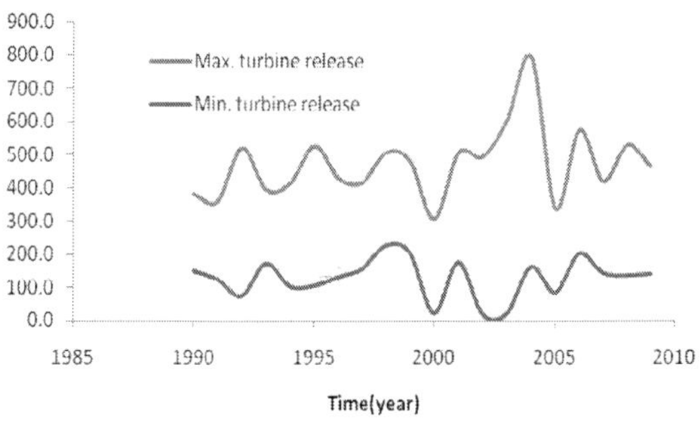

Fig. 3.8 b) Annual variation of maximum and minimum turbine releases at Shororo HEP dam (1990 - 2009).

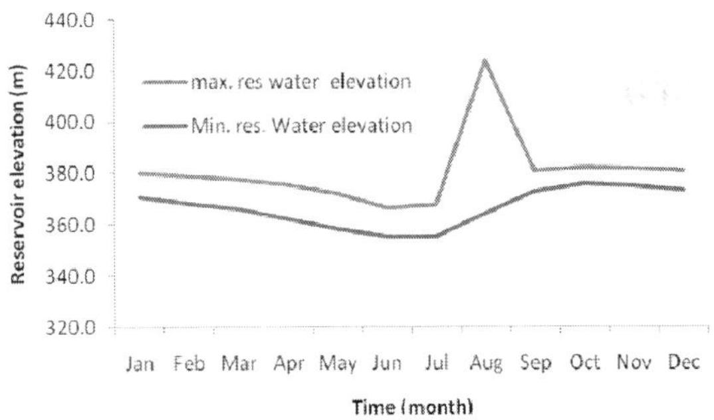

Fig. 3.9 a) Monthly variation of maximum and minimum reservoir elevation at Shiroro HEP dam (1990 - 2009).

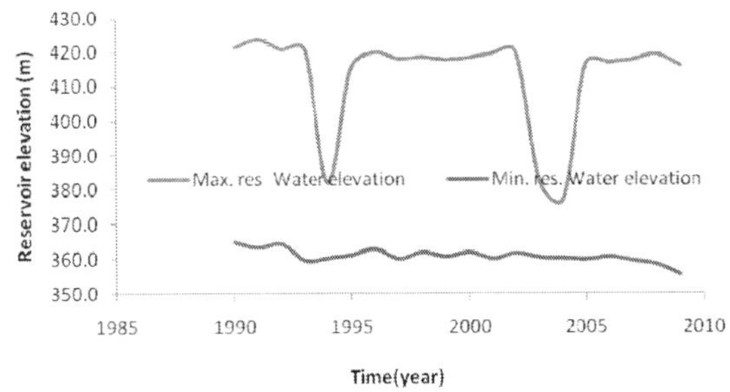

Fig. 3.9 b) Annual variation of maximum and minimum reservoir elevation at Shiroro HEP dam (1990 - 2009).

The trend in Figure 3.9b is that the maximum reservoir elevation has remained relatively constant since 1990, but lowered in 1994, 2003, and 2004. This might be connected to low reservoir inflow has indicated in Figure 3.7b. The minimum elevation was raised to 364.57 in 1990, and was subsequently lowered to 355.33 m in 2009. This might be as a result of over drawn of water for energy generation and to allow for more water to be stored in the dam in case of high flow from River Kaduna and its major tributaries contributing flow to the reservoir.

3.4 Flow duration curve (FDC)

The flow duration analysis was carried out in accordance to the method established by Oregon State University in 2002 to 2005, (http://water.oregonstate.edu/streamflow/). The method involves establishment of relationship between discharge and percent of time that the indicated discharge is equaled or exceeded (exceedence probability). The FDC can be used to determine dependable flow of various reliabilities such as 50%, 60%, 75%, 90%, 95% and 99% of the power output that can be guaranteed at various levels of reliability while ensuring that flooding is either eliminated or reduced. The flow duration curve analysis was carried out for reservoir inflow and turbine discharge at the three hydropower dams. The dependable reservoir inflow and turbine discharge at 50% exceedence probability was obtained as 700 m³/s and 760 m³/s for Kainji, 1000 m³/s for Jebba and 160 m³/s and 300 m³/s for Shiroro respectively. The dependable reservoir inflow and turbine discharges of reliabilities such as 50%, 60%, 75%, 90%, 95% and 99% from Figure 3.10a - 3.12b and the estimated power output that correspond to various reliabilities are presented in Table 3.4a.

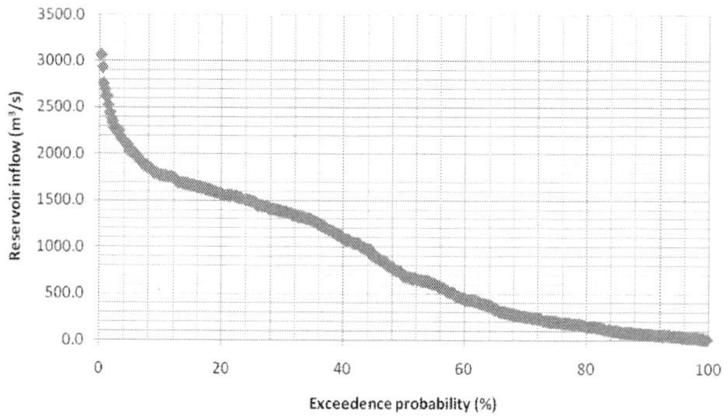

Fig. 3.10 a) Flow duration curve for resrevoir inflow at Kainji H.P dam (1970-2010).

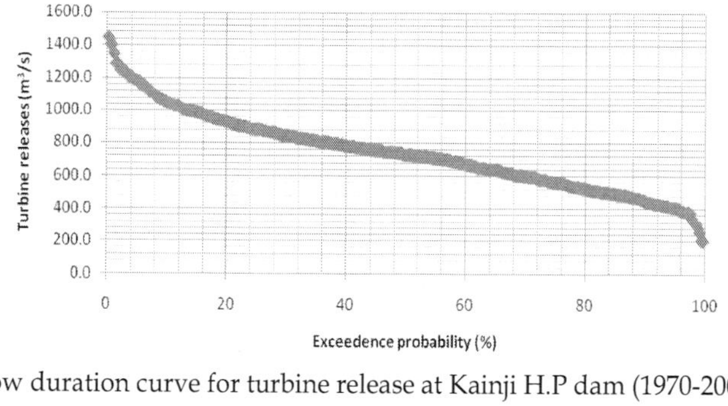

Fig. 3.10 b) Flow duration curve for turbine release at Kainji H.P dam (1970-2009).

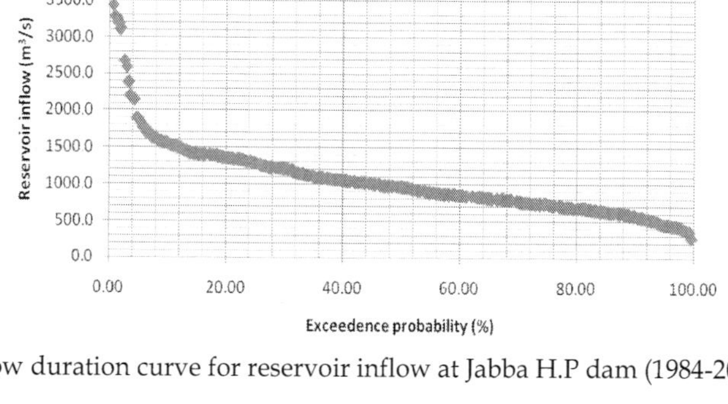

Fig. 3.11 a) Flow duration curve for reservoir inflow at Jabba H.P dam (1984-2010).

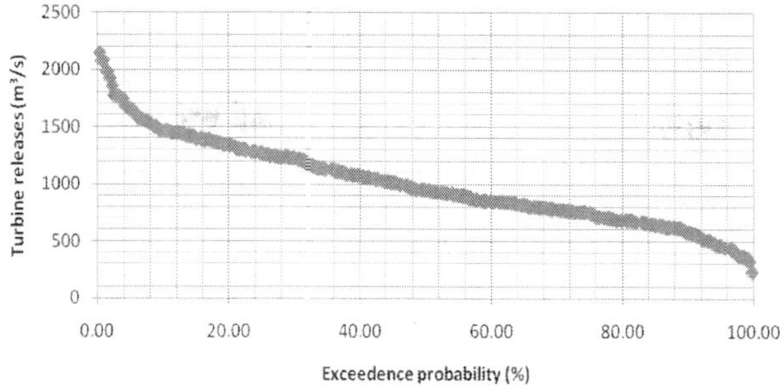

Fig. 3.11 b) Flow duration curve for turbine releases at Jabba H.P dam (1984–2009).

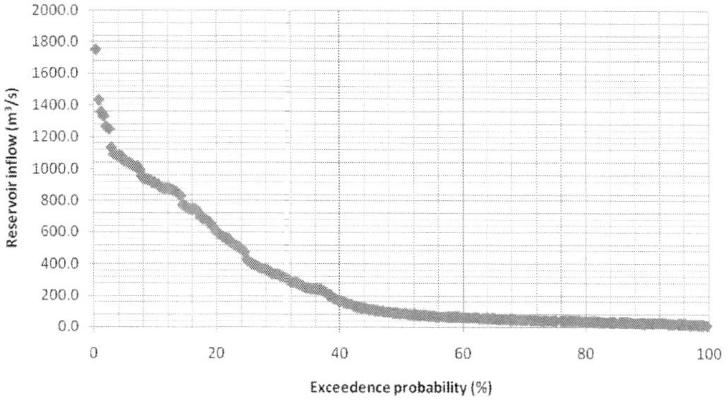

Fig. 3.12 a) Flow duration curve for reservoir inflow at Shiroro H.P dam (1990-2010).

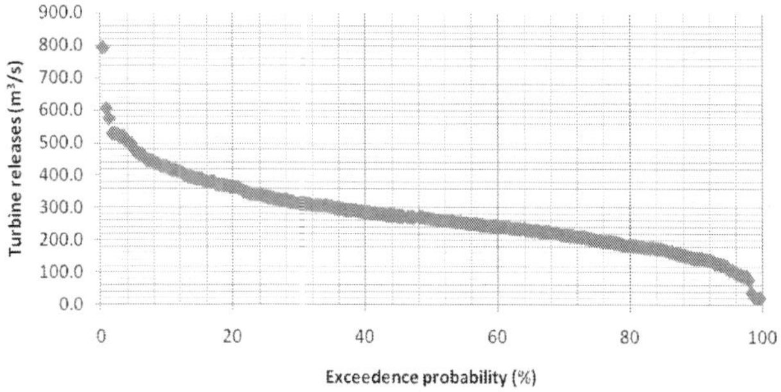

Fig. 3.12 b) Flow duration curve for turbine releases at Shiroro H.P dam (1990-2009).

Table 3.4 shows discharges of various reliabilities and the power output that can be guaranteed at Kainji, Jebba and Shiroro hydropower dams. For example 920 m^3/s can be drawn in 75% of the time and can drive one unit of turbine to generate about 200 MW of power every day. Also 540 m^3/s can be drawn in 90% of the time and can drive one unit of turbine to generate about 120 MW of power every day, while 500 m^3/s can be drawn in 95% of the time and can drive one unit of turbine to generate about 100 MW of power every day.

This implies that the more the amount of energy to be generated the more the quantity of water to be discharged. Care must be taken to avoid downstream flooding.

Kainji Dam Optimum head used H = 42.2m	Level of reliabilities (%)	50	60	75	90	95	99
	Turbine discharge (m^3/s)	760	720	600	480	440	200
	Energy output (MW)	252	239	199	159	146	66
Jebba Dam Optimum head used H = 29.3m	Level of reliabilities (%)	50	60	75	90	95	99
	Turbine discharge (m^3/s)	1000	960	920	540	500	200
	Energy output (MW)	230	221	212	124	115	46
Shiroro Dam Optimum head used H = 97m	Level of reliabilities (%)	50	60	75	90	95	99
	Turbine discharge (m^3/s)	300	280	220	160	120	10
	Energy output (MW)	228	213	168	122	92	8

Table 3.4 Turbine discharge and corresponding power output for the three Stations.

Table 3.4 show discharges of various reliabilities and the power output that can be guaranteed at three hydropower dams. For example 600 m^3/s can be drawn from Kainji in 75% of the time and can drive one unit of turbine to generate about 190 MW of power every day. Also, 480 m^3/s can be drawn in 90% of the time and can drive one unit of turbine to generate about 150 MW of power every day, while 440 m^3/s can be drawn in 95% of the time and can drive one unit of turbine to generate about 140 MW of power every day. This implies that the more the amount of energy to be generated the more the quantity of water to be discharged. Care must be taken to avoid downstream flooding.

3.5 Probability distribution analysis

The Gumbel extreme value type 1 (EV1) probability distribution function was used in fitting the low and high reservoir inflow and turbine discharge at the hydropower dams in order to predict values for various return periods. The probability functions of the form presented in equation (1) was obtained for high and low values of the variables for flood and low flow prediction respectively (Raghunath, 2008).

$$\begin{aligned} Q_T &= \bar{Q} + \sigma(0.78 y_T - 0.45) \\ y_T &= -\ln(-\ln(1-p)) \end{aligned} \tag{1}$$

$$\begin{aligned} Q_T &= \bar{Q} + \sigma(0.78 y_T - 0.45) \\ y_T &= -\ln(-\ln(p)) \end{aligned} \tag{2}$$

where :

Q_T = inflow value having T return period ; $\bar{Q} = Mean \quad value$
σ= Standard deviation; y_T = Reduced Variate; P = Probability

3.5.1 Reservoir inflows and turbine discharges fitted with Gumbel probability function

In order to show how best the Gumbel probability distribution function fits the peak and low values of the variables under consideration, the observed and predicted values were

plotted. The peak and low values of reservoir inflow and turbine discharge at Kainji, Jebba and Shiroro HEP dams were fitted with Gumbel probability distribution and the following functions were obtained as presented in Table 3.5. The result in Table 3.5 shows that the statistical goodness of fit tests such as probability plot coefficient of correlation (r) and coefficient of determination (R^2) were very high for the entire variable; hence the Gumbel probability distribution function can be adequately used to predict both peak and low values of the variables. Based on this fact, values of the variables for different return periods are predicted and presented in Table 3 .6.

3.5.2 Estimation of return period

The analysis of the historical data revealed that the lowest and the peak reservoir inflow rates are 9.40m^3/s and 3065.0m^3/s for Kainji, 378.0 m^3/s and 3636.0 m^3/s for Jebba, and 9.87 m^3/s and 1752.51 m^3/s respectively, while the lowest and the peak turbine discharges are 198.27 m^3/s and 1445.34 m^3/s for Kainji, 232.0 m^3/s and 2143.0 m^3/s for Jebba, and 20.80 m^3/s and 792.45 m^3/s for Shiroro respectively. However, the return periods of these parameters were determined based on the Gumbel probability distribution function developed for each variable. The results are presented in Table 3.6.

Table 3.6 implies that peak reservoir inflow will occur every 33 years while low inflow will be expected every 13 years and the peak turbine discharge (flooding) may occur on the average of every 19 years while the lowest turbine discharge will be rear for Kainji Dam. The peak reservoir inflow will occur every 25 years for Jebba Dam while low inflow will be expected every 20 years. The peak turbine discharge (flooding) may occur on the average of every 18 years while the lowest turbine discharge will also be rear. This also indicates that the peak reservoir inflow will occur at interval of 65 years for Shiroro Dam while low inflow will be expected every 25 year. The peak turbine discharge (flooding) may occur on the average of every 78 years while the lowest turbine discharge will be rear.

4. Result and discussion

The chapter involves collection of data and information concerning the reservoir inflow, turbine discharge and reservoir elevation data from the hydrological unit of the three hydropower stations in Nigeria. The relationships between the observed and predicted values of the peak and low reservoir inflows and turbine discharges are presented in Figures 3.15a and 3.18b. The result in Table 3.5 shows that the statistical goodness of fit tests such as probability plot coefficient of correlation (r) and coefficient of determination (R^2) were very high for the entire variable; hence the Gumbel probability distribution function can be adequately used to predict both peak and low values of the variables. Based on this fact, values of the variables for different return periods were predicted and presented in Table 3.6.

Hydropower dams	Hydrological data		Developed equations	r	R
Kainji Station	Reservoir inflow	Peak	$Q_T = 1798.8 + 562.18(0.78y_T - 0.45)$ $y_T = -\ln(-\ln(1-p))$	0.99	0.98
		Low	$Q_T = 67.9 + 49.72(0.78y_T - 0.45)$ $y_T = -\ln(-\ln(p))$	0.96	0.92
	Turbine Discharge	Peak	$Q_T = 990.3 + 252.88(0.78y_T - 0.45)$ $y_T = -\ln(-\ln(1-p))$	0.98	0.95
		Low	$Q_T = 494.7 + 142.82(0.78y_T - 0.45)$ $y_T = -\ln(-\ln(p))$	0.97	0.94
Jebba Station	Reservoir inflow	Peak	$Q_T = 1893.00 + 859.89(0.78y_T - 0.45)$ $y_T = -\ln(-\ln(1-p))$	0.98	0.95
		Low	$Q_T = 604.40 + 172.99(0.78y_T - 0.45)$ $y_T = -\ln(-\ln(p))$	0.98	0.95
	Turbine Discharge	Peak	$Q_T = 1480.40 + 374.56(0.78y_T - 0.45)$ $y_T = -\ln(-\ln(1-p))$	0.98	0.95
		Low	$Q_T = 570.40 + 180.10(0.78y_T - 0.45)$ $y_T = -\ln(-\ln(p))$	0.98	0.95
Shiroro Station	Reservoir inflow	Peak	$Q_T = 1080.70 + 240.70(0.78y_T - 0.45)$ $y_T = -\ln(-\ln(1-p))$	0.98	0.96
		Low	$Q_T = 20.30 + 7.71(0.78y_T - 0.45)$ $y_T = -\ln(-\ln(p))$	0.99	0.98
	Turbine Discharge	Peak	$Q_T = 473.10 + 108.59(0.78y_T - 0.45)$ $y_T = -\ln(-\ln(1-p))$	0.97	0.94
		Low	$Q_T = 129.10 + 60.09(0.78y_T - 0.45)$ $y_T = -\ln(-\ln(p))$	0.95	0.90

Table 3.5 Summary of the developed equation for the prediction of the hydrological variables.

Reservoir inflow (m^3/s)						Turbine discharge (m^3/s)					
Peak			Low			Peak			Low		
Kainji	Jebba	Shiroro	Kainji	Jebba	Shiroro	Kainji	Jebba	Shiroro	Kainji	Jebba	Shiroro
3065	3636	1753	9	378	10	1445	2143	80	198	232	21
33	25	65	13	20	25	19	18	78	3093	520	288

Table 3.6 Return periods of peak and lowest variables for the three Hydropower Dams respectively.

Statistical analysis carried out on the reservoir inflow data showed that inflow to Kainji Dam is related to the control release from upstream reservoir in the Niger Republic and this implies that the flow hydrology at Kainji is mostly controlled by the upstream reservoir in another region. The same observation can be extended to Jebba reservoir in that the inflow depends on the control releases at Kainji. The reservoir inflow data revealed that spillway releases occur at Kainji throughout the year (monthly), while spillage occurs at Jebba reservoir between the months of August through to December. Peak discharges occur during the month of October at the two reservoirs and is indicated as the month of high flood level downstream of the reservoirs. Analysis of water management options for different level of reliabilities for turbine discharge and energy output performed using probabilistic reservoir inflow of 50%, 60%, 75%, 90%, 95% and 99% reliabilities for the various scenarios were presented in Table 3.5. Values of energy generation were estimated for each value of turbine discharge selected. The developed equations in Table 3.5 relating hydrological variables for energy generation can be used to predict flood limits for any desired amount of energy.

5. Conclusion

The main cause of flood in the downstream regime was identified to be the sudden release of water from the hydropower dams located upstream of the study area. The study revealed that the sudden release of flood water is not due to normal operation at the hydropower stations in Nigeria, but is due to sudden discharges at the reservoirs located in the Niger Republic and the Republic of Mali. This leads to excess releases at Kainji in order to create enough space for the incoming flood water. This automatically forces the release of water at Jebba and thus creating flood problem downstream. The model results revealed that the problem at the Bacita sugar plantation is due to the location of their water abstraction level and the persistence flood problem at the downstream areas is because the flood wall could no longer serve the intended purpose. The flood wall was put in place before the construction of the dams.

5.1 Recommendations

Water management at the three reservoirs needs improvement so that the energy generation output can be improved on and the management of PHCN at Kainji, Jebba and Shiroro should provide information to communities and local authorities on the release of water especially during the months of August to October. It is recommended that joint release policy be established between Nigeria, Niger Republic and Mali in order to alleviate persistence flood problem in the country since the analysis of the reservoir inflow at Kainji revealed that it is a control release from upstream reservoirs in another region. It also recommended that a study should be carried out to assess the state of facilities at the hydroelectric power stations and identify the components that are malfunctioning in order to recommend required improvement.

6. References

Aribisala, J. O. & Sule, B. F. (1998). Seasonal operation of a Reservoir Hydropower system, Technical Transactions, Nigerian Society of Engineers, NSE, Vol. 33 No 2: 1-14

Bolaji, T. (1999). Sacked by the Almighty River Niger, Newspaper Publication, the Guardian on Sunday, October 23, pp 4 -5

Jebba hydro-electric Power Station (2010). Hydrology, Meteorology and Reservoir operational data. Hydrology section. Jebba, Niger State, Nigeria

Kainji hydro-electric Power Station (2010). Hydrology, Meteorology and Reservoir operational data. Hydrology section. Kainji, Niger State, Nigeria

Lawal, A. F. & Nagya, M. T. (1999). Socio-Economic effects of flood disaster on farmers in Mokwa Local Government Area, Niger State, Nigeria, A paper presented at the 33rd Annual Conference of Agricultural Society of Nigeria held at National Cereal Research Institute, Badeggi, Niger State, Nigeria. Pp 1 - 10

Olofintoye, O. O.; Sule, B. F. & Salami, A. W. (2009). Best-fit Probability Distribution model for peak daily rainfall of selected Cities in Nigeria. New York Science Journal, 2(3), ISSN 1554-020, (http://www.sciencepub.net/newyork) (sciencepub@gmail.com)

Olukanni, D. O. & Salami, A. W. (2008). Fitting probability distribution functions to reservoir inflow at hydropower dams in Nigeria. Journal of Environmental Hydrology, Vol. 16 Paper 35, pp. 1-7, 2607 Hopeton Drive San Antonio, TX 78230, USA

Oregon State University (2002). Flow Duration Curve (FDC). Accessed 25th, April, 2011, Available from (http://water.oregonstate.edu/stream)

Raghunath, H.M (2008). Hydrology: Principles, Analysis, Design. New Age International Limited, Publishers, New Delhi, Revised Second Edition. www.newagepublishers.com

Shiroro hydro-electric Power Station (2010). Hydrology, Meteorology and Reservoir operational data. Hydrology section. Shiroro, Niger State, Nigeria

Sule, B. F (2003). Water Security: Now and the future, 65th Inaugural lecture, University of Ilorin, Ilorin, Nigeria. 70 pp, Civil Engineering Department, University of Ilorin, Nigeria

5

Discharge Measurement Techniques in Hydropower Systems with Emphasis on the Pressure-Time Method

Adam Adamkowski
The Szewalski Institute of Fluid-Flow Machinery of the Polish Academy of Sciences, Poland

1. Introduction

Volumetric flow rate (discharge), as the volume of liquid flowing in time unit, belongs to the group of a few basic quantities needed to determine the hydraulic performance characteristics of hydraulic turbines and pumps. Discharge always represents the most difficult quantity to measure and accuracy (uncertainty) of its measurement is worse and very difficult to estimate in comparison with the power and head (or specific hydraulic energy). Despite immense progress in discharge measurement techniques, this part of the hydraulic machine performance tests is often a major challenge even for experienced test teams.

Besides the volumetric method, there are only a few primary methods of the absolute discharge measurement in hydropower plants, generally: (1) velocity-area method (local velocity distribution method), (2) pressure-time (Gibson) method, (3) tracer method, (4) ultrasonic method (ultrasonic flow meters), and (5) electromagnetic method (electromagnetic flow meters). Three first methods belong to the group of traditional (classic) methods, while the fourth one, the ultrasonic method, is relatively new and has been recently the object of numerous research activities oriented on its improvement and validation. This method has not yet reached the proper acceptance among the specialists. However, its basic advantage is the possibility of using it for continuous flow rate measurement. Electromagnetic method reveals also this advantage, however application of electromagnetic flow meters is limited to the conduits of small diameters, for instance up to 2 m. Because of some other advantages, electromagnetic flow meters are widely used in auxiliary water and oil systems of the hydropower plants.

After brief review of discharge measurement techniques, later on the recent important achievements in developing and utilizing the pressure-time method, as one of the primary method for measuring the discharge through the hydraulic machines, is presented. Nowadays, the pressure-time method is more and more frequently used in the hydropower plants equipped with penstocks longer than 15-20 m.

2. Background on flow rate measurement

2.1 Discharge measurement in small closed conduits

Measurement of fluid flow rate in conduits with small diameters, for instance up to 1 m, is not a very difficult task. Typical, standard measuring devices, such as constriction flow meters (measuring orifice plates - Fig 1, flow nozzle - Fig 2 or Venturi tube - Fig 3) and calibrated pipe elbows (bends) - Fig. 4, are used in such conduits. These devices are usually mounted in a properly prepared measuring segment of the conduit. They belong to the group of measuring devices using pressure difference Δp which nowadays can be measured with high accuracy. The flow rate is expressed as a function of the measured pressure difference:

$$Q = f(\Delta p) \tag{1}$$

These devices are highly reliable, however they are very sensitive to flow pattern and cause relatively high pressure drop due to hydraulic resistance, especially in cases of measuring orifices plates.

The principle of the constriction flow meter operation is based upon the Bernoulli equation, which can be written in the form:

$$p_1 + 0.5\rho V_1^2 + \rho g z_1 = p_2 + 0.5\rho V_2^2 + \rho g z_2 \tag{2}$$

and the following continuity equation:

$$V_1 A_1 = V_2 A_2 = Q \tag{3}$$

where p means the pressure, V - flow velocity, ρ - liquid density, g - gravity acceleration, z - elevation, Q - flow rate, A - flow area (A_1=0.25πD^2, A_2=0.25πd^2 - see Figs. 1, 2, 3) and the indexes "1" and "2" denote the cross sections before and within the considered constriction, respectively. Combining Eq. (2) and Eq. (3) after introducing the static pressure difference before and within the constriction (after relating both to the same elevation), $\Delta p=p_2+\rho g z_2-p_1-\rho g z_1$, we can get the following formula:

$$Q_{ideal} = \left(\frac{2\Delta p}{\rho}\right)^{0.5} \frac{A_2}{\left(1-\left(A_2/A_1\right)^2\right)} \tag{4}$$

The above formula applies only to flows of the so called perfect (ideal) liquid. For real liquid flow, viscosity and turbulence are present and act to convert kinetic flow energy into heat. In order to take this effect into account, a discharge coefficient C_d is introduced to reduce the flow rate

$$Q = C_d Q_{ideal} \tag{5}$$

The discharge coefficient C_d has different values depending on the kind of the measuring device and its geometry. The value of C_d has to be experimentally determined. For the standard Venturi tubes the discharge coefficient C_d ranges from 0.93 to 0.97, however, for the standard orifice plates its value is close to 0.6. The flow nozzles are shortened versions of the Venturi tube, with lower pressure drops than orifice plates. It means that the orifice plates, causing much higher energy losses than the Venturi tubes and flow nozzles, should not be widely used for measuring the flow rate in pipelines, especially in large ones.

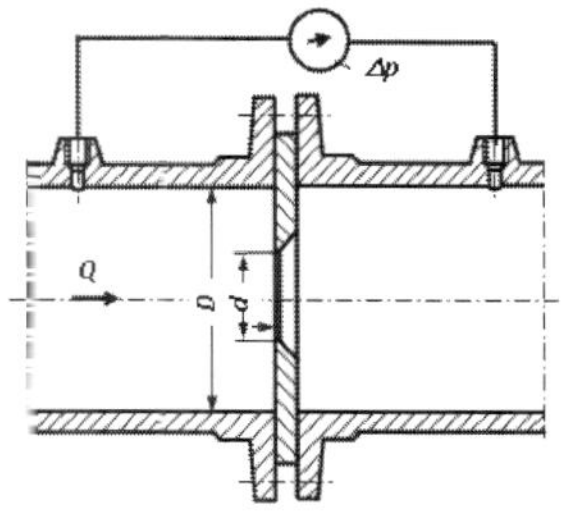

Fig. 1. Orifice plate.

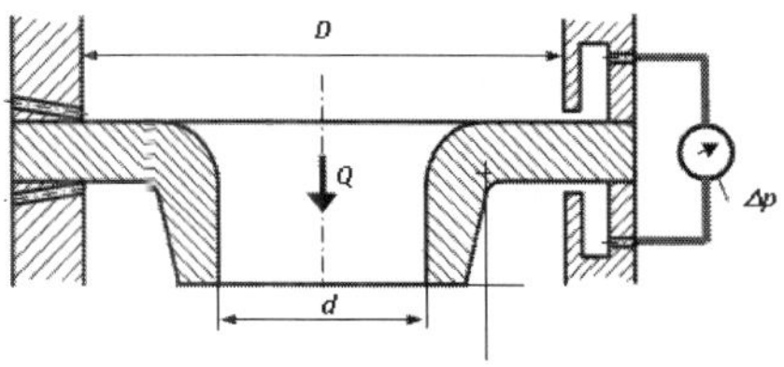

Fig. 2. Flow nozzle.

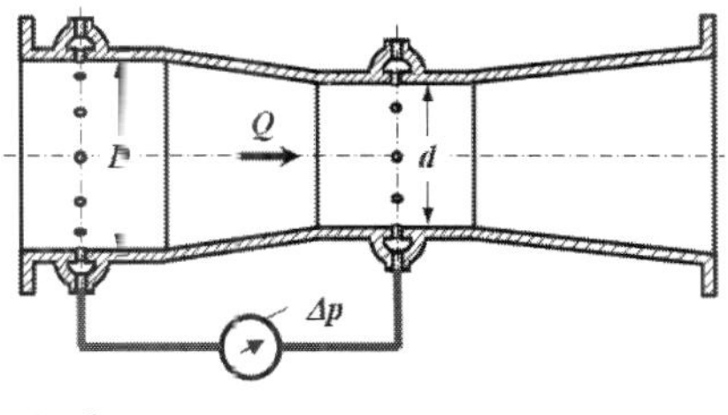

Fig. 3. Conical-type Venturi tube.

The calibrated elbows belong also to the group of flow meters based on pressure difference measurement. When liquid flows through the pipe elbow, the centrifugal forces cause a pressure difference between the outer and inner sides of the elbow, $\Delta p = p_2 - p_1$ – Fig. 4. This pressure difference is used to calculate the pipe flow velocity V or flow rate Q. The simplified relationship between V and Δp can be written in the following form:

$$\Delta p = \rho \frac{D}{R} V^2 \tag{6}$$

The pressure difference generated by the elbow flow meters is smaller than that caused by other pressure differential flow meters, presented above, but the elbow flow meters have less obstruction to the flow.

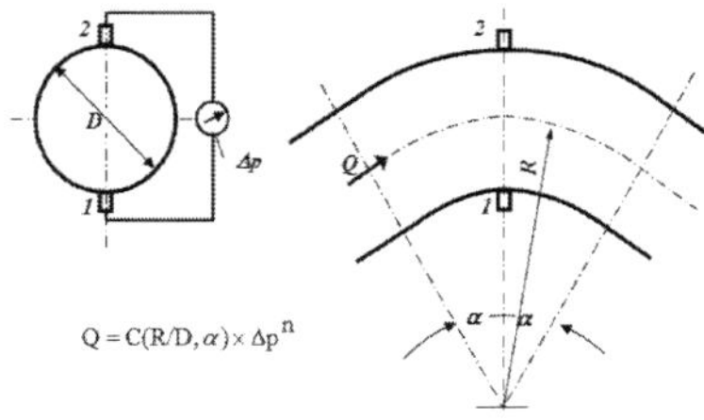

Fig. 4. Elbow flow meter.

The measuring devices briefly presented above can be easily used for continues measurements of flow rate, that is very important in different kind of monitoring hydraulic systems. The relevant pressure transducers have to be applied to achieve such goal.

Other devices are also available, such as electromagnetic flow meters (Fig. 5) and different kind of ultrasonic flow meters that are shortly described in the farther part of this chapter.

The electromagnetic flow meters are based upon Faraday's law of induction, e.g. use the phenomenon of inducing electromagnetic force caused by flow of conducting fluid. The phenomenon can be described in the form of following relationship:

$$U_E = VBL \tag{7}$$

in which U_E means the induced voltage (voltage generated in a conducting liquid), V - the average liquid velocity, B – the magnetic field strength, L - the distance between the electrodes – see also Fig. 5.

The main advantages of the electromagnetic flow meters can be expressed as follows: They do not disturb the flow pattern and do not cause the pressure losses (drops), and also, are not very sensitive to the wear, and can be easily used for continues measurements of flow rate. Such instruments are produced as measuring segments that are installed into pipelines of different diameters, up to 2 meters. Nowadays, due to their many advantages, they are very often used in hydropower plants, mainly in auxiliary water and oil systems. Unfortunately, the author of this chapter does not know any case of electromagnetic flow meter application in turbine penstocks of several meter diameter.

The international standards ISO 6817 and ISO 9104 relate to the flow rate measurements using the electromagnetic meters.

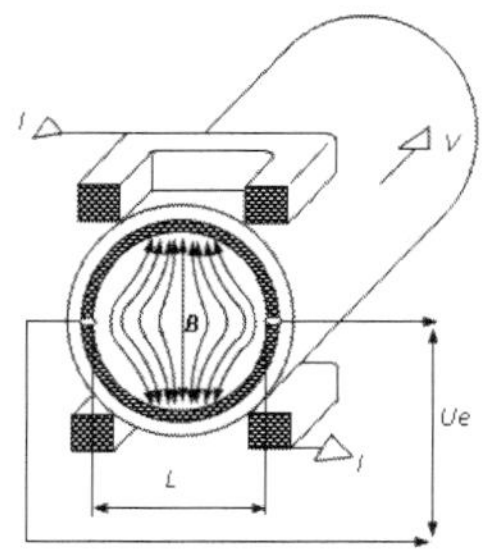

Fig. 5. Electromagnetic flow meter.

2.2 Flow rate measurement in small open conduits (channels) – free-flow meters

The free-flow meters (Fig. 6) consist in raising the head of the liquid stream and relating the discharge to the head accordingly to the theoretical or experimental formulas. The flow rate (Q) is expressed as a function of the head over the weir (h):

$$Q = f(h) \tag{8}$$

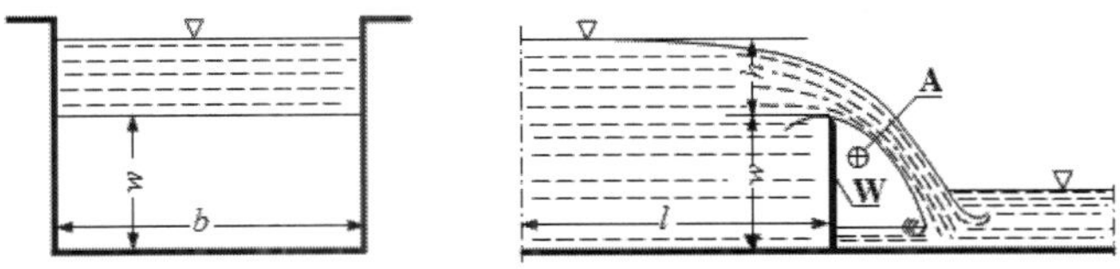

Fig. 6. Measuring weir. W – weir plate, A – ventilation duct.

Rectangular weirs (Figs. 7 and 8) are the most commonly used weirs in hydrometric practice. We can distinguish rectangular weirs without end contractions (Fig. 7) from rectangular weirs with end contractions (Fig. 8), where the width of the crest is less than width of the approach channel $b<B$. There are also free-flow meters with variable-flow area, i.e. the ratio of their throat cross-section A_c to their inlet cross-section A change with the flow rate (A_c/A varies with discharge Q).

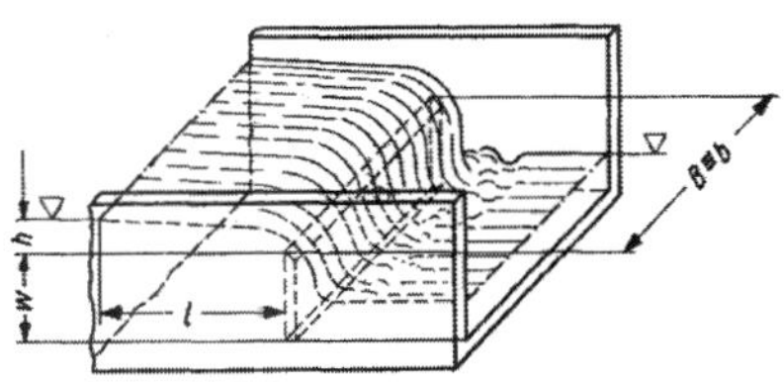

Fig. 7. Rectangular weir (without end contractions) - Hansen weir.

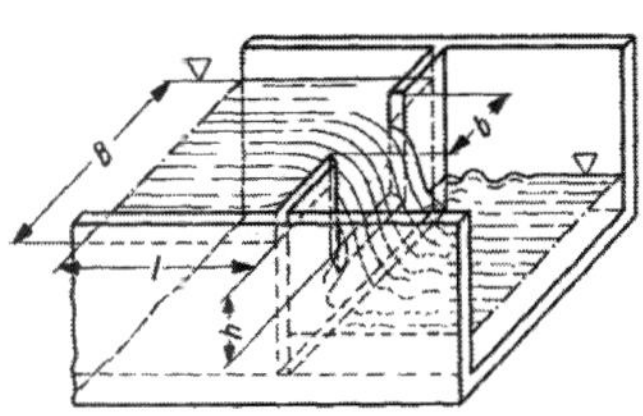

Fig. 8. Rectangular weir with double end contractions - Poncelet weir.

The accuracy of these measuring devices is dependent mainly on the error of measurement of the head over the weir. There are some other factors having important influence on the accuracy, e.g. quality of weir plate and walls, and flow velocity distribution at the inflow side.

The standard ISO 1438-1 relates to the water flow rate measurements in open channel using weirs.

3. Brief review of primary methods for discharge measurement in hydropower systems

The devices presented above are not generally suitable for water discharge measurements in large-dimension conduits, having the diameter of an order of several and more meters. Measurements of discharge in these conduits, frequently used in hydropower engineering, are very difficult and expensive. Preparation and performance of high accuracy measurements is even more difficult by (1) hydro-technical conditions of the power plants (very large real objects, during erection of which the aspects of the use of measuring methods are often neglected) and (2) high requirements referring to the operating conditions of the hydro-units in the electric power system (for instance, very high costs of possible machine stoppage to install the measuring systems - the machine standstill cost is often higher than the cost of preparation and performance of measurements). Additionally, it is still expected to measure water discharge in hydropower with higher precision, for instance, at present within systematic uncertainty of +- 1%.

There are only a few primary methods of the absolute discharge measurement in hydropower plants:

1. The **velocity-area method** [ISO 3354, ISO 7194, ISO 3966] utilizes the distribution of local flow velocities, measured using mainly propeller current meters (Fig. 9) - in cases of large conduits - or impact pressure velocity meters (Pitot static tubes, Fig. 10) - for smaller conduits and clean water. The volumetric flow rate is determined by integrating this distribution over the entire area of the measuring section.

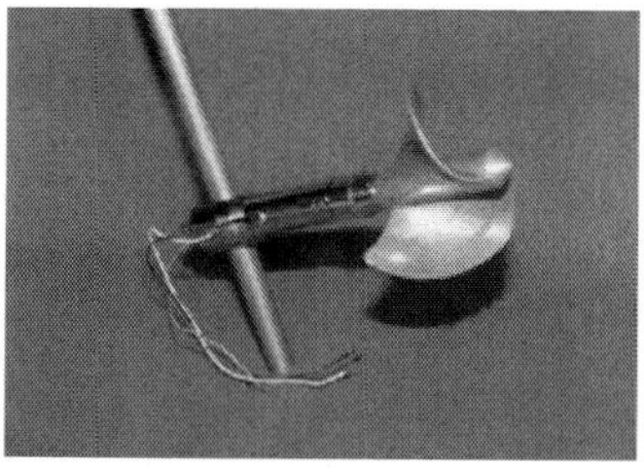

Fig. 9. Propeller-type current meter. (The local flow velocity component is measured based on the current meter rotor revolutions counted in a given time period, and the experimentally determined relationship between the current meter rotation speed and local flow velocity).

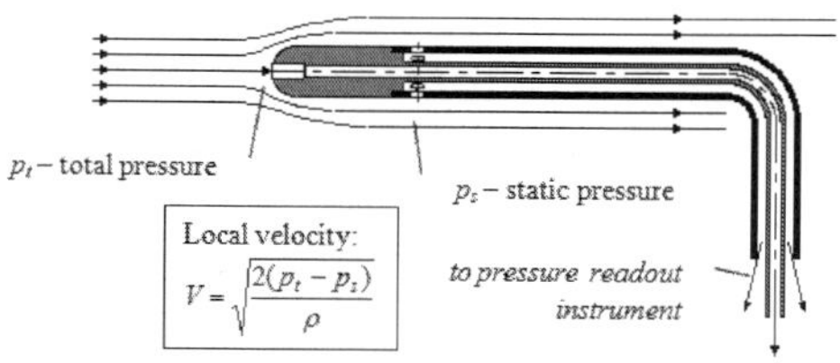

Fig. 10. Pitot static (Prandtl) tube.(The local velocity is resulted from the dynamic pressure measured as the difference between the total (stagnation) pressure and the static pressure).

2. The **pressure time method** (water hammer or Gibson method) [IEC 60041, IEC 62006, ASME PTC 18] is based on measuring the time-history of pressure difference changes between two hydrometric sections of the closed conduit during a complete stop of the fluid flow by means of the shut-off device. The volumetric flow rate of the liquid in the initial conditions, before the stream has been stopped, is determined from relevant integration of the measured pressure difference change occurring while stopping the liquid stream.
3. The **tracer method** [IEC 60041] consists in measurements of the passing time, or concentration, of radioactive or non-radioactive substance (salt, for instance) between two sections of the penstock. The method requires long conduits and conditions for good mixing of the marker.
4. The **ultrasonic** method [IEC 60041] uses the principles of ultrasound and is based on vector summation of the ultrasound (acoustic) wave propagation velocity and the average flow velocity - the use of difference in frequencies or passing times of the emitted and received acoustic signal. Ultrasonic transducers with special software are used for realization of discharge measurement according to rules of this method. The basic advantage of the method is the possibility of its use for continuous discharge measurement. Other primary methods do not reveal this advantage.

The electromagnetic method, mentioned as a fifth one in the introduction, is not considered here due to its application limited to flow measurement in small pipelines.

Three first aforementioned methods belong to the group of traditional (classic) methods, while the fourth method, the ultrasonic method is still the object of numerous research activities oriented on its improvement (Gruber, 2008, Gruber at al., 2010; Hulse at al., 2006; Llobet at al., 2008; Nichtawitz at al., 2004; Strunz at al., 2004; Tresch at al., 2006). Ultrasonic flowmeters are affected by the distribution of flow velocity, turbulence, the temperature, density and viscosity of the flowing medium and the presence of gas bulbs. There is still main question: Is the velocity measured by the ultrasonic meter equal to the average flow velocity along the path of an emitted beam of ultrasound? The relevant integration methods are still developed (Gruber at al., 2010). The ultrasonic method has not reached proper acceptance among the specialists yet. The IEC 60041 standard suggests conditional use of this method, i.e. in cases of explicit agreement of all contracting parties.

Author's own experiences based on multiple verification tests of ultrasonic flow meters, as well as their application, confirm the existence of the above mentioned problems. These problems in many cases make it impossible to obtain measurement results of desired accuracy and reliability using ultrasonic flow meters. On the other hand, basing on many publications in this topic, it can be confirmed that in specially prepared conditions, mainly in the laboratory, tested ultrasonic flow meters give a surprisingly good agreement comparing with other flow measurement techniques. Very often the differences do not exceed (0.5-0.7)%. These divergent assessments point to the need for scientific research of the ultrasonic method of flow measurements, which should be conducted in various real conditions.

The tracer method is the least popular among the above mentioned classic methods. It requires very long measuring segments and special, additional conditions facilitating the mixing process of the introduced marker, the use of turbulizers, for instance.

It should be stressed here that at present the primary methods of absolute discharge measurement in turbine penstocks include the velocity-area method and the pressure-time

(Gibson) method. It is also noteworthy that the velocity-area method, mainly making use of propeller current meters and most frequently used in waterpower engineering in the past, is nowadays being replaced by the pressure-time method in the hydropower plants equipped with penstock longer than 15-20 m. The reason for this is much lower cost of preparation and performance of the measurement based on the pressure-time method. Additionally, the development of computer techniques in recent twenty years has facilitated the measurements and provided opportunities for obtaining higher accuracy of the results using this method.

The different situation is for the low and very low head power plants with short intakes of water turbines. Up to now, generally only the velocity-area method is available in such kind of plants. Local velocities can be measured using various instruments, including current-meters, Pitot static tubes, and recently electromagnetic and ultrasonic meters. The discharge measurements using the velocity-area method is still unprofitable. This reflects that in the past the measurements have been performed only in the minority of low head power plants. Also, an alternative acoustic scintillation method, which allows to scan the flow measurement section, is still expensive and very sensitive to flow disturbances (Llobet at al., 2008).

Discharge measurements in low-head installations are linked with lack of sufficiently long straight flow channel, allowing for parallel streamlines, perpendicular to the hydrometric section. Usually, due to various reasons, the plane of stoplog hollows (Fig. 11) is considered as the optimum one. On the one hand side, such a location enables mounting current-meters on a traversing frame that results in high density of the local measurement points. On the other hand problems linked with deviation of the streamlines from direction perpendicular to the hydrometric section plane occur. Under such circumstances the use of CFD flow analysis in order to determine the streamline deviation angles is of essential significance.

Fig. 11. Frame with fixed the current meters for traversing the hydrometric (measuring) section (author's application).

Application of CFD calculations is highly recommended also in cases of measuring discharge with current-meters installed in short intake penstocks of hydraulic turbines - Fig. 12-13. Figure 14 shows a framework with current-meters situated along the calculated streamlines.

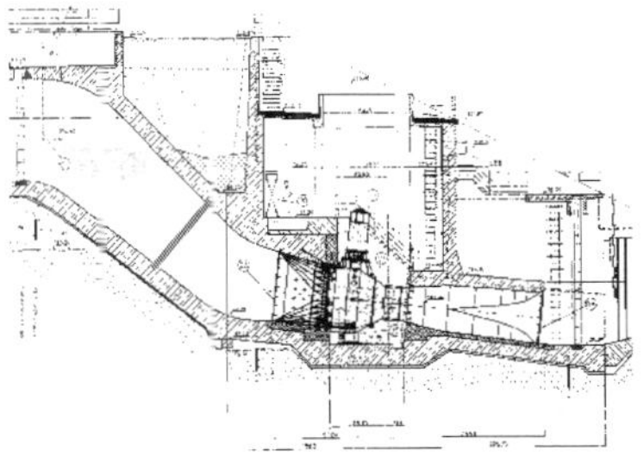

Fig. 12. Vertical section of the short penstock of the bulb turbine with hydrometric section indicated (own application).

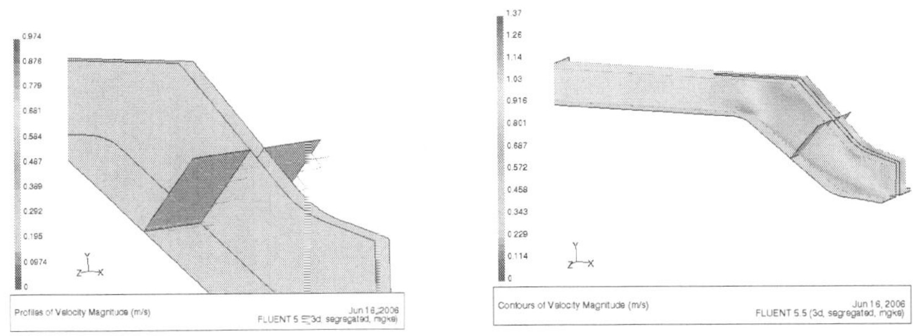

Fig. 13. Results of CFD calculations of velocity distributions in the considered case (own application).

Fig. 14. Hydrometric section with the supporting frame and installed current-meters (own application).

Generally, other methods and devices, like the standardized weirs, the differential pressure devices and the electromagnetic flow meters, described earlier, as well as the volumetric method, are also possible to use in hydropower plants for absolute discharge measurement [IEC 60041]. However, these methods have very limited area of application - they can be used mainly for small hydropower plants, not for large-scale discharge measurements. The volumetric method, conventionally confined to low flow rates, because of the size of the reservoirs required, can be used in some large hydropower plants, particularly in pumped-storage plants with big artificial reservoirs but only when the characteristic (geometry) of these reservoirs are exactly known.

Discharge measurement using the volumetric method consists in determining the rise of water volume ΔV in the head or tail water reservoirs during the measured period of time, Δt, as it can be written in the following form:

$$Q = \frac{\Delta V}{\Delta t} = \frac{V(z_0 + \Delta z(t)) - V(z_0)}{\Delta t} \tag{9}$$

where z_0 means the water level at the initial conditions, at the test beginning.

The water volume rise ΔV is determined by means of measurement of water level increment Δz during the test and the function $V(z)$ known from the geometry of reservoir.

Some experiences of the author of this chapter concerning application of the volumetric method allow to indicate a few important problems (issues) relating to the accuracy of the relationship between water level and reservoir volume and the accuracy of measurement of water level rise (Adamkowski, 2001, Adamkowski at al., 2006). The reservoir volume should be determined by precise standard geometrical measurement or by photogrammetry. Practically, the required accuracy of such measurement can be only obtained for the concrete artificial reservoirs. Water level cannot be accurately measured using the instruments commonly used in the control system of the plants because very often they have much too high range and much too low precision class, additionally also various disturbing effects do not ensure the required accuracy of the measurements. Therefore the change in water level has to be determined using the special method [ISO 60041].

Fig. 15 presents an effective technique developed and applied by the author of this chapter for precise measurement of water level rise (Adamkowski, 2001). The technique was successfully utilized during the efficiency tests with the aid of the volumetric method in the Polish pump-storage power plants. Water level rise Δz is determined by measuring the difference between pressure in the hydropower plant water reservoir and the constant pressure in the auxiliary tank underhung at constant height during the tests. For this purpose, the differential pressure transducer of high precision class has to be connected to the plant reservoir and auxiliary tank.

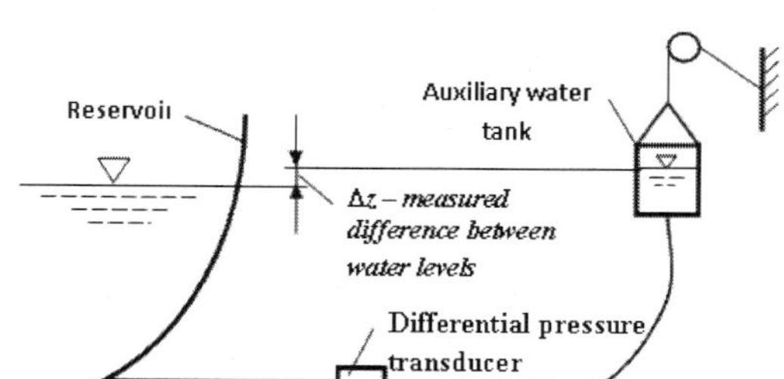

Fig. 15. Technique of water level rise measurement applied in the volumetric method in a few Polish pumped-storage power plants (own application).

Wave motion of water in the reservoir is one of factors that absolutely should be taken into consideration as the effect most affecting the measurement results. Computer data acquisition system and regression line (function calculated using the least squares method) applied to the recorded water level values can allow to eliminate the water waves disadvantageous effect - Fig. 16. Traditional readings realized in such kind of method do not give any chance for obtaining the required accuracy of discharge measurements.

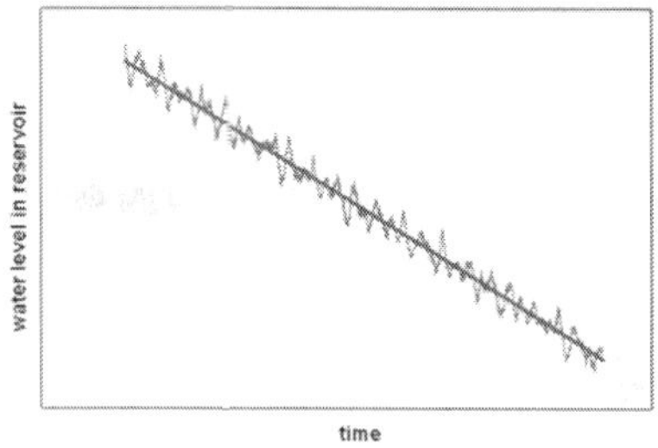

Fig. 16. Example of recorded water level signal and its regression line (own application).

4. Pressure-time (Gibson) method of discharge measurement

4.1 The physical principles of the pressure-time method

The pressure-time method (popularly called "Gibson method") utilizes the inertia force effect which manifests in the pressure rise during deceleration of liquid mass flowing in a closed conduit (penstock in hydropower plant). The method is based on recording the time-history (time diagram) of pressure difference changes between two hydrometric (measuring) cross sections of the conduit during a complete stop of the fluid flow realized by means of the shut-off device. Fig. 17 shows a scheme of turbine penstock with marked measuring sections relevant to pressure-time method use. The volumetric flow rate of the liquid in the initial conditions, before full stop of the stream, is determined from proper integration of the measured pressure difference change during closing up the shut-off device installed in the penstock - pressure difference caused by inertia force. It can be proved that the area between the pressure difference time-history curve recorded during the transient state and the curve representing the hydraulic loss in the conduit segment (and the dynamic pressure difference between the end sections of this segment) is proportional to the change of the volumetric flow rate between the initial and final conditions - Fig. 18.

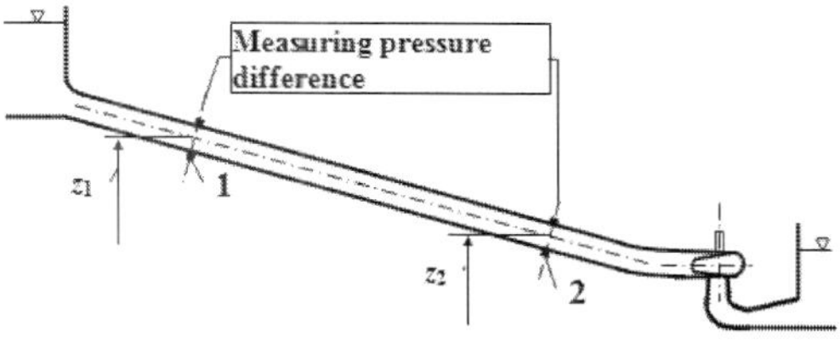

Fig. 17. Schematic layout of turbine penstock with marked measuring cross sections applicable for use in the pressure - time method.

4.2 Basic information about the pressure-time method

The method was introduced in the first half of the 20th century (year 1923) by Norman R. Gibson (Gibson, 1923, 1954). (That is why it is called, after its author, the Gibson method). This method has been mainly used for measurements of discharge in the penstocks of water turbines and pump-turbines. It is recommended by the international standards IEC 60041 and IEC 62006 and the American standard ASME PTC 18. It is assumed that in the conditions consistent with the recommendations of these standards, the measurement accuracy is not worse than +/-(1.5-2.0)% and does not differ from that represented by other primary methods, the current meter method, for instance.

The Gibson method was more frequently used in the North America than in Europe and in other parts of the world, especially when the optical techniques were employed to record pressure changes and combined with the manual graphical integration. Nowadays, the increased accuracy of pressure measurement instruments, along with the availability of the hardware for computer acquisition and processing of the measured data are the reasons why this method is becoming more attractive also all over the world. For example, during the latest two decades the method was extensively and successfully used in the performance tests of hydrounits that were carried out at many hydropower plants in Poland (Zarnowiec, Solina, Dychow, Zydowo, Niedzica, Koronowo, Zur, Pilchowice), as well as in Mexican hydropower plants (Angostura, Infiernillo, Chicoasen, Aquamilpa, Temascal, Villita, Infiernillo, Villita, Cobano, Novillo, Santa Rosa, Bacurato). In this period, the author of this chapter actively participated in developing process of the method in different kind of its aspects (Adamkowski & Kwapisz, 2000; Adamkowski at al., 2006a, 2006b and 2006c, 2007, 2008, 2009; Adamkowski & Janicki, 2007, 2008, 2009, 2010; Urquiza at al., 2007).

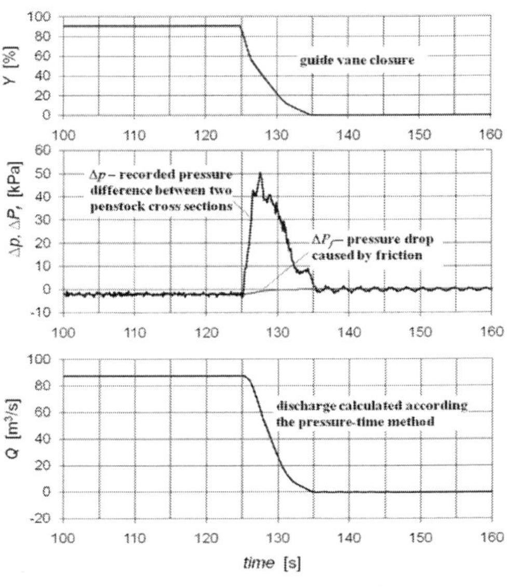

Fig. 18. Example of measurement of discharge through a turbine using the pressure-time method (own experiment).

4.3 Theoretical principles of the pressure-time method

In order to derive the relationship for calculating flow rate Q let us consider the closed conduit with the cross section area A changing along its longitudinal axis – see Fig. 19. In this conduit the stream of liquid is stopped. Let us extract from this conduit a segment of a length L between cross sections 1-1 and 2-2.

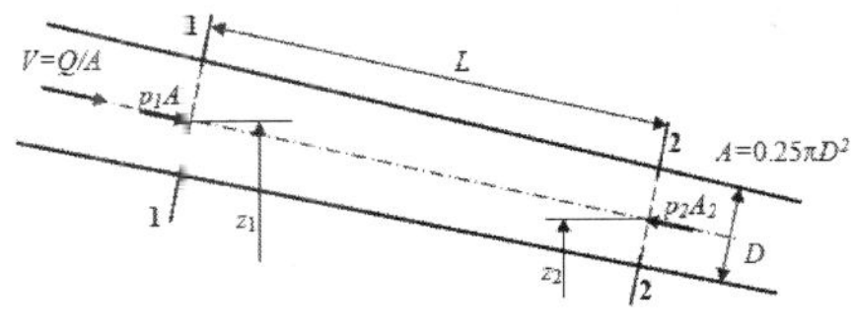

Fig. 19. Segment of a closed conduit with marks needed to explain the theoretical basis of the pressure-time method.

Let us farther assume that the liquid velocity and pressure distributions are constant in these sections, and that the liquid density and the flow section area do not change due to pressure variations.

According to these assumptions, the dependence between the parameters of the one-dimension unsteady flow in two selected cross sections of the conduit can be described using the energy balance equation, well known from the literature in the form:

$$\alpha_1 \frac{\rho Q^2}{2A_1^2} + p_1 + \rho g z_1 = \alpha_2 \frac{\rho Q^2}{2A_2^2} + p_2 + \rho g z_2 + \Delta P_f + \rho \int_0^L \frac{dx}{A(x)} \cdot \frac{dQ}{dt} \tag{10}$$

where ρ denotes liquid density, p_1 and p_2 present static pressures in pipeline cross sections 1-1 and 2-2, respectively (see Fig. 17), z_1 and z_2 are levels of 1-1 and 2-2 hydrometric pipeline section weight centres, α_1, α_2 are the Coriolis coefficients (kinetic energy correction coefficients) for 1-1 and 2-2 sections, respectively, Q is the volumetric flow rate (discharge), g means gravity acceleration and, finally, ΔP_f is the pressure drop due to friction losses between 1-1 and 2-2 cross sections.

The last but one term on the right-hand side of the above equation represents hydraulic (friction) losses. In the discussed method the pressure drop due to these losses is determined from the following square discharge function:

$$\Delta P_f(t) = K_f Q(t)\left|Q(t)\right| \tag{11}$$

in which the constant K_f is calculated basing on the measured values of the initial flow conditions as follows:

$$K_f = K_{f0} = \frac{\Delta P_{f0}}{Q_0 \left|Q_0\right|} \tag{12}$$

The last term in the Eq. (10) is the unsteady term, depending on the rate of change of the discharge $Q = VA$. This term represents the effect of liquid inertia in the considered conduit segment. For the steady-state flow this term is equal to *0*, and then the above equation takes the form of the Bernoulli equation for the flow of a real liquid.

For simplification of notation let us introduce the following quantities:

- Static pressure difference between conduit sections 2-2 and 1-1 related to a reference level:

$$\Delta p = p_2 + \rho g z_2 - p_1 - \rho g z_1 \tag{13}$$

- Dynamic pressure difference between conduit sections 2-2 and 1-1:

$$\Delta p_d(t) = K_d\ [Q(t)]\ \text{, where}\ K_d = \frac{\alpha_2 \rho}{2A_2^2} - \frac{\alpha_1 \rho}{2A_1^2} \tag{14}$$

- Geometrical factor of the examined conduit segment of L length and A cross section area:

$$C = \int_0^L \frac{dx}{A(x)} \tag{15}$$

Then we get the differential equation (Eq. (10)) in the form:

$$\rho C \frac{dQ}{dt} = -\Delta p - \Delta p_d - \Delta P_f \tag{16}$$

After integrating this equation over the time interval (t_0, t_e), in which the flow conditions change from initial to final stage, we obtain the discharge difference between these conditions. If we assume that we already know the discharge value in the final conditions (q_l), i.e. after fully closing the shut-off device, we get the following formula for discharge in the initial conditions (before the water flow stopping was initiated):

$$Q_0 = \frac{1}{\rho C} \int_{t_0}^{t_f} \left(\Delta p(t) + \Delta p_d(t) + \Delta P_f(t) \right) dt + q_l \tag{17}$$

The discharge in the final conditions, if different from zero due to leakages through the shut-off device, has to be measured or assessed using a separate method.

The above integral formula reveals that in order to determine the discharge, the pressure drop ΔP_f due to hydraulic loss in the examined conduit segment and the dynamic pressure difference Δp_d in the hydrometric cross sections of the conduit should be extracted from the measured static pressure difference Δp. The need for calculating these quantities, using their simplified dependence on the square of the flow rate (Eqs. 5, 6 and 8), unfavorably affects the accuracy of the measurement method.

4.4 Versions of the pressure-time method

In practice, various versions (different variants) of the pressure-time method are used. The most important of them include - Fig. 20:

1. The classic version based on direct measurement of pressure difference between two hydrometric sections of the conduit using a pressure differential transducer.
2. The version making use of separate measurements of pressure variations in two hydrometric sections of the conduit
3. The version based on measurement of pressure changes in one hydrometric section of the conduit and relating these changes to the pressure in the open liquid reservoir, to which the conduit is directly connected.

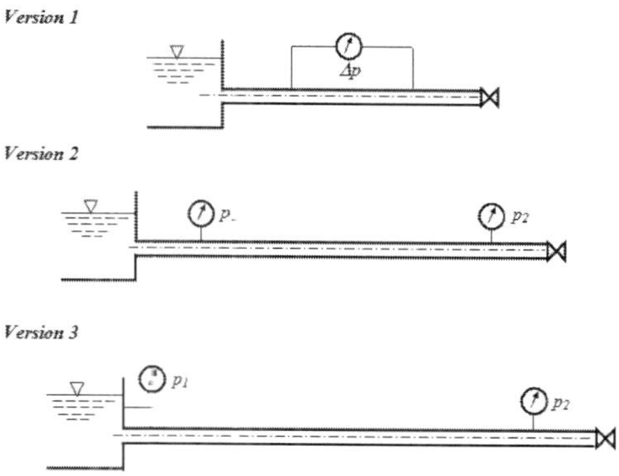

Fig. 20. Various versions of the pressure-time method.

The classic version based on the straight penstock measuring segment with constant diameter is mostly recommended in the standards. The results of measurements obtained using this version have the lowest uncertainty. Then, the version 3 belongs to the simplest and cheapest in using, that has significant importance from the economical point of view for small hydropower plants.

4.5 Simplifying assumptions

When using the pressure-time method, one should be aware of differences between the real flow in conduits and its theoretical model taking into account certain simplifications. Along with the inaccuracy of the measuring devices used and numerical calculations applied, those simplifications can be the source of inaccuracy of the measurement method. The point is that the effect of those simplifications should not be too excessive and should not provoke significant errors in discharge measurement.

Basic simplifying assumptions of the pressure-time method refer to:

1. pressure independence of the liquid density and flow section area of the conduit,
2. negligible effect of residual (free) pressure oscillations in the conduit,
3. required constant pressure distribution in the measuring sections of the conduit,
4. method used for calculating friction loss in the measuring segment of the conduit,
5. method used for calculating dynamic pressure difference in cases of conduits with section area A changing with their longitudinal axis.

The effect of these simplifications is briefly discussed below.

Ad.1. The relative change of water density and the area of the penstock cross section in relation to the pressure increase Δp can be evaluated using the following formulas:

$$\frac{\Delta\rho}{\rho}=\frac{1}{E_w}\Delta p \tag{18}$$

$$\frac{\Delta A}{A}=\frac{D}{eE}\Delta p \tag{19}$$

where: ρ - liquid density, E_w - liquid bulk module, p - pressure, A - area of penstock cross-section, D - internal diameter of penstock, e - thickness of penstock wall, E - elasticity (Young) module of penstock material.

In the majority of cases of practical implementation of the pressure-time method to the penstocks of hydraulic machines, relative changes of the above quantities are extremely limited. For the water turbine flow systems with steel or concrete penstocks considered by the author of this chapter, the relative changes of those quantities did not exceed 0.1%.

$$\frac{\Delta\rho}{\rho}<0.1\% \quad \frac{\Delta A}{A}<0.1\% \tag{20}$$

Remark: The method can be used for cases in which the liquid density change and the pipeline wall deformation resulting from the pressure rise caused by stopping the liquid stream are negligibly small. These requirements are completely fulfilled for steel or concrete penstocks and for water considered as low-compressible fluid.

Ad.2. Despite the fact that the liquid density and the conduit diameter changes are negligibly small, their effect can be observed in the time-histories of the pressure changes in the conduit, mainly in the form of diminishing residual (free) pressure oscillations after full closure of the shut-off device. These oscillations around the equilibrium state do not affect the measured flow rate considerably, of course provided that they are properly considered. The way that can be used in this purpose is presented in the farther part of this chapter.

Ad.3. In large-dimension penstocks, keeping the pressure distribution constant is not physically possible. In order to obtain the best possible conditions, one should make sure that the locations of the measuring sections are properly selected, and the pressure measurements are performed using the manifolds that collect pressure from a several number of points uniformly distributed along the penstock perimeter. The IEC 60041 standard recommends measuring the static pressure difference between particular taps - it is important for the measured pressure differences not to exceed a certain limiting value, equal to 20% of the dynamic pressure.

Ad.4. A questionable issue in the method is calculating the pressure drop due to the friction between the measuring sections of the conduit - see Eq. (11). Since the nature of the water flow in large-dimension penstocks is strongly turbulent, then the use of these functions returns good results, especially for steady-state and for slowly changing unsteady flows. For such kind of flow, this is confirmed by the difference in the friction losses values (see Fig. 21) determined using the following formula:

$$\Delta P_f = f\frac{L}{D}\frac{V|V|}{2} \tag{21}$$

for cases when the friction coefficient f is considered as Re-dependent, calculated in accordance with the well-known empirical Colebrook-White formula (Cengel at al., 2006) and for cases when the friction coefficient f is constant $f = f_0$, determined for the initial flow conditions. The maximum relative difference between the losses calculated by these two ways does not exceed 2.5% for the conditions most frequently existed in hydro power plants. This value can be recognized as negligible when considering the small share of the friction losses in the pressure increment Δp - see below.

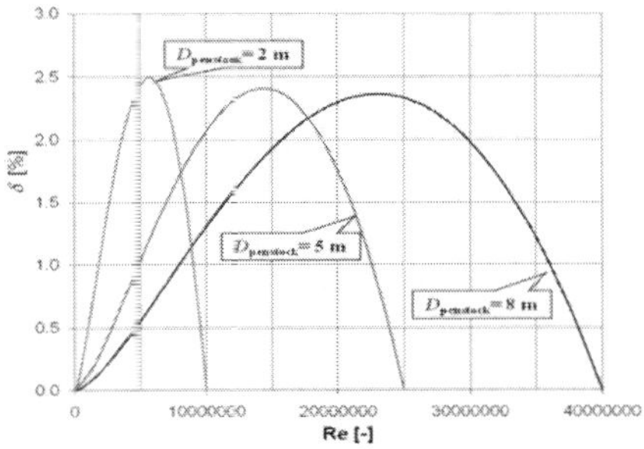

Fig. 21. Relative difference δ between the friction losses in the pipeline segment determined using a variable friction factor f calculated from the Colebrook-White empirical formula and a constant coefficient f_0 specified for the initial flow conditions (author's own calculations).

Moreover, references (Brunone at al., 1991a; Bughazem at al., 2000) show that dissipation of mechanical energy during flow deceleration (taking place when the pressure-time method is applied) is slightly less than that obtained from the quasi-steady hypothesis, which is in opposite to accelerating flow where energy dissipation is much larger. Some of unsteady friction losses models in the closed conduits use these features (Brunone at al., 1991b. These models have been confirmed experimentally - there is a high conformity between experimental and numerical results of water hammer course (Adamkowski & Lewandowski, 2006).

In this context, it is of great importance for achieving good accuracy of the measurement performed using the considered method, that the contribution of the pressure difference attributed to hydraulic loss would be possibly the smallest and would not exceed a certain limit. This requirement can be written in the form of the following inequality, referring to the ratio of the pressure drop caused by the resistance in the initial conditions to the average value of the static pressure difference measured between the sections during stream stopping:

$$\left|\frac{\Delta P_{f0}}{\Delta p_m}\right| \le \varepsilon_f \tag{22}$$

According to IEC 60041 standard, the value of ε_f is equal to 0.2 (20%) in cases of the classic version of the method (there is no difference of dynamic pressures between penstock measuring sections).

The consideration presented below shows that this requirement is sufficient when the discharge measurement accuracy is analyzed. For this purpose let us assume that discharge varies linearly during flow cut-off. This assumption is consistent with the conditions mostly occurring in the measurement practice. For such conditions, the ΔP_{f0} value does not exceed 6.7% (33%*20% = ~6.7%) of the Δp_m value. Taking into account the results presented in Fig. 21 (2.5% inaccuracy of friction loses calculation using Eq. (11) and (12), the uncertainty of the discharge measurement coming from the calculation of the friction loss between the measuring sections of the conduit is very small, not higher than about 0.2% (6.7%*2.5% = ~0.17%).

Remark 1: Highly important for the realization of the requirement presented in the Eq. (22) is to select speed of closure of the shut-off device properly. The higher speed of the closure process causes the higher rise of the measured pressure difference change (Δp_m), for the same pressure drop relating to the hydraulic loss in the initial conditions (ΔP_{f0}).

Remark 2: The theoretical description of the pressure-time method presented above is valid for both turbine and pump modes of operation. However, the IEC 60041 standard recommends using the method only in cases of turbine operation mode. Author's experiences indicate on the possibility to utilize the method also in cases of pumping operation mode. One of the necessary requirements in such cases is correct calculation of the hydraulic losses - pressure drop caused by the friction loss between the hydrometric cross sections of a pipeline. Typical calculation procedures, including the presented in the IEC 60041 and ASME PTC 18 standards, assume the hydraulic losses to be dependent on the square of the discharge value, as in the following equation:

$$\Delta P_f = K_f Q^2 \tag{23}$$

The hydraulic losses calculated in accordance with Eq. (23) do not depend on flow direction (both are always of the same sign) – as contrary to the Eq. (11). So, following this type of calculation may lead to the generation of additional error while determining the discharge value in the pressure-time method. It results from the fact that under some conditions, particularly in cases of pump-turbine tests, the significant temporary change of liquid flow direction takes place. The calculation procedure, based on Eq. (11) enables to account for actual flow direction and to increase measurement accuracy, particularly under pump mode operating conditions for which, in general, the temporary flow reversion during the pump shut-down required by the method occurs. The tests performed in the laboratory and *in situ* confirm the advisability of calculation of hydraulic losses from Eq. (11), particularly in cases when the pressure-time method is used to investigate hydraulic machinery characteristics under pumping regime (Adamkowski & Janicki, 2010).

Ad.5. Like for the friction loss calculations, a questionable issue is the dynamic pressure change Δp_d (t) calculations from the Eq. (14) in the cases of conduits with different hydrometric section areas. It is a well-known fact that the values of the Coriolis coefficients (α_1 and α_2) change in relation to the nature of the flow, in particular to the velocity distribution in the hydrometric sections of the conduit. For turbulent flows, these values are within the limits (1.04 - 1.1) (Cengel, 2006). Therefore, like in the case of the friction loss, assuring that the contribution of the dynamic pressure difference in the measured pressure difference is possibly the smallest and does not exceed certain limit is of high importance for achieving good accuracy of the measurements performed using the pressure-time method. This requirement can be written in the form of the ratio of the dynamic pressure difference in the initial conditions (Δp_{dmax}) to the average value of the static pressure difference measured between the measuring sections during stopping the liquid stream (Δp_m):

$$\left. \frac{\Delta p_{d\max}}{\Delta p_m} \right| \leq \varepsilon_d \tag{24}$$

Obviously, to reduce the effect of the dynamic pressure difference on the result of the measurement, the proper selection of the hydrometric sections is highly important. They should be selected in such a way that their areas would least differ from each other. Moreover, it is recommended to tend to shorten the closure time of the shut-off device, and to keep the pressure distribution in the hydrometric sections not very disturbed, close to constant.

Besides simplifications discussed above, other main sources of inaccuracy of the considered discharge measurement method include the inaccuracy of the measuring devices used (Adamkowski & Janicki, 2007), the numerical calculations applied (Adamkowski & Janicki, 2010) and determination of the C factor in Eq. (17) (Adamkowski at al., 2009). Some selected issues are presented and discussed in the next part of this chapter.

4.6 Some chosen problems to be solved during applications of the pressure-time method

4.6.1 Determination of the upper integration limit of the pressure-time curve

Calculation of the initial value of discharge Q_0 using the Eq. (17) requires to specify the time integration limits, t_0 and t_f. These values should determine the time interval in which the flow is completely cut-off. Contrary to t_0 time (lower limit of integration), the determination of t_f time (upper limit of integration) presents difficulties. Even precise synchronization of recording of the flow cut-off device run with measurement of the pressure rise does not ensure the exact determination of t_f time value. The reason for this is often the lack of the strict relation between the time moment at which the closing device movement is stopped and the time moment of flow cut-off finish - in some cases despite the termination of flow cut-off run, the closing device is still in motion, e.g. in result of elastic strain.

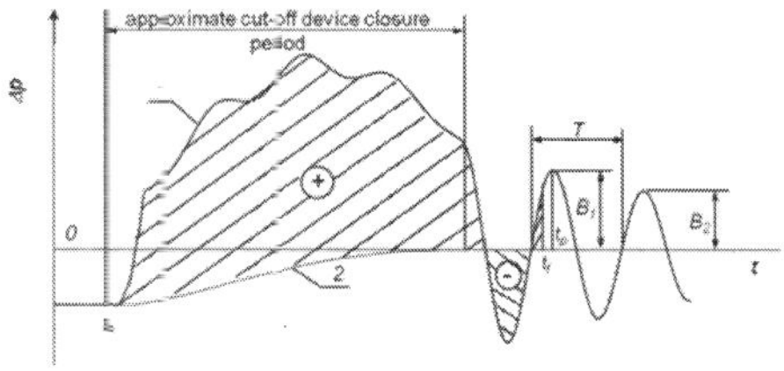

Fig. 22. Typical character of changes of pressure difference between the measuring penstock sections during flow stoppage and the notation applied.

Therefore, to determine the upper integration limit t_f the character of free pressure oscillations is presumed - Fig. 22. These oscillations remain in the penstock after the flow cut-off, as a result of interaction between inertial effects and effects associated with liquid compressibility and deformability of the penstock shells. One of the procedures relating to the upper integration limit calculation in the pressure-time method is given in the IEC 60041 standard. However, it includes mathematical inaccuracy - the author of this chapter has proved that it does not ensure to set a zero-value integral of free pressure difference oscillations that intent to eliminate their influence on the discharge measurement (Adamkowski & Kwapisz, 2000; Adamkowski & Janicki, 2010). The consideration regarding the relevant explanation and the procedure improvement are presented below.

Let us assume that free pressure oscillations after the termination of flow cut-off may be described by the following function (Fig. 23):

$$\Delta p(t) = B_o e^{-ht} \cos(\omega t) \tag{25}$$

with $\omega = 2\pi/T$ denoting the circumferential wave frequency, $h = (1/T) \ln(B_i / B_{i+1})$ - oscillation damping coefficient (reciprocal of the relaxation time), $\ln(B_i / B_{i+1})$ - logarithmic damping decrement, T - pressure wave period.

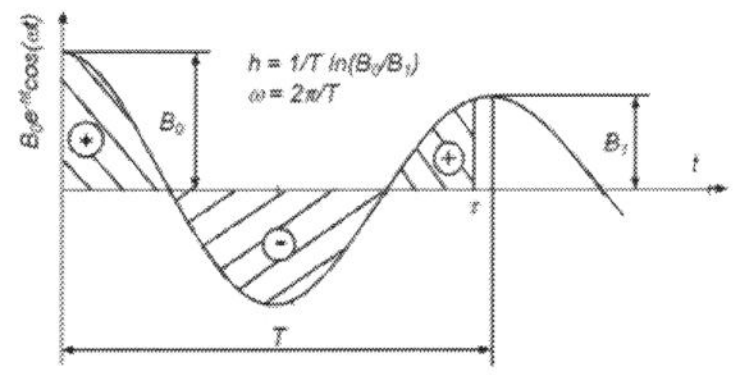

Fig. 23. Free pressure oscillation run including the notation applied.

In order to avoid the influence of diminished free pressure oscillations on the discharge value to be determined, the time point τ, fulfilling the condition

$$\int_0^\tau B_o e^{-ht} \cos(\omega t)\, dt = 0 \tag{26}$$

is sought for. Eq. (26) represents the condition of equal total fields (areas) defined by the diminished pressure wave curve below and over the time axis - Fig. 26.

Basing on the analysis performed, it has been stated that the procedure of determining the τ time, as presented in the IEC 60041 standard does not lead to a strict solution of Eq. (26). It can be proved that this procedure follows from the solution of an equation based on the indefinite integral of Eq. (26) which can be determined analytically. By solving the following equation:

$$\int B_o e^{-ht} \cos(\omega t)\, dt = B_o \frac{e^{-ht}}{h^2 + \omega^2}\left[-h\cos(\omega t) + \omega \sin(\omega t)\right] = 0 \tag{27}$$

in respect to time t, one derives an analytical expression which is consistent with the presented in the IEC 60041 standard and used to determine the end of the integration interval. The precise (strict) solution should be based on the definite integral of Eq. (26) which can be written down in the following form:

$$\int_0^{\tau} B_o e^{-ht} \cos(\omega t)\, dt = \frac{B_o}{h^2 + \omega^2}\left\{e^{-h\tau}\left[-h\cos(\omega\tau) + \omega\sin(\omega\tau)\right] + h\right\} = 0 \tag{28}$$

Equation (28) cannot be solved analytically. By comparing equations (27) and (28) it can be seen that the term:

$$\text{Eq. (28) - Eq. (27)} = B_o h / (\omega^2 + h^2) \tag{29}$$

has not been taken into account by the IEC 60041 standard procedure.

From these considerations it is evident that one of the possibilities to eliminate the influence of free pressure oscillations on the calculated discharge is to subtract the value $B_o h/(\omega^2 + h^2)$ from the integral value calculated from the recorded pressure difference diagram as proposed in the IEC 60041 standard The value resulting from Eq. (29) can be easily calculated basing on the recorded pressure-time diagram.

Another approach aimed at eliminating the effect of free pressure oscillations is the numerical determination of solution of the integral given by Eq. (26) (Adamkowski & Kwapisz, 2000).

The influence of free pressure oscillations generated in the pipeline after the flow cut-off on the discharge measurement realized by means of the pressure-time method increases as the free oscillation amplitude increases. It appears from the author's experience that in some cases it may achieve 0.5% of determined discharge.

4.6.2 The influence of a curved penstock application on discharge measurement

Following the classical approach (version 1), the pressure-time method applicability is limited to straight cylindrical pipelines with constant diameters. However, the IEC 60041 standard does not exclude application of this method to more complex geometries, i.e. curved penstock (with an elbow). It is obvious that a curved pipeline causes deformation of the uniform velocity field in its cross-sections, which subsequently causes aggravation of the accuracy of the pressure-time method flow rate measurement results - Fig. 24. So, the influence of a curved penstock application on discharge measurements by means of the considered method should be taken into account. The author of this chapter and his co-workers have developed the special calculation procedure for solving that problem (Adamkowski at al., 2009). The procedure is based on the CFD *(Computational Fluid Dynamic)* simulation - Fig. 25. It allows calculating the equivalent value of the geometry factor C (see Eq. (15)) for a measuring penstock segment with an elbow (or elbows). The value can improve the discharge measurement results of the standard pressure-time method without curved penstock correction. As an example, the systematic uncertainty caused by neglecting the effect of the two elbows on measured discharge values has been estimated - Fig. 26 (Adamkowski at al, 2009). In the considered case, the average value of the quantity Δf = 0.45% was taken to correct the discharge values following from the calculation carried out for the C factor obtained only from the geometry of measuring penstock segment.

Similar approach, based on the *CFD* simulation, is needed for determining the value of C factor for whole penstock while the pressure time method in the version 3 is used. The author of this chapter strives to develop a special procedure for this purpose. It is assumed that this procedure should allow for applying the version 3 of pressure-time method in penstocks shorter than it follows from the current requirements of the IEC 60041 standard - in addition to ensure increasing measurement accuracy of this method.

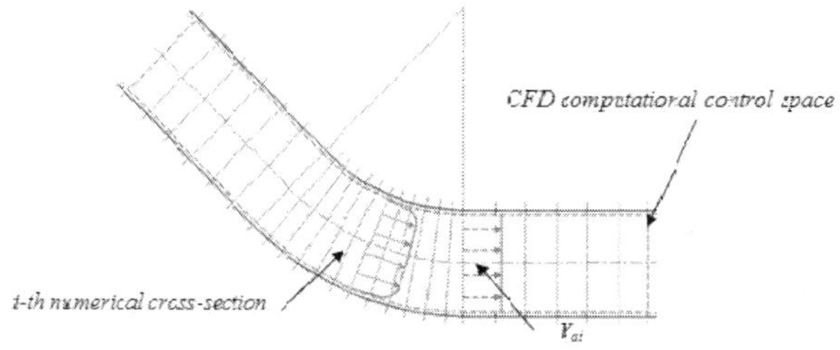

Fig. 24. A penstock elbow with marked computational space.

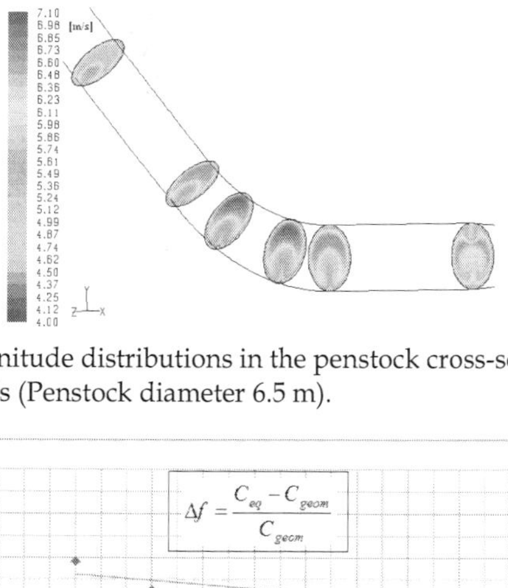

Fig. 25. The velocity magnitude distributions in the penstock cross-sections calculated for discharge of Q = 200 m³/s (Penstock diameter 6.5 m).

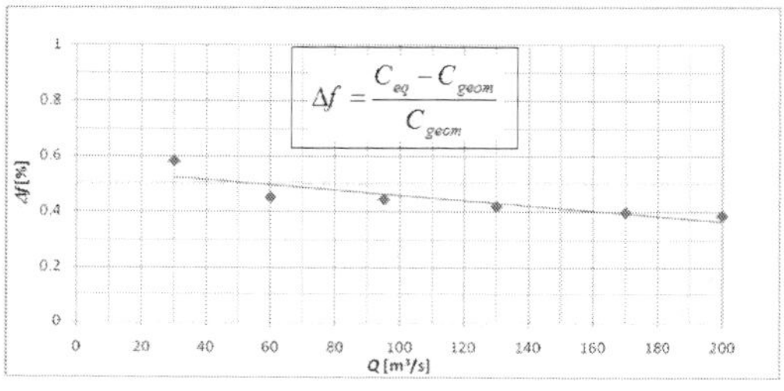

Fig. 26. The values of Δf deviation factor C determined for the assumed discharge values for the considered case of measuring penstock segment with two elbows (C_{geom} is the value of C factor determined only from the geometry of the measuring penstock segment and C_{eq} is the equivalent value of C factor determined with correction obtained using the proposed procedure).

4.6.3 Using the pressure-time method based on special instrumentation installed inside penstocks

Discharge measurement using the pressure-time method typically requires mounting instrumentation on the outside of the penstock. In the case of a hydropower plant where the penstock is embedded in concrete traditional way for using that method is not possible. Therefore, an approach that involved installing discharge measurement instrumentation inside the penstock has been recently developed (Adamkowski at al., 2006b, 2008; Adamkowski & Janicki, 2008). One of the first applications of such kind of techniques is presented below.

In the considered case, the discharge was measured using the pressure-time method in the version based on separate pressure difference measurements in two hydrometric cross sections of the penstock, 1-1 and 2-2 – Fig 27.

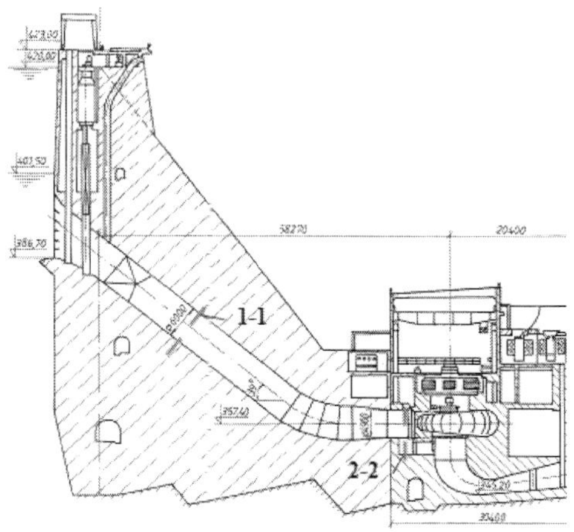

Fig. 27. A scheme of the flow system in one of the Polish hydropower plants with marked cross-sections used for discharge measurement.

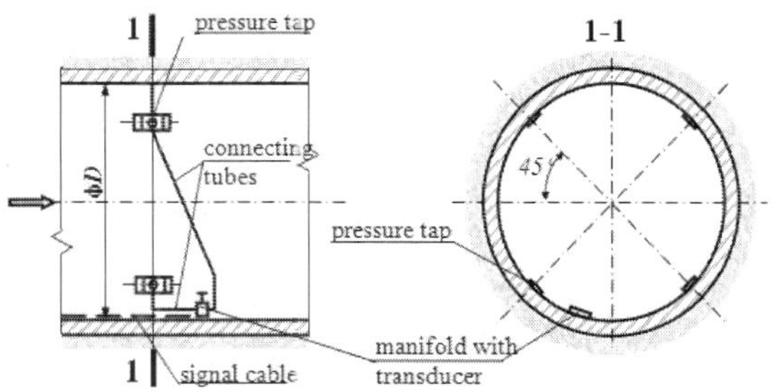

Fig. 28. Distribution of pressure reception holes (taps) in measuring section 1-1 and their connection to the hermetic manifold with absolute pressure transducer installed inside.

In each of those sections, 4 pressure taps (or pressure reception holes) were prepared and connected using small tubes to the manifold and pressure transducer. A typical manifold was used in the lower penstock section 2-2. There was possibility to prepare the whole system of collecting and measuring pressure from the taps, having the access from outside.

Since there was no access from outside to the upper penstock section 1-1, a special internal installation was prepared for pressure reception and measurement - see Fig. 28, 29 and 30. The Fig. 29 shows a flat bar with the specially drilled hole for the pressure reception. The hole diameter was equal to 3 mm. The flat bar was welded to the penstock shell, respectively to the flow direction. Four pressure reception holes were connected, using copper tubes of 10 mm diameter, to the manifold with absolute pressure transducer of precision class 0.1 mounted hermetically inside it - Fig. 30.

It is worth stressing that preparing the pressure measuring system inside a penstock of 6 meters in diameter and inclined by an angle of 40 degrees was an extremely difficult task.

Figure 31 shows: the time-histories of the static pressures measured in the selected and prepared sections 1-1 and 2-2, the static pressure difference determined from them, and the discharge calculated according to the pressure-time method.

After preparing and mounting the entire instrumentation inside the penstock, filling this with water and full deaeration was of high importance task. On the contrary to easy way of the deaeration performed in typical, external installation of pressure measurement, in the presented case for the 1-1 hydrometric section a special procedure had to be applied (description of this procedure goes beyond the scope of this chapter).

Fig. 29. View of the plate with a pressure reception hole.

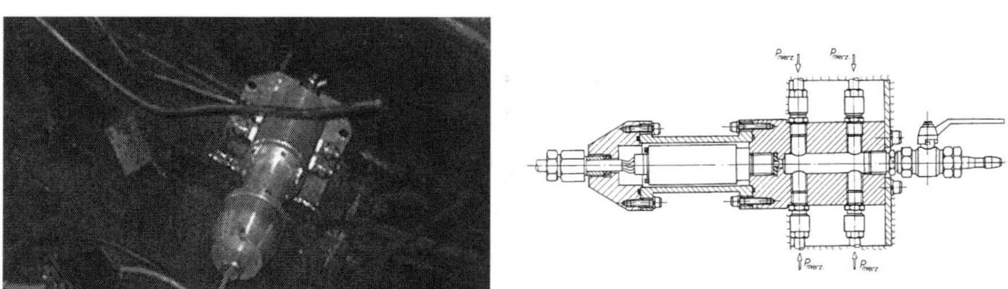

Fig. 30. Manifold with absolute pressure transducer hermetically mounted (own application).

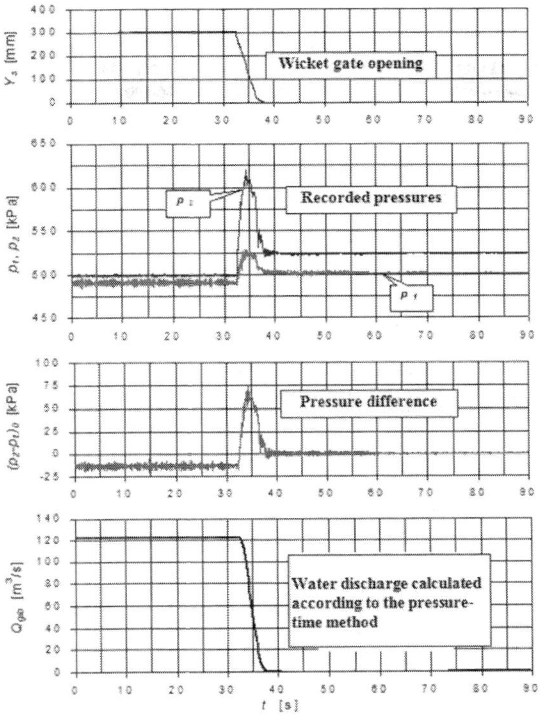

Fig. 31. Pressure changes measured in the measuring sections of the classical turbine penstock, and the discharge determined from them (own application).

4.6.4 Measurement of leakage flow through closed hydraulic turbine wicket gates

As it was mentioned above, applying the pressure-time method for discharge measurement requires determination of the rate of leak flow through gaps of closed shut-off devices, for instance wicket gates of hydraulic machines. The technique mostly used for this purpose is based on time-record of water level decrease in cylindrical segments of a penstock or surge tanks during turbine standstill. This technique allows to determine the leakage rate through closed devices in cases where the inflow due to leakage past closed intake gates has to be measured or estimated using an additional, separate method. The separate measurement of the rate of such inflow is rather complicated and usually burdened with significant uncertainty. Therefore, the extension of the aforementioned technique aimed at eliminating separate measurement of leakage through closed intake gates has been developed and presented in (Adamkowski & Janicki, 2009). The proposed technique is based on simultaneous measurement of pressure on both sides of a closed guide vane apparatus and on applying the square root dependence between determined leakage flow rate and recorded pressure difference between the measuring points. Short description of this technique is presented below.

According to the principle of mass conservation it can be written (see Fig. 32):

$$q - \frac{dV}{dt} = q_l \tag{30}$$

where q means the flow rate of leakage through the turbine intake gates closure (flow into the control water volume V), q_l – flow rate of leakage through the closed turbine guide vanes and dV/dt – time derivative of water volume inside the penstock segment.

The derivative dV/dt can be calculated numerically according to formula:

$$\frac{\Delta V}{\Delta t} = \frac{A}{\sin\beta}\frac{\Delta z}{\Delta t} = \frac{A}{\rho g \sin\beta}\frac{\Delta p_t}{\Delta t} \tag{31}$$

where A is the area of constant pipeline cross-section, z – water level inside the penstock, p_t - pressure before the closed shut-off device, for instance, in the turbine spiral case when guide vane is a shut-off device, ρ – water density, g – acceleration of gravity and β - slope angle of penstock axis.

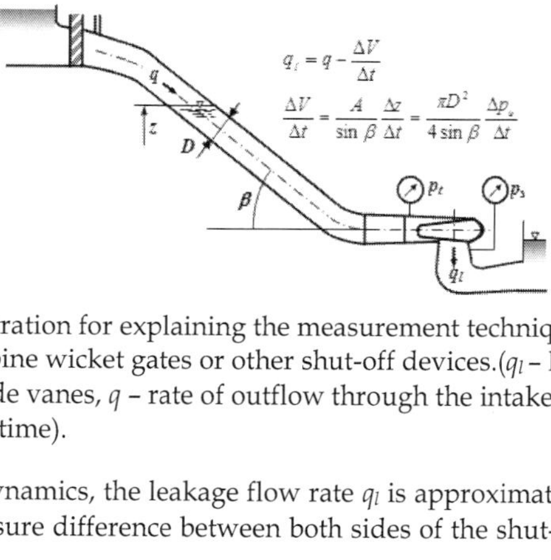

Fig. 32. Schematic illustration for explaining the measurement technique of leakage flow rate through the closed turbine wicket gates or other shut-off devices.(q_l – leakage flow rate through the closed guide vanes, q – rate of outflow through the intake gate, V – volume of water in a penstock, t - time).

On the basis of fluid dynamics, the leakage flow rate q_l is approximately proportional to the square root of the pressure difference between both sides of the shut-off device and may be expressed as follows:

$$q_l = a(p_t - p_s)^{0.5} \tag{32}$$

The proportionality coefficient a can be determined on the basis of recorded pressures in time $p_t(t)$ i $p_s(t)$ – Fig. 33 – and the calculations of time derivative of water volume change inside the pipe $\Delta V/\Delta t$ in accordance with Eq. (31). (The pressures p_t and p_s have to be referred to the same level.) The way of determination of leakage flow rates q_l and q is presented graphically in Fig. 34. The results obtained from the test and calculation are presented in the coordinate system of $x = (p_t - p_s)^{0.5}$ and $y = \Delta V/\Delta t$. Then, they are approximated by means of the least square method using linear function $y = ax+c$, where constants a i c stand for, respectively, the value of coefficient a in Eq. (32) and the rate of the outflow through the intake gate, $q = -c$.

Applying Eq. (32) will enable to recalculate the leakage flow rate q_l through the closed devices for the conditions existing after the turbine shutdown during discharge measurement test by means of the pressure-time method. (Also this technique enables to determine value of leakage flow rate during the turbine standstill for the pressure difference equal to the gross head.)

Two considered cases concerning the measurement of leakage flow rate through the hydraulic turbines are presented in (Adamkowski & Janicki, 2009).

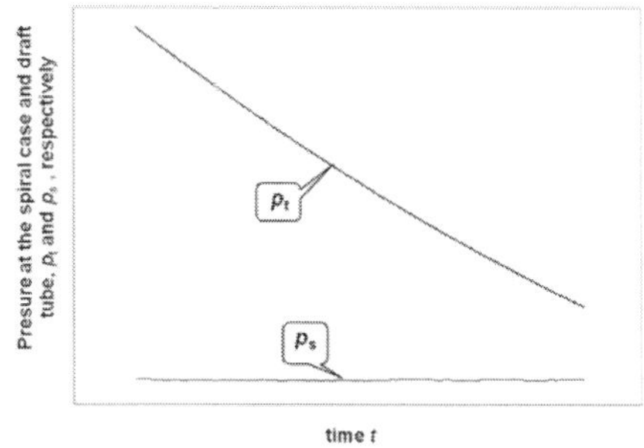

Fig. 33. Example of pressure variations in a spiral case and a draft tube recorded during leakage test (own application (Adamkowski & Janicki, 2009)).

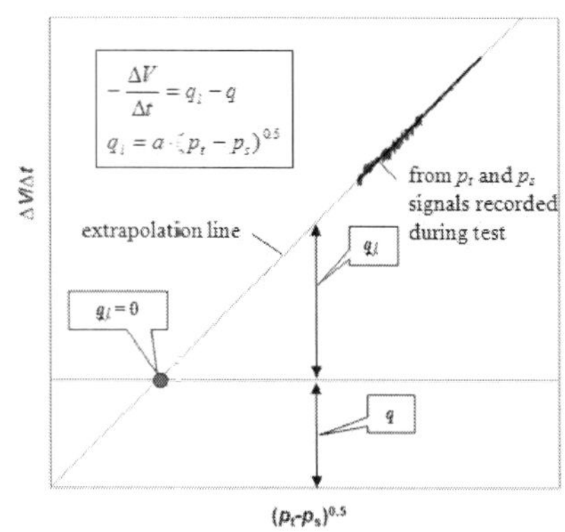

Fig. 34. Example of the results obtained from the leakage test conducted in a Francis turbine system - rate in water volume ΔV change in penstock control area was determined as a square root function of pressure difference between two sides of the turbine guide vanes, p_t i p_s. (own application (Adamkowski & Janicki, 2009)).

5. Conclusions

1. Standard measuring devices, such as constriction flow meters (orifice plates, flow nozzle or Venturi tube), calibrated pipe elbows, electromagnetic and ultrasonic flow meters can be used for flow rate measurements in conduits of small diameters, for instance up to 1 m. Such kind of devices are mounted in a properly prepared measuring

segment of the conduit and can be used for continues discharge measurement as an important part of monitoring of hydraulic systems.

2. Large-scale discharge measurements in hydropower systems are the most complicated, costly ineffective and having higher level of uncertainty with comparison to measurement of other quantities (head and power) needed to determine hydraulic performance of water turbines. The correct application of discharge measurement techniques in large-scale objects is usually a major challenge, despite recent immense progress in those techniques.
3. Among a few primary methods for absolute discharge measurements in hydropower systems (velocity-area method, pressure-time method, tracer method, ultrasonic method and volumetric method), the most attractive from practical point of view is the velocity-area method and the pressure-time method. The tracer method is less popular, mainly, due to requirement of a very long measuring segment of flow conduits and special, additional conditions facilitating the mixing process of the introduced marker. The volumetric method has very limited area of application - generally it can be used in hydropower plants equipped with artificial reservoirs, particularly in pumped-storage plants. The ultrasonic method is still the object of numerous research activities oriented on its improvement and has still not reached proper acceptance among the most specialists.
4. The velocity-area (propeller current meter method, for instance) is specially recommended and available for discharge measurement in low and very low head water power plants with short intakes of turbines, although further progress in acoustic scintillation techniques may change this situation.
5. The pressure-time method is a very attractive one for discharge measurements in the hydropower plants equipped with penstocks longer then 15-20 m. Recently, the increased accuracy of the devices used for pressure measurements and the use of computer techniques for collecting recorded data and their numerical processing make this method more attractive than earlier when for its outdated versions the optical techniques to record pressure changes combined with the manual graphical integration was used.
6. The recent progress and improvement in discharge measurements using the pressure-time method refers, mainly, to: (1) successful development of the special instrumentation installed inside penstocks in the frequent practical cases where there is no access to the penstock from outside, (2) changes of the discharge calculation procedure based on the recorded pressure-time diagram, and (3) development of the original way for measuring of leakage flow through the closed turbine guide vanes.
7. The special numerical procedure for evaluating the influence of an elbow (or elbows) on discharge values measured using the pressure-time method was recently developed. The procedure, based on the CFD simulation, allows to determine the equivalent value of the geometry factor for a measuring penstock segment with elbows. It is recommended to use for correction of discharge measurement results for cases of penstocks with elbows.
8. It is expected to use the pressure-time method in the simplified version - the version 3 based on measurement of pressure changes in one chosen hydrometric section of the penstock and relating these changes to pressure in the water reservoir supplying the penstock directly. Such kind of an application is very important for small hydropower

plants concerning economical issues. It seems that the special procedure based on the *CFD* simulation should allow to use this version of pressure-time method in penstocks shorter than it follows from the IEC 60041 standard requirements, and, in addition, to increase accuracy of discharge measurement in such hydraulic systems.

9. Not all theoretical issues of the pressure-time method have been solved comprehensively. Ones of those unsolved issues concern calculation of the friction loss during the unsteady flow of liquid in closed conduits and calculation of dynamic pressure difference between the measuring conduit sections. The method adopted for calculating these quantities, justified for the steady-state flows, should be verified in unsteady conditions.
10. To achieve the expected accuracy of the measurement, the selection of conditions in which the pressure-time method is used, along with the measuring devices applied, should be based on the analysis of the measurement uncertainty, recommendations given in relevant standards, as well as personal experiences gained.

6. References

ASME PTC 18:2002, American National Standard: *Hydraulic Turbines and Pump–Turbine, Performance Test Codes* (Consolidation of ASME PTC 18-1992 and ASME PTC 18.1-1978).

IEC 60041: 1991, International Standard: *Field acceptance tests to determine the hydraulic performance of hydraulic turbines, storage pumps and pump-turbines,* European equivalent: EN 60041: 1999.

IEC 62006: 2009, *Hydraulic machines - acceptance tests of small hydroelectric installations.*

ISO 5168: 1978, *Measurement of fluid flow - Estimation of uncertainty of a flow-rate measurement.*

ISO 1438-1: 1980, *Water flow measurement in open channels using weirs and Venturi flumes - Part 1: Thin-plate weirs.*

ISO 3354:1988, *Measurement of clean water flow in closed conduits - Velocity-area method using current-meters in full conduits and under regular flow conditions.*

ISO 9104: 1991, *Measurement of fluid flow in closed conduits - Methods of evaluating the performance of electromagnetic flow-meters for liquids.*

ISO 5167-1: 1991, *Measurement of fluid flow by means of pressure differential devices - Part 1: Orifice plates, nozzles and Venturi tubes inserted in circular cross-section conduits running full.*

ISO 6817: 1992, *Measurement of conductive liquid flow in closed conduits - Method using electromagnetic flowmeters.*

ISO 748:1997, *Measurement of liquid flow in open channels - Velocity-area methods.*

ISO 7194: 2008, *Measurement of fluid flow in closed conduits - Velocity-area methods of flow measurement in swirling or asymmetric flow conditions in circular ducts by means of current-meters or Pitot static tubes.*

Adamkowski, A. & Kwapisz, L. (2000). Determination of the integration limits in the GIBSON method for measuring the flow rate in closed conduits, *Proceedings of International Conference HYDROFORUM'2000,* pp. 287-298, Czorsztyn, Poland, 18-20.10.2000

Adamkowski, A. (2001) Flow rate measurement in operation conditions of hydraulic turbines. *Scientific-Technical Monthly Journal POMIARY AUTOMATYKA KONTROLA*, (June 2001), pp. 10-13

Adamkowski, A. & Lewandowski, M. (2006) Experimental examination of unsteady friction models for transients pipe flow simulation, ASME Journal of Fluid Engineering, , Vol. 128, (November 2006), pp.1351-1363

Adamkowski, A.; Lewandowski, M.; Lewandowski, St. & Cicholski, W. (2006). Calculation of the cycle efficiency coefficient of pumped storage power plants units basing on measurements of water level in the head (tail) water reservoir, *DAM ENGINEERING*, Vol. XVI, Issue 4, (February 2006), pp. 247-258 (International papers of technical excellence affiliated by International Water Power and Dam Construction)

Adamkowski, A.; Janicki, W.; Kubiak, J.; Urquiza, B., G.; Sierra, E.; F. & Fernandez, D., J., M. (2006). Water Turbine Efficiency Measurements Using the Gibson Method Based On Special Instrumentation Installed Inside Penstocks, Proceedings of 6th International Conference on Innovation in Hydraulic Efficiency Measurements, pp. 1-12, Portland, Oregon, USA, July 30 - August 1 2006

Adamkowski, A. & Janicki, (2007) W.: Influence of some components of pressure-time method instrumentation on flow rate measurement results, Proceedings of International Conference HYDRO2007, Granada, Spain, October 15-17, 2007

Adamkowski, A.; Janicki, W.; Urquiza, B. G.; Kubiak, J. & Basurto, M. (2007) Water turbine tests using the classical pressure-time method with measuring instrumentation installed inside a penstock, Proceedings of International Conference HYDRO2007, Granada, Spain, October 15-17, 2007

Adamkowski, A. & Janicki, W. (2008). A new approach to using the classic Gibson method to measure discharge, Proceedings of 15th International Seminar on Hydropower Plants, pp. 511-522, Vienna, November 26-28, 2008

Adamkowski, A.; Kubiak, J.; Sierra Fernando, Z.; Urquiza, G.; Janicki, W. & Fernández, J. M. (2008). An Innovative Approach to Applying the Pressure-Time Method to Measure Flow, *HRW (Hydraulic Review Worldwide)*, Vol.16, No.6, (December 2008), pp. 40-49

Adamkowski, A.; Krzemianowski, Z. & Janicki, W. (2009) Improved discharge measurement using the pressure-time method in a hydropower plant curved penstock, *ASME Journal of Engineering for Gas Turbines and Power*, Vol. 131, (September 2009), pp. 053003-1 - 053003-6

Adamkowski, A. & Janicki, W. (2009). A method for measurement of leakage through closed turbine wicket gates, *Proceedings of the HYDRO2009, International Conference and Exhibition*, pp. 1-8,Lyon, France October 26-28 2009, CD

Adamkowski, A. & Janicki, W. (2010) Selected problems in calculation procedures for the Gibson discharge measurement method, *Proceedings of 8th International Conference on Hydraulic Efficiency Measurement - IGHEM 2010*, pp. 73-80, Rookie, India, 2010

Brunone, B.; Golia, U. M. & Greco M. (1991) Some Remarks on the Momentum Equations for Fast Transients, International, *Proceedings of Meeting on Hydraulic Transients with Column Separation, 9th Round Table, IAHR*, pp. 201-209, Valencia, Spain, 1991

Brunone, B.; Golia, U. M. & Greco, M.(1991) Modelling of Fast Transients by Numerical Methods, *Proceedings of Meeting on Hydraulic Transients with Column Separation, 9th Round Table, IAHR*, , pp. 273-280 Valencia, Spain, 1991

Bergant, A. & Simpson, A. (1994) Estimating Unsteady Friction in Transient Cavitating Pipe Flow, *Proceedings of the 2nd International Conference on Water Pipeline Systems*, pp. 3-16, Edinburgh, Scotland, 1994.

Bughazem, M. B. & Anderson, A. (2000) Investigation of Unsteady Fiction Model for Waterhammer and Column Separation, Proceedings of the BHR Group 8th International Conference on Pressure Surges, pp. 483-498, Hague, The Netherlands, 2000.

Cengel, Y. A. & Cimbala, J. M. (2006) *Fluid Mechanics. Fundamentals and Applications*, McGraw-Hill International Edition, New York.

Gibson, N. R. (1923) The Gibson method and apparatus for measuring the flow of water in closed conduits, *ASME Power Division*, pp. 343-392.

Gibson N. R. (1959) Experience in the use of the Gibson method of water measurement for efficiency tests of hydraulic turbines, *ASME Journal of Basic Engineering*, pp. 455-487, 1959.

Gruber, P. (2008) Upgrading the IEC41 code with respect to acoustic discharge measurement, Proceedings of the 7th International Conference on Hydraulic Efficiency Measurements IGHEM'2008, pp.1-8, Milan, 2008.

Gruber, P.; Staubli, T.; Tresch, T. & Wermenlinger, F. (2010) Optimization of the ADM by adaptive weighting for the Gaussian quadrature integration, Proceedings of the 8th International Conference on Hydraulic Efficiency Measurement , IGHEM 2010, pp. 14-24, Rookie, India, 2010.

Hulse, D.; Miller, G. & Walsh, J. (2006) Comparing Integration Uncertainty of an 8 and 18-path Flowmeter at Grand Coulee Dam, Proceedings of the 6th International Conference on Innovation in Hydraulic Efficiency Measurements, pp.1-8, Portland, Oregon, USA, 2006.

Kubiak, J.; Urquiza, B. G.; Adamkowski, A.; Sierra E. F.; Janicki, W. & Rangel R. (2005) Special Instrumentation and Hydraulic Turbine Flow Measurements using a Pressure-Time Method, Proceedings of ASME Fluids Engineering Division Summer Meeting and Exhibition, FEDSM2005-77394, Houston, TX, USA, June 19-23, 2005.

Llobet, J.V.; Lemon, D.D.; Buermans, J. & Billenness, D. (2008) Union Fenosa Generación's field experience with Acoustic Scintillation Flow Measurement, Proceedings of IGHEM'2008, the 7th International Conference on Hydraulic Efficiency Measurements, Milan, 2008.

Nichtawitz, A.; Grafenberger, P.; Tormann, J.; Gaisbauer, R. & Kepler, J. (2004) Discussion on Acoustic Flow Measurement Method, Proceedings of IGHEM'2004, the 5th International Conference on Hydraulic Efficiency Measurements, pp. 1-8, Lucerne, 2004.

Strunz, T.; Wiest, A.; Fleury, A. & Fröhlich, T. (2004) Influence of turbulence on ultrasonic flow measurements, Proceedings of IGHEM'2004, the 5th International Conference on Hydraulic Efficiency Measurements, pp. 1-6, Lucerne, 2004

Tresch, T.; Staubli, T. & Gruber, P. (2006) Comparison of integration methods for multipath acoustic discharge measurements, Proceedings of IGHEM'2006, the 6th International Conference on Innovation in Hydraulic Efficiency Measurements, pp.1-16, Portland, Oregon, USA, 2006.

Troskolański, A. (1960). *Hydrometry*, Pergamon Press Ltd., Oxford, London, New York, Paris

Urquiza, G.; Adamkowski, A.; Kubiak, J.; Janicki, W. & Fernandez, J.M.(2007) Medicion del flujo de 8una turbina hidraulica de 170 MW utilizando en metodo Gibson, *Ingenieria hidraulica en Mexico*, Vol. XXII, No. 3, (Julio-septiembre de 2007), pp. 125-137.

6

Sediment Management in Hydropower Dam (Case Study – Dez Dam Project)

H. Samadi Boroujeni
Shahrekord University,
Iran

1. Introduction

Hydropower reservoirs are loosing their capacity due to sedimentation processes, and are therefore seriously threatened in their performance. Without any mitigating measures the viability of many reservoirs in the worldwide is questionable, as the impacts and losses are not balanced by the profits It is apparent that for mastering the reservoir sedimentation issues the use of strategies for controlling reservoir sedimentation becomes increasingly important because of sustainable development issues. Basic principles in sustainable development and use of reservoirs are:

- Human beings are at the center of concerns for sustainable development and use.
- Humans are entitled to a healthy and productive life in harmony with nature.
- Along with the right to develop and use reservoirs comes the responsibility to meet the needs of present and future generations.

To achieve sustainable development and use of reservoirs and a higher quality of life for all people, we should gradually reduce and eliminate unsustainable patterns of development and use subject to social, environmental, and economic considerations. Reservoir sedimentation shortens the useful life of reservoirs. Systematic and thorough consideration of technical, social, environmental, and economic factors should be made to prolong the useful life of reservoirs.

Approximately 1% of the storage volume of the world's reservoir is lost annually due to sediment deposition [Morris and Fan, 1998]. In some developing countries, where watershed management measures are not carried out effectively, reservoir storage is being lost at much larger rates. Although the reduction of sediment yield via a watershed management program is the best option for reducing the rate of reservoir sedimentation, flushing may be one of the most economic methods which offer recovering of lost storage without incurring the expenditure of dredging or other mechanical means of removing sediment. Flushing is the scouring out of deposited sediments from reservoirs through the use of low level outlets in a dam by lowering water levels, and thus increasing the flow velocities in the reservoir. The technique is not widely practiced because of the damages caused by the injection of high sediment concentrations to the downstream river system; involving large volumes of water being passed through the dam; being usually only effective in narrow reservoirs and requiring the reservoir to be emptied.

Mangahao reservoir in New Zealand 59% of the original operating storage had been lost by 1958, 34 years after the reservoir was first impounded. The reservoir was flushed in 1969, when 75% of the accumulated sediment was removed in a month [Jowett, 1984]. Another examples of reservoirs that have been successfully flushed are Baira (in India), Gebidem (in Switzerland), Gmund (in Austria), Irengshan, Honglingyin and Naodehai (in China)[Atkinson, 1996]. Another technique for decreasing Reservoir sedimentation is the passing of turbidity currents through the reservoir and low-level sluices in the dam. Turbidity currents occur when sediment-laden water enters an impoundment, plunges beneath the clear water, and travels downstream along the submerged thalweg. As the current travels downstream, it will generally deposit the coarser part of its sediment load along the bottom, and if the current reaches the dam, it can be vented through low-level outlets. For example, the records of turbidity current releases from a reservoir were made in July 1919 at Elephant Butte reservoir in the United States, where the inflow suspended sediment concentration was 72 g/l and discharge from the low-level outlet at the dam was 41 g/l [Lane, 1954]. Also in this subject a literature survey on different methods and their geomorphologic and sediment transport effects has been carried out by Brandt [Brandt, 2000].

It is well accepted that reservoir sedimentation poses a serious threat to available storage. The annual loss of storage in reservoirs is roughly 1% corresponding to a about 50 km worldwide (Mahmood, 1987). Some reservoirs have a much higher storage loss, e.g., the Sanmenxia Reservoir in China looses about 1.7% yearly. In the meantime significant transformations can occur in the stream basin due to the redistribution of sediments and discharges. Sloff (1991) reviewed these phenomena by means of a survey of the scattered literature in order to find the remaining gaps in the applied theory. Theoretical approaches are here desired to estimate the sedimentation threat and even to reconsider the design. In the past highly empirical models were used for this purpose, but often resulted in an underestimation of the actual sedimentation rate. This can be ascribed to failing theory as well as to a lack of data. For instance sedimentation rates of the Sefid-Rud reservoir in north-west Iran can be estimated with a 60 years old highly empirical approach to be about 35 million m^3/yr (Tolouie et al., 1993). However, after construction (in 1962) the measured rate was about 45 million m^3/yr causing a storage loss of over 30% in 1980. The original predicted useful reservoir life of one century based on old data was found to be actually about 30 years (Pazwash, 1982). Not until 1980 flushing operations were started which were able to regain about 7% of the lost capacity. In Dez Dam reservoir after 40 years from starting operation of the dam, the height of sediment surface behind face of the dam was reached near power intake level.

When dealing with reservoir-sedimentation problems engineers are challenged by the difficult questions emerging. How to incorporate reservoir problems in feasibility studies (cost-benefit analyses) including environmental and technical effects, limitations on benefit and possible measures? Or what is the impact of sedimentation on the reservoir performance such as power generation and water supply, and what is the impact of the reservoir on the river-system morphology? Obviously a good prediction of the processes and trying to better understanding of the reservoir behavior is essential to master the reservoir-sedimentation issues.

2. Reservoir sedimentation problems

During the 1997 19th Congress of the International Commission on Large Dams (ICOLD), the Sedimentation Committee (Basson, 2002) passed a resolution encouraging all member countries to the following measures:

1. Develop methods for the prediction of the surface erosion rate based on rainfall and soil properties.
2. Develop computer models for the simulation and prediction of reservoir sedimentation processes

Extensive literature exists on the subject of reservoir sedimentation. The book by Morris and Fan (1997), entitled *Reservoir Sedimentation Handbook* is an excellent reference and provides an extensive list of references.

At the first time we introduce the principle processes involved with sedimentation in a storage reservoir in Fig. 1, as treated in Sloff (1991 and 1997).

The most important distribution principles of these sediments in the reservoir can be subdivided into the following groups: be subdivided into the following groups:

- **Coarse sediment deltaic deposits:** mainly the coarse sediment fractions are deposited in the head of the reservoir by backwater effects during high discharges, forming a delta. The delta proceeds into the reservoir while the foreset slope can be considered as an area of instability and slumping.

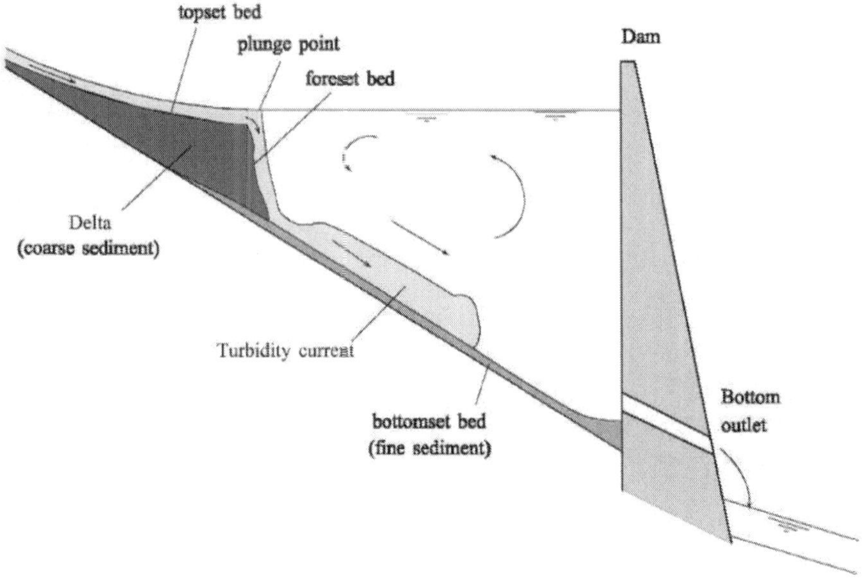

Fig. 1. Schematic presentation of principle sedimentation processes in river-fed storage reservoirs (Sloff, 1997).

- **Fine sediments in homogeneous flow:** A large part of the fine sediments transported in suspension or as wash load are transported beyond the delta after which they settle out to form the bottom set bed. They are more evenly spread than coarse sediment, but there distribution is highly dependent on reservoir circulation and stratification, for instance generated by river inflow and wind shear, or precluded by an ice cover. Also for this type of deposition the quantification methods still yield rough predictions.
- **Turbidity currents:** another important transport mode for fine sediments, i.e., silt and clay, is the turbidity current. It is formed when the turbid river inflow plunges below the clear reservoir water and continues as a density underflow. Also other processes can generate them, such as underwater slides (slumping of delta front) or coastal erosion. Turbidity currents are driven by an excess gravity force (negative buoyancy) due to the presence of sediment-laden water in a clear ambient fluid.

3. Sediment problems in hydropower plant

The sediment management is challenging discipline in civil engineering especially in the many regions. The storage capacity of reservoirs decreases due to accumulation of sediment. Dealing with reservoir-sedimentation problems, engineers are challenged by the difficult questions emerging. How to incorporate reservoir problems in feasibility studies (cost-benefit analyses) including environmental and technical effects, limitations on benefit and possible measures? Or what is the impact of silting and desilting of the reservoir?

By focusing on sediment impacts in hydropower plant the following impacts may be notified:

1. The effect of reservoir sedimentation on regulating water resources and its impact on power generation
2. The effect of sediment inflow to power intakes and its impacts on turbine system and other components of the hydropower plant.

The erosion of turbine component depend on: (i) eroding particles - size, shape, hardness, (ii) substrates–chemistry, elastic properties, surface hardness, surface morphology, and (iii) operating conditions - velocity, impingement angle, and concentration and like that. Depending on the gradient of the river and distance traversed by the sand particles, the shape and size of sediment particles vary at different locations of the same river system, whereas mineral content is dependent on the geological formation of the river course and its catchments area.

Run-of-river projects are constructed to utilize the available water throughout the year without having any storage. These projects usually consist of a small diversion weir or dam across a river to diver the river flow into the water conveyance system for power production. Therefore, these projects do not have room to store sediments but should be able to bypass the incoming bed loads to the river downstream. The suspended sediments will follow the diverted water to the conveyance system. Settling basins are constructed close to the intake to trap certain fractions of the suspended sediment (Thapa and Dahlhaug, 2003).

4. Sediment management measures for reservoirs

Sediment management practices for reservoirs are often as different as their physical and technical conditions and social-economic and environmental aspects. Based on literatures and existence experiences, a tentative long-list of alternatives for sediment control of dam reservoirs can be found. The list is sub-divided into four general categories as follows:

i. watershed rehabilitation (Structural and non- Structural Measures)
ii. sediment flushing
iii. sediment routing
iv. sediment removal and disposal

Based on the above general categories some of the measures commonly used to reduce reservoir sedimentation are summarized in the following sections.

4.1 Soil conservation

This strategy focuses on reducing sediment inflow to dam reservoir. In the upstream watershed of a reservoir, three basic patterns of soil conservation measures are commonly taken to reduce sediment load entering the reservoir: structural measures, vegetative measures, and tillage practice. Structural measures include terraced farmlands, flood interception and diversion works, gully head protection works, bank protection works, check dams, and silt trapping dams. Vegetative measures include growing soil and water conservation forests, closing off hillsides, and reforestation. Tillage practice includes contour farming, ridge and furrow farming, pit planting, rotation cropping of grain and grass, deep ploughing, intercropping and interplanting, and no-tillage farming. For a large watershed with poor natural conditions, soil conservation can hardly be effective in the short term.

4.2 Bypass of incoming sediment

Rivers carry most of the annual sediment load during the flood season. Bypassing heavily sediment-laden flows through a channel or tunnel may avoid serious reservoir sedimentation. The bypassed flows may be used for warping, where possible. Such a combination may bring about high efficiency in sediment management. When heavily sediment-laden flows are bypassed through a tunnel or channel, reservoir sedimentation may be alleviated to some extent. In this method, however, the construction cost of such a facility may be high.

4.3 Sediment diverting

Sediment diverting (Warping) has been used around the world. It has a history of more than 1,000 years in China as a means of filling low land and improving the quality of salinized land. Now, this practice may have a dual role, not only improving the land but also reducing sediment load entering reservoirs. Warping is commonly carried out in flood seasons, when the sediment load is mainly concentrated, especially in sediment-laden rivers. Warping can also be used downstream from dams when hyperconcentrated flow is flushed out of reservoirs.

4.4 Joint operation of reservoirs

Joint operation of reservoirs is a rational scheme to fully use the water resources of a river with cascade development. For sedimentation management of reservoirs built on sediment-laden rivers, such an operation may also be beneficial to mitigate reservoir sedimentation and to fully use the water and sediment resources, provided a reasonable sequence of cascade development is made. There are various patterns of joint operation of reservoirs built in semi-arid and arid areas. The idea is to use the upper reservoir to impound floods and trap sediment and to use the lower reservoir to impound clear water for water supply.

Another idea is to use the upper reservoir for flood detention and the lower reservoir for flood impoundment. Irrigation water in the lower reservoir is used first; when it is exhausted, the water in the upper reservoir is used. The released water from the upper reservoir may not only erode the deposits in the lower reservoir, but also cause warping by the sediment-saturated water

4.5 Drawdown flushing

Drawdown flushing is a commonly used method of recovering lost storage of reservoirs. It may be adopted in both large and small reservoirs. The efficiency of drawdown flushing depends on the configuration of the reservoir, the characteristics of the outlet, the incoming and outgoing discharges, sediment concentrations, and other factors. Sometimes reservoir emptying operations may be used for increasing efficiency of the flushing. In the process of reservoir emptying, three types of sediment flushing occur: retrogressive erosion and longitudinal erosion, sediment flushing during detention by the base flow, and density current venting. Environmental impacts are the most constraints for drawdown flushing.

4.6 Venting density current

Density currents have been observed in many reservoirs around the world. A density current may carry a large amount of sediment and pass a long distance along a reservoir bed without mixing with surrounding clear water. The conditions necessary to form a density current, and allow it to reach the dam and be vented out if the outlet is opened in time, have been studied extensively, both from the data of field measurements and laboratory tests. Venting of density currents is one of the key measures for discharging sediment from several reservoirs in the world wide. Density current venting may be carried out under the condition of impoundment, thus maintaining the high benefit of the reservoirs.

4.7 Lateral erosion

The technique of lateral erosion is to break the flood plain deposits and flush them out by the combined actions of scouring and gravitational erosion caused by the great transverse gradient of the flood plains. In so doing, it is necessary to build a low dam at the upstream end of a reservoir to divert water into diversion canals along the perimeter of the reservoir. The flow is collected in trenches on the flood plains. During lateral erosion, because the surface slope of the flood plain is steep and the flow has a high undercutting capability, intensive caving-in occurs at both sides of the collecting trench. The sediment concentration of the flow may be as high as 250 kg/m3. This technique has the advantage of high efficiency and low cost, and no power or machines are required.

4.8 Siphoning dredging

Siphoning dredging makes use of the head difference between the upstream and downstream levels of the dam as the source of power for the suction of deposits from the reservoir to the downstream side of the dam. Siphoning dredging has a wide range of applications in small and medium-size reservoirs. Such an application is valuable to solve reservoir sedimentation and to fulfill the demand of irrigation if the head difference is adequate and the distance between upstream and downstream ends of the siphon is not too great.

4.9 Dredging by dredgers

Dredging is used to remove reservoir deposits when other measures are not suitable for various reasons. In general, dredging is an expensive measure. However, when the dredged material may be used as construction material, it may be cost effective.

5. Overview of hydropower potential in Iran

Iran is a vast country with an area exceeding 106 MKm^2 with two major Zagross and Alborz mountain ranges. Based on the mean annual precipitation of the country about 250 mm and its processes of transformations into water resources, the annual average precipitation is 417 Billion Cubic Meters (BCM), average annual evaporation is 299 BCM, surface water is 92 BCM and seepage to alluvial aquifers is 26 BCM. Further about 72 percent of precipitation is not accessible due to evaporation and transpiration and about 22 percent of precipitation flows on surface water resources and nearly 6 percent of precipitation that falls within the limits of the country in used for direct recharge of alluvial aquifers. Accounting the 13 BCM surface flows enter into the country from across its borders, the total water resources potential is 130 BCM of water that available for various usage and also developing hydropower plants. In Iran 310 large dams (42 concrete dams and 268 embankment dams) were constructed and 81 dams at least 60 m in height are under constructing and 172 dams are under study (Molanezhad, 2008).

The comparative study of dam built in Iran in last one decade to earlier three past decade indicated of increasing (56.5 %) of developing new water structure activities and increasing of (73.6%) of completed new dams through today.

While the capacity of hydroelectric power in the National Power Industry in Iran is about 15% of the total installed power capacity, they have an important role to peak electricity production and they play a key role in stability of the National Power network. despite the potential capacity of hydropower development in Iran may be reach up to 20,000 MW, nowadays total installed hydropower capacity in Iran is about 8500 MW. So the hydropower can be a key energy sector in the National Power Industry.

6. Case study: dez hydropower dam

6.1 Dez dam project background

The Dez Reservoir is located in the Zagros Mountains in the Southwest Iran and was created by the construction in 1963 of the 203m high Dez Dam. An underground powerhouse contains eight 65 MW units for a total installed capacity of 520 MW which has generated an average of 2400 GWh/year energy production over an operating period of 45 years. The minimum and maximum water level of the reservoir operation is 300m and 352m from sea level respectively. Flow releases are through the spillway, power tunnels and three low-level irrigation outlets. The original reservoir volume was 3315 million m3 and over an operating period of 40 years the storage volume was reduced to 2600 million m3 by sedimentation. Fig.2. shows a photo of Dez Dam Project and a plan view of Dez reservoir.

a.

b.

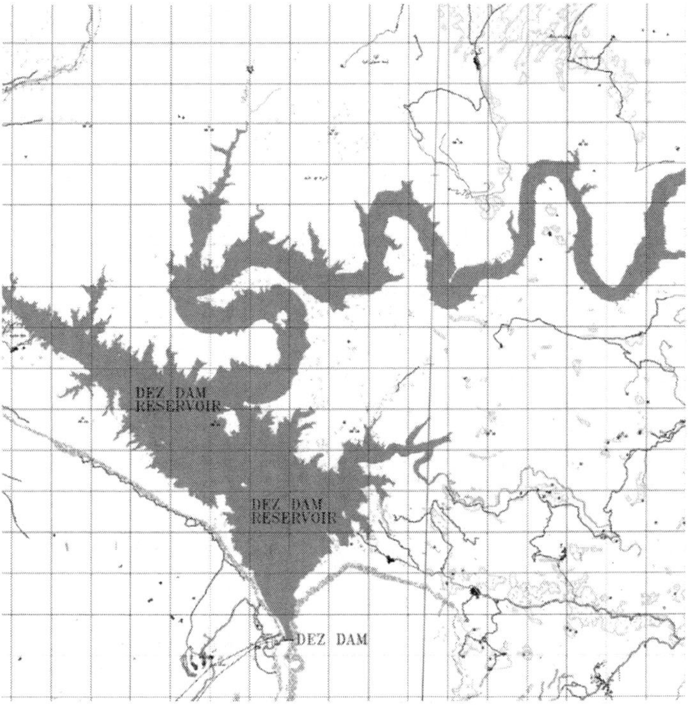

Fig. 2. (a) Photo of Dez Dam Project, (b) A plan view of Dez reservoir

6.2 Dez reservoir sedimentation

Because of importance of sedimentation issues in Dez reservoir, in recent years many field surveys and measurements were conducted by financial supporting of Khuzestan Water and Power Authority (KWPA). These measurements and field investigation is included the following terms:

1. Physical Properties of deposited sediments
2. Mechanical Properties of deposited sediments
3. Volume of deposited sediment
4. Profile and pattern of deposited Sediment in Dez reservoir

6.3 Physical properties of deposited sediments

Field measurement was conducted to obtain the undisturbed and disturbed samples from deposited sediment of Dez reservoir close to power intakes (KWPA, 2004). Sediments deposited in Dez reservoir near the dam body have always been as submerged sediments and thickness of water upon the sediments has been more than 50 meters over an operating period of 45 years. In order to survey physical properties of sediment deposited in front of the power intakes, two deep boreholes which they are called boreholes A and B, were dug near power intakes located at 55 and 100 meter upstream of face of the dam, respectively. Digger was installed on a barge and it was fixed by strong wires which they were anchored to abutments as shown in Fig.3. A piston-type core sampler used to obtain undisturbed samples. The sampler is operated by lowering it until the digger weight touches the sediment surface. With the digger weight resting on the bottom, further lowering of the sampler causes the digger arm to rise and release the coring head. As the cutting sloe is just about to penetrate the sediment, the sampler is penetrated in sediment by hydraulically force so the sampler drives the coring tube into the sediment deposit. The piston remains fixed as the outside tube moves past and serves to fold the undisturbed sample in the tube as it is withdrawn. For boreholes A and B the depth of sampling was 33 and 63 meter respectively in maximum state. During the sampling the height of water above the sediment surface for boreholes A and B was 79 and 76 meter respectively. Height of water for the boreholes was different because time of digging the boreholes was different.

In borehole A, sediment obtained up to depth of 18 meter was very loose so that obtaining undisturbed sample was impossible by piston-type core sampler. From depth of 18 to 33 meter sediments was reported less loss so that obtaining undisturbed sample was possible. In the other hand in borehole A one can see two overall different layers. One layer which it is very loose (soft) sediment is started from sediment surface of reservoir and is continued up to depth of 18 meter Another layer is more dense sediment and it is which it is started from depth of 18 meter and is continued up to original bed reservoir.

In borehole B, sediment obtained up to depth of 30 meter was very loose so that obtaining undisturbed sample was impossible. From depth of 30 to 63 meter sediments was reported denser so that obtaining undisturbed sample was possible. It should be noted that two overall different layers were found in borehole B as same as for borehole A.

Fig. 3. Photos of barge and digger located near Dez power intakes.

The results of field investigation show that loose sediment region for two boreholes are different. It refers to different location for the boreholes. The borehole A is located in front of irrigation outlets (in the river thalweg path) but the borehole B is closed to left abutment of reservoir. As it can be seen a part of sediments in boreholes A and B are very loose. This is because during the past years part of the sediment deposited in the reservoir has been flushed via irrigation outlets forming a wedge in the upstream face of the dam. Years after the wedge have been refilled by new sediment. Therefore sediment deposited in the wedge flushing region has not had much time for consolidating and the sediments have remained is loose stages.

The bulk density of sediments was measured using ASTMD854 procedure and percent of saturated water content were also measured by weighting method. The void ratio of the sediments was calculated based on the following relationship:

$$e = \frac{\rho_w G_s - \rho_{sat}}{\rho_{sat} - \rho_w} \tag{7}$$

Where ρ_{sat} is saturated density of the sediments, ρ_w density of water, G_s specific gravity of sediment solid. Based on the result of the experiments of the samples obtained from boreholes A and B, the relative density of the sediments (G_s) was averagely measured at a value of 2.69.

The results of measured bulk density are presented in Fig.4. The grain size distribution of obtained sediment samples from boreholes **A** and **B** are shown in Fig.5.

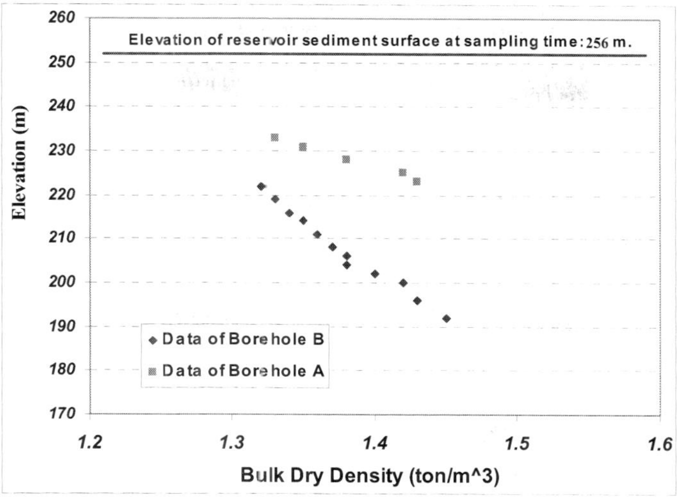

Fig. 4. Measured bulk dry density versus Elevation of sampling for obtained samples from boreholes **A** and **B**.

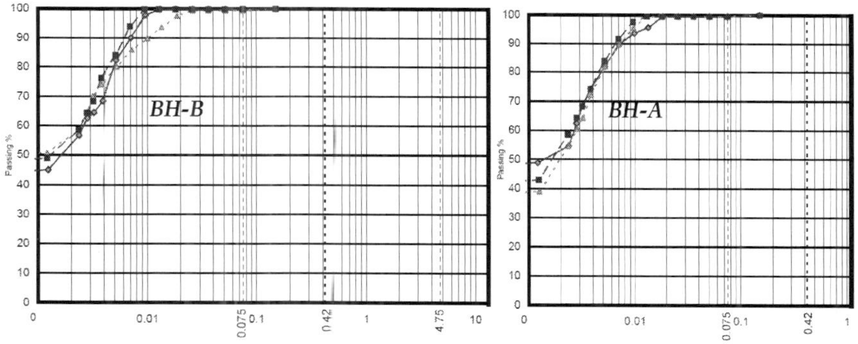

Fig. 5. Grain size distribution curve for obtained samples from boreholes **A** and **B**.

Field measured data obtained from the Dez deep reservoir were applied to evaluate the bulk density of deposited sediment using empirical methods discussed previously. From Fig.5, the percent of clay and silt content in the sediments samples are 55 and 45 percent, respectively. From the Lane's equation, the initial bulk density of the sediments is obtained by 732.8 kg/m^3. This value is an estimated of the initial bulk density for reservoir surface sediment layer which their deposition time is less than 1 year. The measurement conducted of unit weight of sediment at the reservoir bed surface shows a value of 825 kg/m^3(Samadi Boroujeni, 2005). Comparing of measured and calculated initial bulk density shows that Lara and Pemberton's empirical method predict the initial bulk density 11% less than that of measured value. This shows that the accuracy of the empirical formula is acceptable.

Also the Lane's, Miller's and power relationship's methods were applied for various points of A and B boreholes. Obtained results showed that the accuracy of the above empirical methods weren't acceptable, therefor based on the the field data of A and B boreholes, the following relations was developed (Samadi Boroujeni , 2010):

Modified Lane's equation for Dez reservoir:

$$\rho_t = \rho_1 + 1.52\ B\left[\frac{t}{t-1}(\ln t) - 1\right] \tag{9}$$

Modified Miller's equation for Dez reservoir:

$$\rho_t = \rho_1 + 2.57 \times B.\log t \tag{10}$$

Modified Power 's equation for Dez reservoir:

$$\rho_t = 1.35\,\rho_1.t^{0.5\log(1+\frac{2B}{\rho_1})} \tag{11}$$

6.4 Progressive deposition in dez reservoir

The original estimate for sediment accumulation in the Dez reservoir was for a 50 yr volume of 840 million cubic meters (million m3), i.e., equivalent to filling the dead storage within the reservoir to el 290 m. This estimate was made on the basis of there being upstream sediment retention structures and a reforestation program. However, these programs were not carried out. Nevertheless out of an initial reservoir volume of 3,315 × 10^6 m^3 (at elevation 350 m), the available storage in 2002/2003 was found to be 2,698 × 10^6 m^3 (based on the latest bathymetric in 2002/2003) which corresponded to a volume loss of about 615 × 10^6 m^3or 19% of original volume of the reservoir. This result shows that the sediment incoming into the reservoir have been annually 15.7 × 10^6 m^3. Much of the sediment drops out along the upper reaches to form a delta, which is slowly progressing to the dam, as shown in Fig. 6, while about 11% reaches the dam in the form of relatively fine sediment transported by turbidity currents. Size distribution analysis of the sediment samples which was taken near the dam showed that the median particle of the sediments was less than 0.01 mm and it was included 60% silt and 40% clay. Although it will take some time for the reservoir to reach its 50% life, a level normally accepted as the end of effective life, the effect of present sediment inflow on key elements of the project is already of concern. For example the reservoir bed at the face of the power intakes has risen from an original elevation of 180m to 256m in 2003 at average rate of 2 m/year over the past operation period, which is only 14 m below the invert level of the power intake at elevation 270 m as shown in Fig.7 (Acres and Dezab, 2004). There is now concern that sediment will begin to be drawn into the power tunnels within a decade with potential to damage the turbine runners.

The volume elevation curve for the reservoir when it reaches the end of life condition of 50% depletion is given in the Fig.8. This figure was obtained by using the empirical area reduction method to derive the volume elevation relationship. The derivation and application of this method is discussed in more detail in the American Society of Civil Engineers Transaction (1960). The methodology was calibrated against the results of the

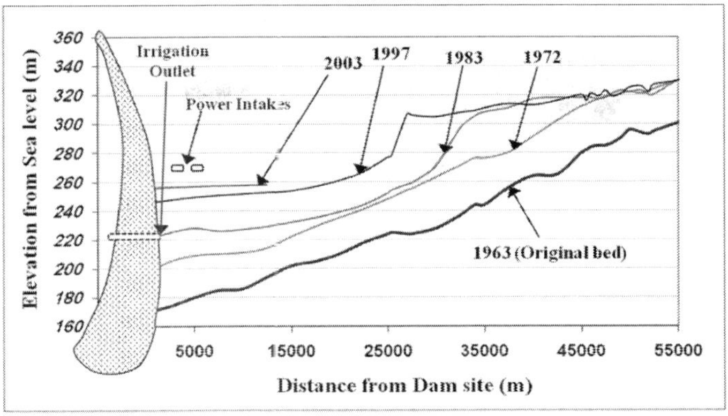

Fig. 6. Sediment longitudinal profile of Dez Reservoir.

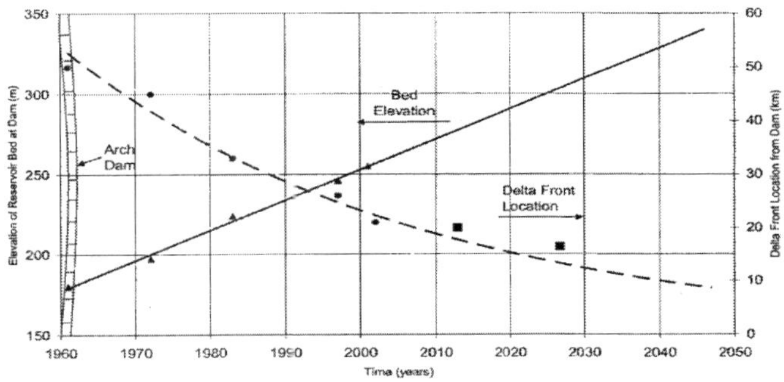

Fig. 7. Progressing Rate of delta front and increasing rate of bed elevation (near Dez dam) versus time based on Longitude profiles of Dez reservoir thalweg in different years.

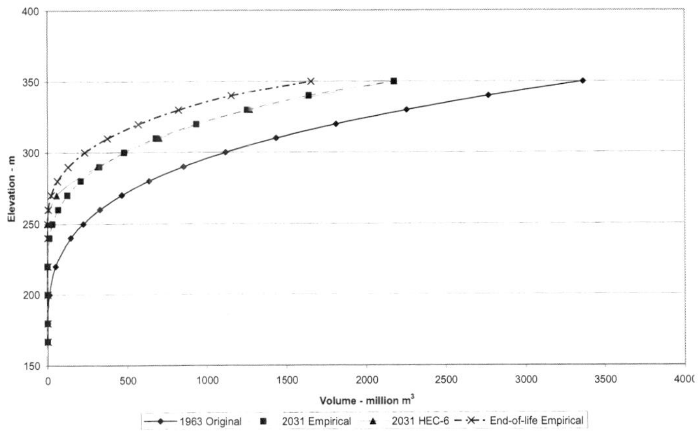

Fig. 8. End of Dez reservoir life volume elevation curve.

HEC-6 model for the period from 2003 to 20031 (Acres and DezAb, 2004). The Figure shows the comparison of the HEC-6 modeling with the empirical method. The agreement is reasonably good except near the lowest elevations where the configuration of the actual reservoir will differ from the empirical method due to the presence of turbidity current. Also shown is the anticipated volume elevation curve for the reservoir after it is 50% depleted resulting in a total volume of 1658 million m^3.

6.5 Impacts of dez reservoir sedimentation

Generally with particular regard to the elements affected by sedimentation these include:

- the reservoir, its remaining live storage and its ability to provide the degree of regulation originally envisaged for the project;
- elements already affected by the passage of water and sediment, including the irrigation outlets, and which may in the future be affected, including turbine equipment, particularly runners and wicket gates.
- Dezful re-regulating dam sedimentation. Since the main function of the re-regulating dam is to make regular the released water from Dez dam hydropower plant in downstream river, the useful volume of storage is, therefore, very important for this dam.

An overview of the reservoir sedimentation situation, its effects on the elements noted above, an indication of the impacts on the project, if remedial work is not implemented, and in this regards, one of the measures proposed to address the sediment management issue is cone flushing of sediment through the dam irrigation outlets. Flushing of sediment has been conducted over the last decade to augment irrigation flows as well as to remove the sediment that has accumulated immediately upstream of the irrigation outlets. This issue has prompted the Khuzestan Water and Power Authority to consider performing the Dez dam flushing operations on the agenda. Predictions indicate that between 1-1.5 million cubic meters of fine-grained sediment in the Dez dam Reservoir will be discharged into the downstream river, annually, in the years to come, by performing flushing operations through the Dez dam irrigation outlets (Samadi Boroujeni and Galay, 2004). It should be noted that here are various important constraints during sediment flushing operation via the irrigation outlets such as environmental problems downstream, loss of reservoir water storage, possibility of irrigation outlets clogging by large boulders, the effect of sediment spray around of the power house, and decreasing of power generation, and Dezful re-regulating dam reservoir sedimentation. Since the sedimentation in the re-regulating dam reservoir play an important role in Dez dam flushing downstream effects, it is essential that the sedimentation processes in the re-regulating dam reservoir to be simulated for surveying the negative effects of Dez dam flushing operation on downstream. It should be noted that sediment released from Dez irrigation outlets are mostly cohesive sediments therefore this matter should be considered for modeling of sediment transport processes in downstream river.

6.5.1 Irrigation outlets

The Dez Dam is provided with three irrigation outlets located at el 222.7 m within the central blocks of the dam body, some 129 m below the full supply level of the reservoir. The original function of these outlets was to provide irrigation water to the downstream area, partial flood control and a measure of emergency release.

Normal operation of the irrigation outlets can be assumed to have occurred for the first 23 years of the project life. However, KWPA operating staff at the dam site advise that flushing of sediment through the irrigation outlets first occurred some 17 years ago and has occurred relatively frequently since then. In early June 2003 sediment had accumulated at the face of the dam to el 256 m, which is over 30 m above the centerline of the irrigation outlets. Typically flushing of sediment now occurs two to three times each winter.

As a consequence of sediment and trash flushing, the trashracks at the upstream entrance have been damaged to the extent that some of the trashrack bars are missing, the guard valves are leaking at the seats/seals and erosion of the valve stem on all three valves has occurred. Major leakage is evident on the Hollow-cone discharge valves to varying degrees on each outlet. It has also been reported that because of high vibration, the valves are not operated beyond 50% open (Samadi Boroujeni, 2003).

Fig. 9. Photo of Dez Dam Irrigation Outlet, Flushing in June 17, 2003.

6.5.2 Turbine equipment

The plant comprises eight 65-MW units with vertical Francis turbines. The maximum operating head for the turbine, given on the nameplate, is 180 m. Each turbine has a butterfly-type turbine inlet valve. Studies are presently underway to upgrade the generating equipment. Given this investment by KWPA it will be important to manage the reservoir sedimentation process with the objective of protecting the units from damaging sediment-laden flows. It will be important to manage the reservoir sedimentation process with the objective of protecting the units from damaging sediment-laden flows.

6.5.3 The effect of dez reservoir sedimentation on power generation

Sedimentation in dam reservoir reduces the reservoir storage and it may cause the ability of the reservoir and therefore power generation to be decreased. Samadi-Boroujeni and Kalali (2012), used WEAP model to assess Dez dam reservoir sedimentation effects on hydropower generation. The results, as seen in Fig.10, showed that without Bakhtiari dam (see section 6.6.2), energy production would be decreased by an average rate of 0.08% per year and in scenario of with Bakhtiari dam, the average rate would be 0.06% per year. In the other hand when the reservoir reach to its half-life, the amount of power generation in comparing with base scenario (by assuming no sedimentation), to be decreased by 9.1 %.

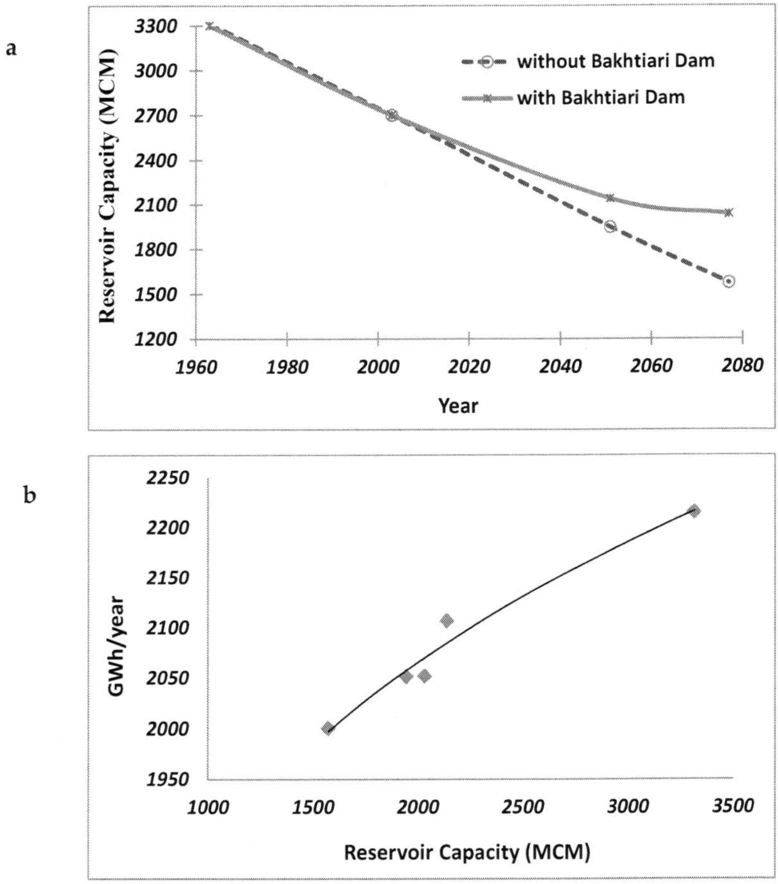

Fig. 10. results from WEAP model: a. reservoir capacity versus time; b. energy production versus reservoir capacity of Dez reservoir.

6.6 Reservoir sedimentation management alternatives

Based on literatures and existence experiences, a tentative long-list of alternatives for sediment control of dam reservoirs can be found. The list is sub-divided into four general categories as follows:

i. watershed rehabilitation (Structural and non- Structural Measures)
ii. sediment flushing
iii. sediment routing

iv. sediment removal and disposal
v. Construction of New power intake

The categories will be described and applicability of each one for Dez reservoir will be discussed. It should be noted that the downstream impacts of Dez dam flushing and proposal of appropriate hydraulic conditions to decrease the negative effects of Dez dam flushing operation on downstream is one of the most important issues.

In anticipation of the need to manage the sedimentation process KWPA has proactively commenced reservoir studies. A tentative long-list of alternatives for sediment control and low-level flushing was initially formulated. The list was sub-divided into the above four general categories. These general concepts were expanded into some 17 alternatives along with several additional sub-alternatives (Acres and Dezab, 2004). A more manageable short-list of most likely suitable candidate alternatives was then selected including:

- Dam on the Bakhtiari River
- Irrigation outlet rehabilitation and conversion to flushing facility
- Bypass flushing tunnels within the dam abutments
- Excavation in the dry of sediment in the delta
- Dredging near the dam
- Watershed management
- Dam raising.

Acres and Dezab (2004) assessed power system benefits on a system wide basis considering both the effects on the Dez and Karun river systems. Differences in power and energy benefits were determined using Acres Reservoir Simulation program (ARSP model) which has been used extensively in the Khuzestan Province for over more than 20 years. As part of these analyses, irrigation, domestic and industrial water demands on the Dez River were satisfied with varying priorities. The value of irrigation water deficits were calculated for each alternative using a water value of Rials 712 per m^3. This water value was derived from information provided by KWPA regarding the weighted average value of a composite crop and amount of water used to grow this crop. The various sediment management alternatives considered the incremental water, power and energy benefits of each alternative compared to a case of no intervention.

6.6.1 Watershed management

The reduction of sediment yield from a watershed is generally accomplished by measures such as erosion check structures, terracing and tree planting. From overflights and inspection of satellite images it appears that measures would be difficult to construct in the Bakhtiari watershed because of steep canyons and no access roads. In addition measures in the upper watershed of the Tireh and Marloreh Rivers would not serve any purpose because sediment yields are very low. Therefore, the study of erosion reduction measures was confined to the Northwest portion of the watershed, namely the Keshvar and Girit Rivers. From brief studies by DRC (1972), by DezAb and Acres (2004 and 2005) and with due regards to the status of the watershed management activities over the past 40 years, approximate estimates show that the reduction in sediment yield would range from 10 to 15 percent of the annual yield for the entire basin. However, this alternative would not result in immediate reduction because implementation of check structures and road construction would take many years. Therefore, this alternative was ranked as having a low status.

6.6.2 Dam on the Bakhtiari river

Several dam sites are being studied on the Bakhtiari River and preliminary plans indicate that the first dam could be completed by 2018. The Bakhtiari reservoir could trap about 55 percent of the annual sediment load of the entire watershed and would double the active life of the Dez Reservoir. Although the power and energy economics of this alternative were not evaluated, it would contribute to significantly reducing the sediment accumulation in the Dez reservoir and is therefore ranked high on the list.

6.6.3 Flushing sediment through the irrigation outlets

Flushing of sediment has been conducted over the last decade to augment irrigation flows as well as to remove the sediment that has accumulated immediately upstream of the irrigation outlets. The level of sediment adjacent to the irrigation outlet had reached el. 256 m in 2003. At that time a pilot flushing program was conducted from April to June 2003 to assess the headcutting process upstream from the dam and the subsequent concentrations downstream. The findings from the pilot project were as follows:

i. The flushing of the sediment on June 17, 2003 caused a wedge to be headcut upstream from the dam, having a length of 400 m, a width of 90 m and a maximum depth of 26 m at the dam face (Fig. 11). This results in approximately 470,000 m³ of fine sediment being flushed in about 4 hours.
ii. The flushing was undertaken during low flow into the reservoir with no turbidity current arriving at the dam.
iii. The suspended sediment concentration downstream from Dez Dam, at the regulatory structure, rose from a low value of about 100 mg/l to 8,000 mg/l and this high concentration lasted for about one day (Fig.12).

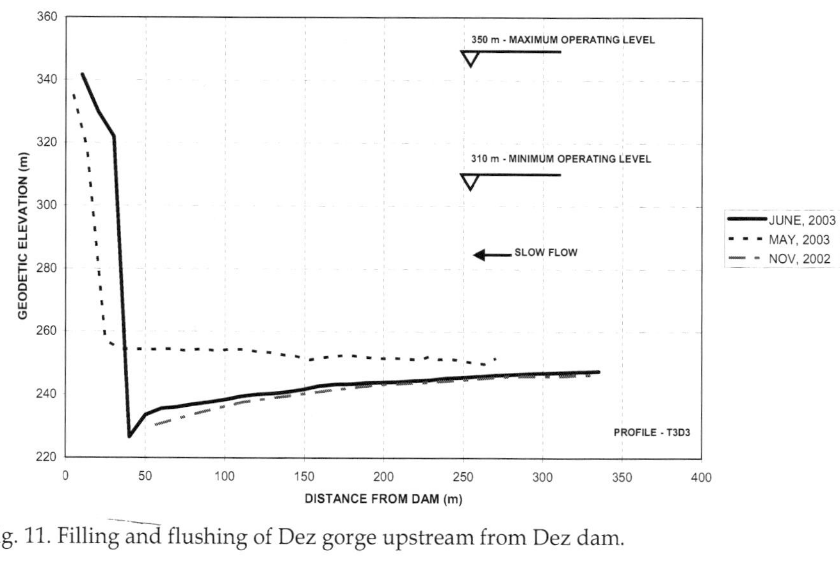

Fig. 11. Filling and flushing of Dez gorge upstream from Dez dam.

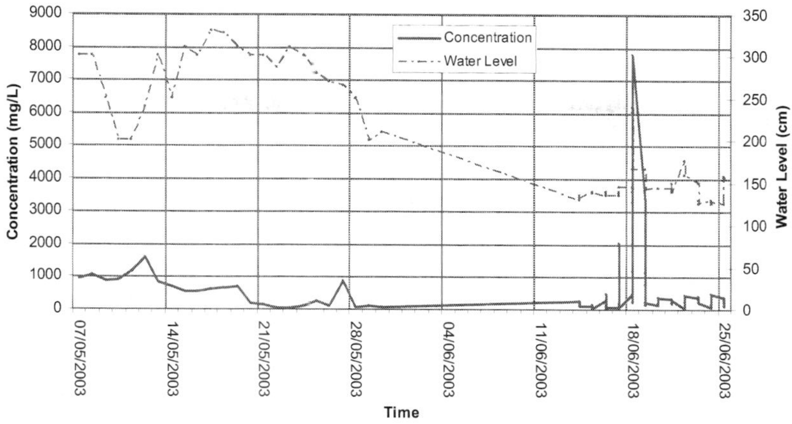

Fig. 12. Sediment concentration and water levels for Dez-Harmaleh.

The pilot project generally indicates that wedges (or slugs) of sediment of about 500,000 m^3 can be flushed from the reservoir if a turbidity current has previously filled the narrow wedge to about elevation 250 m. From analysis of historical peak flows entering the reservoir, it was shown that about 2.2 million m^3 of sediment could be brought to the dam annually by turbidity currents - the balance of the inflowing sediment, about 14 million m^3 would be deposited on the delta and in the lake portion of the reservoir. Therefore, the flushing of sediment should be closely coordinated with the inflow of turbidity currents, but this would remove only about 10 percent of the annual sediment inflow. However, such wedge flushing has the advantage of causing the sediment surface level at the face of the power intakes to be effectively decreased.

Samadi Boroujeni, et al. (2010) modeled the Dez dam flushing through the irrigation outlets by using MIKE11 model. MIKE11 model is able to solve the equation of mass transport and can be used for the simulation of cohesive sediment transport in the reservoir. Results obtained from MIKE11 model showed that manner of carrying out the flushing for the Dez dam sediments influences the trapping efficiency of re-regulating dam reservoir and one can, therefore, reduce the sedimentation in the re-regulating dam reservoir through proper management at the Dez dam flushing. The same results indicate that the less duration of flushing operations at the Dez dam and the more discharge from re-regulating dam will decrease trap efficiency coefficient in the re-regulating dam reservoir. This, however, causes the concentration of outflow from the re-regulating dam to be more than the allowable level of concentration at the downstream river. If all the factors involved in decision-making, such as the loss of Dez reservoir water storage, threat of flooding downstream, sedimentation in the reservoir of re-regulating dam, downstream environment, and risk of opening the Dez dam outlets are considered to have equal degrees of impact, the option that the irrigation outlets (sediment discharge) are opened for two times with a total discharge of 30m^3/s, the duration of performing the flushing operation being 6 hours each time at a minimum 6-hour interval between the two flushing, will gain the lowest impacts and therefore, it will be the most appropriate option. To decide which option is the best flushing manner, the beneficiary administration needs to pay attention to the relevant conditions and to consider the incremental as well as minimal coefficients for each criterion, and to calculate the total scores of each option.

6.6.4 Bypass flushing tunnels within the dam abutments

Consideration was given to re-opening the existing diversion tunnel, which had been plugged following the original construction of the project, and adapting it and an access tunnel to a sediment flushing tunnel. The intake to the diversion tunnel is at elevation 184 m which means that it is under a 50 to 70 m thickness of sediment which may require dredging prior to re-opening the tunnel. Alternatively, the sediment may be removed during the initial blast, resulting in rapid headcutting and a slug of sediment in the downstream river with concentrations as high as 500,000 mg/l for possibly 6 hours. This sediment slug would be damaging to downstream fish habitat and recreation and therefore it would be required that the sediment concentration be decreased by spillway and turbine flows.

6.6.5 Venting of turbidity currents

Field measurements of turbidity currents in the Dez Dam reservoir in Jan 2002 to July 2003 showed that turbidity currents occur when the river is flooded (Samadi-Boroujeni and Galay, 2004). The preliminary results indicate the following information:

i. at high reservoir levels, the turbidity current is indicated at the upstream end of the reservoir and as it moves over the delta foreset slop fine sediment is entrained into the current.
ii. tow current were tracked with suspended sediment concentration up to 7000 mg/l and the volume of sediment moving within the current in January 28, 29, 2003 was 44,000 m^3 while larger event of Aprial 23, 24, 2003 was 1,188,500 m^3/year;
iii. after analysis of 26 years of high flow events and using a relationship between peak flow events and turbidity loads determined in 2003, one obtains an annual turbidity current of about 2,243,200 m^3/year;
iv. the estimate of annual turbidity load amounts to approximately 13 percent of the annual sediment inflow to the reservoir.

From this information it appears that the extension of reservoir life by flushing of turbidity currents would not be large, but could be economically viable provide that downstream environment guidelines could be adhered to. This event occurs two to four times per year and if the rehabilitation of the irrigation outlets or the bypass flushing tunnels within Dez dam abutments to be performed, the turbidity current can be vented from the reservoir by opening the outlets or the tunnels.

6.6.6 Excavation in the dry of sediment in the delta

A brief study indicates that this alternative would be rather expensive, with high costs for road construction necessary to truck the excavated sediment to disposal sites and possibly to improve cropland. The excavation site would be somewhere between km 35 and 45, upstream of the dam, and would require a bridge so that excavation could occur on both sides of the active channel. This alternative is, however, difficult to implement as hundreds of trucks would have to be loaded every day to excavate about 10 percent of the annual sediment load which is about 16 million m^3/year. The annual operating cost could be as high as $5 million which with a high initial capital cost makes this alternative unattractive.

6.6.7 Dredging near the dam

Dredging near the dam would require a special dredge capable of operating in water depths of 100 to 120 m which would be expensive. The dredge spoil could be pumped to the spillway tunnels. This alternative was ranked low.

6.6.8 Dam raising

Dam raising would considerably increase the effective volume of the reservoir. Structural analysis of the raised dam shows potential limitations with any large height increase due to the high seismic loads in the region. Raising the dam 10 m shows large increases in joint opening during extreme earthquake (Acres and Dezab, 2004). From preliminary results, it is recommended that the maximum extent of dam height increase be limited to not more than 10 m. In this case the added effective volume after 10 m of dam raising, would amount to about 700 million m^3, which is approximately equal to the sedimentation volume from the start of operation to date. Economic analysis shows that, the present value of costs and benefits, for a 10 m dam raising, will result in a positive net benefit to the project, anticipated to be in the order of 900,000 Million Rials.

Other impacts of the dam rising can be summarized as follows:

- Spillway

For a 10-m raised dam, the maximum spillway discharge capacity would increase from the present capacity of 3000 m^3/s to 4300 m^3/s. This rate of discharge could have serious consequences including potential erosion damage of the gorge wall, cavitation damage potential at inlet and outlet of the spillway tunnels.. Thus it has been recommended to maintain the present discharge rate of the spillway. In a raised reservoir condition, some modifications will be required at the inlet structure. For example the radial gates should be replaced and a lintel beam provided.

- Power Generation

For a 10-m dam raising, the water passages are generally capable of handling the extra loads. Some existing leakage might increase or new leakage may occur, but is expected to be controllable. The provision of more efficient generating equipment would be the major change. Average energy production of Dez hydropower with a 10 m dam raising would be up to 2600 GWh/yr which is about 8% more than that of without dam raising option.

- Access Tunnels

To provide access to the raised dam crest elevation, new access tunnels have to be excavated. These tunnels branch from the existing main access tunnel and are of rather short length. Construction of this part is simultaneous with construction activities on the spillways.

- Reservoir Rim and Environmental Issues

For a 10 m dam raising, three small villages would be flooded and their habitants should be resettled. There is no interference between the raised reservoir and the adjacent railroad.

Inundation of reservoir rim does not cause major change in environmental condition of the reservoir.

6.6.9 Flushing of sediment from upstream

Diversion of sediment from upstream is an approach to mange of Dez reservoir sedimentation. For this purpose many alternatives for a tunnel path were surveyed and ultimately with considering geological conditions, the best path was selected by Emamgholizadeh et al (2005), as seen from the topographic map in Fig.13. This path is shorter than all other options which is approximately 15 km in length, and its outlet is out of the river system, thus negating any environmental impacts downstream. In addition, we can reclaim the gravel land of the Andimeshk region (Outlet in Fig.13) by means of disposed flushed sediment. The proposed method for flushing of the delta sediments of Dez reservoir include erecting a submerged dam at location A-A and a diversion tunnel, as shown in Fig.13. In this method, the flushing is started when the water surface of reservoir is drawn down to the crest of the submerged dam. During the flushing, the water requirement for power generation would be supplied by water storage existing between Dez dam and submerged dam which in this paper will be termed named the main reservoir (M-R), (Fig 13, 14) after flushing operations, the gate of the diversion tunnel will be closed and the area upstream of the submerged dam, which in this paper is termed the upstream reservoir (U.R), (Fig 13, 14) will be refilled and the water surface will rise again up to the crest of the submerged dam.

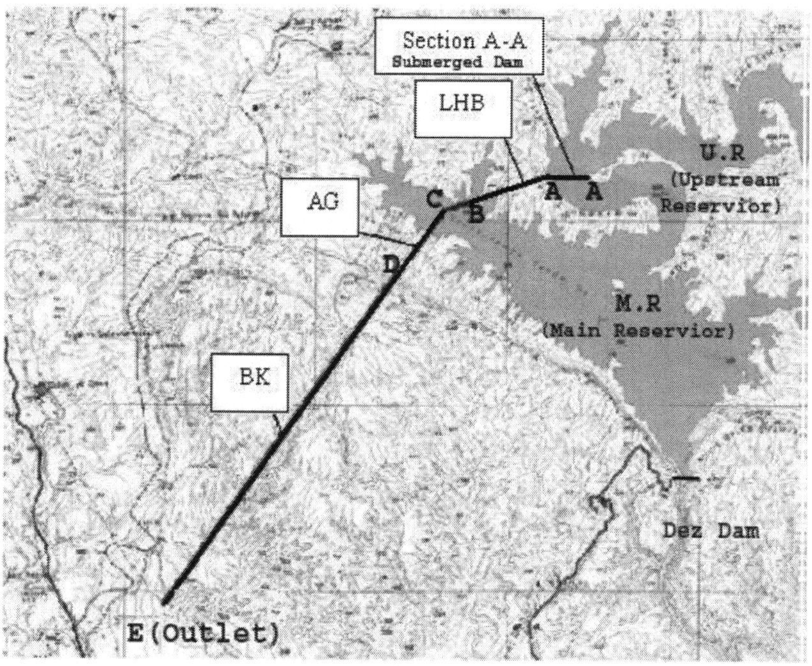

Fig. 13. Layout of diversion tunnel (path ABCDE).

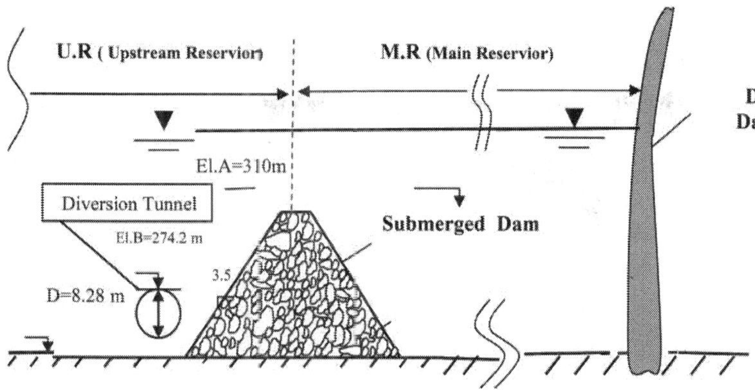

Fig. 14. location of submerged dam (Section A-A from Fig. 13) and outlet of diversion tunnel.

The dimensions of the diversion tunnel and the submerged dam will depend on the hydraulic and hydrological aspects of flushing and economic considerations.

- The most suitable period for flushing operations

Since in normal conditions, flushing operations are used to dispose the deposited sediments from the upstream submerged dam (U.R), the best time to carry out flushing operations is when the reservoir elevation is at minimum operating level. Based on the Dez Dam Rule Curve, this period is during the months of Dec, Jan, and Feb.

- The transporting capacity of flushing flows

The transporting capacity of flushing flows can be estimated by using an empirical equation that is reported in IRTCES (1985). This equation is:

$$Q_S = \psi \frac{Q_f^{1.6}.S^{1.2}}{W^{0.6}}$$

Where Q_S= the sediment transporting capacity (ton/sec); Q_f= the flushing discharge (m^3/sec); S= the bed slope; W = the channel width (m); and Ψ= constant parameter that is dependent on the size of the sediment. With respect to the type of sediments in the Dez Dam (D50>0.1mm), this parameter is selected to be 300. Based on the regime theory, the bed slope should be about 0.0010 m/m.

After flushing begins, the flushing channel created in the sediment deposits will be self-formed, and its width can be controlled principally by the discharge flow, slope and sediment properties. However, the channel formed during flushing has been found to correlate well with sediments derived from flushing discharge (Q_f) alone. The flushing equation which was derived from the data presented by IRTCES (1985) can be applied for the calculating of the channel width (W):

$$W = 12.8 Q_f^{0.5}$$

Thus, with Q_f =200 the channel width (W) is calculated 181 m. Based on this equation, the transporting capacity of flushing flows is calculated as 16 ton/sec.

To estimate the volume of sediments that can be eroded by flushing, it is necessary to determine the geometry of the flushing channels. The channel bed width is calculated as 78 m and the side slope of flushing channel can be determined by the following equation [Teissin, 1991]:

$$\tan\alpha = \frac{36}{5}\rho_d^{4.7}$$

Thus, ρ_d = 0.8 ton/m^3, the side slope of the flushing channel (α) is calculated 65.6^o .Before sediment flushing, the delta sediments bed slope is 0.0005 and after that, it would be 0.0015.With due regards to the information provided, the volume of the flushing channel can be calculated as 9.3 ×10^6 m^3. It is assumed that flushing operation would be carried out once per year.

Criteria for determining whether flushing in a particular reservoir has been successful are required. One of the methods to designate the criteria for successful flushing is the Sediment Balance Ratio (SBR), which is defined as sediment mass flushed annually divided by sediment mass depositing annually [Atkinson, 1996]. If SBR>1 then sediment balances can be obtained. For the proposed flushing method, since the sediment flushed and the sediment deposited have been calculated as 9.3 and 17×10^6 m^3 respectively, the SBR ratio will be 0.55. Another term for the evaluation of flushing is Flushing Efficiency (Fe), which is defined as the ratio of deposited volume eroded to the water volume used during flushing over any time interval [Atkinson, 1996]. An alternative expression for flushing efficiency is the Water Sediment Ratio (WSR) which is the inverse of the Fe. Since the transporting capacity flows and the total of the flushed sediments volume have been calculated as 16 ton/s and 9.3 ×10^6 m^3 respectively, therefore duration of flushing operation is calculated 8.1 day and waste water will be 140 ×10^6 m^3. In this calculation it is assumed that bulk density of the sediment to be 1.2 tons/m^3 .According to the aforementioned information, the WSR ratio for Dez Reservoir will be 15.05.

7. Conclusion

In this chpter reservoir sedimentation and sediment management measures were presented and the sediment problems in Dez dam reservoir, as a case study, was disscused. Also a comprehensive introducing and evaluation of the various alternatives was briefly undertaken by considering the technical and environmental issues of each alternative.

In addition to the technical details of each alternative the effects of various sediment management alternatives on the life of the reservoir were considered. For this purpose a "useful reservoir life" corresponding to 50% of the initial reservoir volume ,which it also call resrvoir half-life, was assumed. The initial volume of the Dez reservoir was estimated to be 3316 million m^3, thus the 50% storage depletion volume is approximately 1658 million m^3. Understandably the major effect on the "useful reservoir life" comes from those alternatives which directly increase the reservoir volume or intercept the sediment entering the Dez reservoir, namely Dez dam raising or construction of the Bakhtiari dam.

Surveys undertaken in the Iranian years 1381 and 1382 (2002 and 2003) show that, after 40 years of operation, some 81% of the original Dez reservoir volume remains. Sediment is entering the reservoir at an average rate of about 15.8 million m^3/yr. Coarse sediment deposition is occurring in a delta area, which is presently about 27 km from the dam, while the finer silt and clay fractions are transported beyond the delta. The delta is progressing downstream at a rate of about 0 5 to 1.0 km/year.

The fine sediments carry about 2.2 million cubic meters which settle out in the main body of the reservoir or are transported to the dam area. These suspended sediments travel in density currents and are reflected by, and deposited near, the dam, increasing the depth of sediment at this location by about 2.0 m/year. Accordingly the hydro-mechanical equipment within the dam body is being affected by the sediment laden flows and without interventions the powerhouse turbine equipment will be similarly affected in the near term. To assess Dez dam reservoir sedimentation effects on hydropower generation WEAP model was used (Samadi-Boroujeni and Kalali, 2012). The results showed that when the reservoir reach to its half-life, the amount of power generation in comparing with base scenario (by assuming no sedimentation), to be decreased by 9.1 %.

It was important to distinguish between sediment management alternatives which consider different issues. Two key categories of alternatives were identified, those that directly address the power intake sediment issue and those that address the wider issue of reservoir sedimentation and storage.

Power intake sedimentation was found to be best addressed by flushing the sediment through rehabilitated irrigation outlets. However, flushing is only effective in transporting sediment from the wedge area in close proximity to the dam, essentially to clear the passage to the irrigation outlets and to protect the power intakes. Subsequently, flushing will remove some or all of the turbidity current as it arrives at the dam and before it deposits. For this purpose an operational procedure is required that involves the passage of flows at various intervals, coordinated with power and spillway flows, to control downstream sediment concentrations and to allow water clearance time, both of which are required to address the present downstream fisheries operations. Initial planning of such an operation has been based on environmental criteria which have yet to be made specific to Iran and the Dez River. Thus operational experience is required to be gained over time as the flushing operations are implemented. Although the rehabilitated irrigation outlet alternative had a limited net present value of benefits it had the added benefit of maintaining reservoir drawdown capabilities and of being implemented in a relatively short period of time.

Sediment removal of the delta material by dredging is not considered to be a viable alternative due to the expected high cost of operation and maintenance. The alternative of machine excavation at times of low reservoir has been evaluated as part of this study, although only about 7% to 10% of the annual sediment inflow could be excavated by this method. Also diversion of sediment from upstream was an approach to mange of Dez reservoir sedimentation that surveys showed that this option is possible by constructing a submerge dam with a height of approximately 80 meters and a tunnel approximately 15 km in length at midrate location of Dez reservoir. By this measure a total sediment volume of 9.3 MCM per year can be discharged from the reservoir. The Sediment Balance Ratio (SBR) for this option was obtained 0.55 and total cost for that was estimated 176 million US$. It is so extensive comparison with irrigation outlet rehabilitation option.

Dam raising would considerably increase the effective volume of the reservoir (approximately 70 million m^3 for each meter of dam raising). In general it is considered that the various technical issues involved in dam heightening can be addressed. However, further refined study is required of the dam body which was found to have significant stress and deformation conditions both under static and dynamic load conditions to the extent that the maximum height increase may be limited to about 10 m.

Construction of the Bakhtiari dam in upstream basin of Dez Dam, perhaps in the period 2014 to 2018 would further reduce the sediment entering the Dez reservoir, but at great cost. It will trap about 50% of total sediment inflow to Dez reservoir so it can be addressed as an efficient measure.

Results obtained from the use of MIKE11 model for flushing option through irrigation outlets showed that if the irrigation outlets (sediment discharge) to be opened for two times with a total discharge of $30m^3/s$, the duration of performing the flushing operation being 6 hours each time at a minimum 6-hour interval between the two flushing, it will gain the lowest impacts and therefore, it will be the most appropriate option.

8. Acknowledgements

I wish to acknowledge the Khuzestan Water and Power Authority and Shahrekord University, Iran for their assistance and guidance provided in the preparation of this chapter.

9. References

[1] Acres and Dezab, (2004), "Dez Dam Rehabilitation Project- Stage1: Sedimentation Study", Report to KWPA, Ahwaz, Iran.

[2] Acres and Dezab, (2004), "Dez Dam Rehabilitation Project- Stage2: Irrigation Outlet Study",Report to Khuzestan Water and Power Authority, Ahwaz, Iran.

[3] Atkinson, E., (1996) "The feasibility of flushing sediments from reservoirs", Report OD 137, HR Wallingford, UK.

[4] Brandt, A., (2002)"A review on reservoir desiltation", International journal of sediment research. 307-320.

[5] Batuca, D, G & Jordan, J. (2000) "Silting and Desilting of Reservoirs". USA: Balkema.

[6] Basson, G. (2002). "Mathematical Modelling of Sediment Transport and Deposition in Reservoirs-Guidelines and Case Studies, " Inter~zational Commission on Large Dams Sedimentation Committee Report.

[7] Chao, P.C. and S. Ahmed. (1985) A mathematical model for reservoir sedimentation planning. Water Power & Dam Construction, Vol. 37, No. 1, Jan. p.45-52.

[8] Chikita, K., N. Yonemitsu and M. Yoshida (1991) Dynamic sedimentation processes in a glacier-fed lake, Peyto Lake, Alberta, Canada. Japanese. J. of Limnology, Vol.52, No.1, p.27-43.

[9] DHI, (2000), Reference Manual MIKE11 Model.

[10] Emamgholizadeh, S., H.Samadi-Boroujeni, and M.Bina, 2005, "The Flushing of the Sediments Near the Power Intakes in the Dez Reservoir", Proceeding, The Third International Conference on River Basin Management, Italy.

[11] Fan, J. and G.L. Morris (1992) Reservoir Sedimentation II: Reservoir desiltation and long-term storage capacity, J. Hydr. Engrg., Vol. 118, No.3, p.370-384.

[12] Garcia, M.H. (1993) Hydraulic jumps in sediment-driven bottom currents. J. Hydr. Engrg., ASCE, Vol.119, No.10, p.1094-1117.

[13] Gebrehawariat, G., and Haile. H. (1999). Reservoir Sedimentation Survey Conducted on Four micro dams in Tigray. Commission for Sustainable Agriculture and Environmental Rehabilitation in Tigray, Ethiopia.

[14] Khusestan water and power Authority. (2003). "Field measurement of turbidity current in Dez Dam reservoir".

[15] Lara, J.M., and Pemberton, E.L. (1963). Initial Unit Weight of deposit sediments. Proceedings of the Federal Inter-Agency Sedimentation Conference, Paper No. 82., USDA.

[16] Lane, E.W., and Koelzer, V A., (1943). "Density of Sediments Deposited in Reservoirs," Report No.9. In A Study of Methods Used in Measurement and Analysis of Sediment Loads in Streams. Hydraulic Lab, Iowa University.

[17] Mahmood, K. (1987) Reservoir sedimentation: Impact, extent, and mitigation. Techn. Paper No.71, The World Bank, Washington D.C., USA.

[18] Morris,G.L. and J.Fan, (1998),"Reservoir Sedimentation Handbook", McGraw-Hill, New York, USA,.

[19] Molanezhad, M., (2008), Development of Eco-Efficient Water Infrastructure In Iran, UNESCAP Workshop10-12 November 2008 , Republic of Korea.

[20] Miller, C.R., (1953), "Determination of the Unit Weight of Sediment for Use in Sediment Volume Compaction", USBR, Denver.

[21] Pazwash, H. (1982) Sedimentation in reservoirs case of Sefidrud dam. Proc. 3rd Congress of the ADP, IAHR, Bandung, Indonesia, Vol. C, Paper Cc7, p.215-223.

[22] Sadeghi, M. A. (2002) "Surveying the Trend of Sedimentation in the Dez Re-regulating Dam Reservoir Using the BRI–STARS Mathematical Model, Faculty of Agriculture, University of Shahid-Chamran, Ahwaz, Iran, Dissertation (M.S. Thesis).

[23] Samadi-Boroujeni, H. (2004), "Prediction of Sediment Concentration Resulting from the Dez Dam Hydraulic Flushing Operations at the Downstream River Along with a Survey of its Environmental Impacts", Report of Research project, Khuzestan Water and Power Authority, Iran.

[24] Samadi-Boroujeni, H. (2005), "Modeling of Sedimentation and Consolidation of Cohesive Sediments", Faculty of Water Engineering, University of Shahid-Chamran, Ahwaz, Iran, Dissertation (Ph.D. Thesis).

[25] Samadi-Boroujeni, H. and V.J. Galay,(2005), "Turbidity Current Measurments Within the Dez Reservoir,Iran", Proceeding, 17th Canadian Hydrotechnical Conference,.

[26] Samadi-Boroujeni, H., M.Fathi-oghaddam, M.Shafaie-Bajestan and H.M.V., Samani. (2005). Modelling of Sedimentation and Self-Weight Consolidation of Cohesive Sediments. Sediment and Ecohydraulics Intercoh2005.1stEdn,Elsevier B.V.Oxford,UK, ISBN: 978-444-53184-1 :165-191.

[27] Samadi-Boroujeni, H., A.H. Haghiabi, and E. Ardalan, (2010), Determination of appropriate hydraulic conditions to decrease the negative impacts of Dez dam flushing operation on the downstream, International Journal of Water Resources and Environmental Engineering, Vol.2 (1), pp. 001-008.

[28] Samadi-Boroujeni,H. and Kalali, M. (2012), Dez dam reservoir sedimentation effects on hydropower generation. Report of Research project, Water Resources Research Center, Shahrekord University, Iran.

[29] Scarlatos, P D., Lin Li, (1997), "Analysis of Fine-grained Sediment Movement in Small Canals", Journal of hydraulic engineering, Vol.123, No.3:200-207.

[30] Shafai Bejestan, M., (2008). Hydraulics of Sediment Transport. Shahid Chamran university press, Ahwaz, Iran, 549p.

[31] Sloff, C.J. (1991) Reservoir Sedimentation: a literature survey. Comm. on hydr. and geotechn. engrg., Report No. 91-2, Delft Univ. of Technology, The Netherlands, 126 pp.

[32] Sloff, C.J. (1994) Modelling turbidity currents in reservoirs. Comm. on hydr. and geotechn. engrg., Report No. 94-5, Delft Univ. of Technology, The Netherlands, 142 pp.

[33] Sloff, C.J. (1997) Sedimentation in Reservoirs, Doctoral Thesis, Delft University of Technology, 270 pp.

[34] Tamene, L., Park, S. J., Dikau, R., and Vlek, P. L. G. (2006), Reservoir siltation in the semi-arid highlands of northern Ethiopia: sediment yield-catchment area relationship and a semi quantitative approach for predicting sediment yield, Earth Surf. Process. Landforms, 31, 1364-1383.

[35] Thapa, B., Dahlhaug, O.G, (2003). Sand erosion in hydraulic turbines and wear rate measurement of turbine materials, Int. Conf. on Hydropower, Hydro Africa -2003, ICH, Arusha.

[36] Tolouie, E., J.R. West, and J. Billam (1993) Sedimentation and desiltation in the Sefid-Rud Reservoir, Iran.

7

Application of Microseismic Monitoring Technique in Hydroelectric Projects

Nuwen Xu, Chun'an Tang, Hong Li and Zhengzhao Liang
Institute of Rock Instability and Seismicity Research,
Dalian University of Technology,
People's Republic of China

1. Introduction

Slope instability and landslides have been always one of the most significant subjects of slope engineering, and also one of the hottest and most difficult topics in geotechnical engineering research all over the world. How to effectively predict and control slope failure hazards and ensure safety of these engineering projects is a significant task that people often probe. Recently, with the rapid and sustainable development of China' economy, mines proceed to ever greater depth and into more and more complex geological settings. Meanwhile, civil engineering projects, particularly large hydroelectric projects (e.g. West-East power transmission) in southwest China, are being advanced at greater buried-depths. These projects are challenged by violent rock mass failure processes due to deep cracks and faults, high stress levels, loose rockmass with low wave velocity, intense weathering, inter-layer extrusion zones and unloading fissures etc. As we know, extensive use of measurement technology, such as GPS (Global Positioning System), SAR (Synthetic Aperture Radar) Interferometry, TDR (Time Domain Reflectometer), multiple position extensometers, convergence meters and surface subsidence monitoring, is currently found to be very useful in surface deformation monitoring of slopes. However, it is unrealistic for them to effectively monitor the occurrence of micro-fractures in deep rock masses prior to the formation of a macroscopic rock fracture outside slope surface. With regard to rock slope, these internal micro-fractures may often lead to macroscopic instability of slope. Consequently, there must be an intrinsical correlation between rock slope macro-instability and its internal micro-fractures (i.e., microseismicity). It is well known that rocks loaded in testing machine and rockmasses that are stressed near underground excavations emit detectable acoustic or seismic signals. If these signals can be captured sufficiently clearly as seismograms by a number of sensors nearby, the origin time of seismic events, its location, another source parameter such as source radius, static stress drop, dynamic stress drop and apparent stress can be estimated (Cai and Kaiser, 2005; Mendecki, 1997). Microseismic monitoring techniques have been thus employed to locate damage in order to identify and delineate the potential hazardous regions in rock engineering practice, which would provide early warning of rock slope instability.

Earthquakes with local magnitude lower than 2.5 are called microearthquakes and are rarely felt. Microseismicity may be induced in mining areas, hydroelectric reservoirs or geothermal sites. Those activities involve changes in stress, pore pressure, volume and load, which can result in sudden shear failure in the subsurface, usually along pre-existing weakness zones such as fault structures. Therefore, long-term microseismic monitoring in related fields above would have a potential to reveal fracture geometry and to investigate the progressive failure of rock masses. During the past two decades, microseismic monitoring technique has been emerged from a pure research means to a mainstream industrial tool for daily safety monitoring at various geotechnical engineering. Plenty of engineering applications of microseismic monitoring technique were carried out in South Africa, Canada, Japan, Australia and North America. Significant achievements have been obtained in open pit slopes (Abdul-Wahed et al., 2006; Lynch et al., 2005), underground mining (Ge, 2005; Hudyma, 2008; Kaiser, 2005; Mendecki, 2000), tunnels (Hirata et al., 2007; Milev et al., 2001), oil and gas exploration and development, and electricity generation by hot dry rock (Tezuka and Niitsuma, 2000) etc. The application and its related theory of microseismic monitoring technique have been also investigated in China. For instance, in order to real-time monitor and analyze microseismicity of the left bank slope at Jinping I Hydropower Station, along Yalong River in Sichuan province, a routine micro-seismic monitoring system was installed in May 2009 and the monitoring results can identify and delineate the potential rock damage regions and sliding surface (Xu et al., 2011c; Xu et al., 2010). Some researchers (Jiang et al., 2006) have studied the relationship between strata fracturing and rock-burst in underground mines and discussed the possibility of forecasting rock-burst based on the rules of strata fracturing monitoring by different microseismic monitoring systems. In order to study the progressive failure of geological structures (faults, karst collapse colums) and predict their microseismic activities associated with water inrush, micro-seismic monitoring was employed in deep mines successfully (Jiang et al., 2006). Tang et al. (Tang et al., 2011.) investigated the tempo-spatial distribution of microseismicity for rockburst prediction of the drainage tunnels with a maximum buried depth of 2 500 m in Jinping II Hydropower Station, which is also situated in Sichuan province, southwestern China. These achievements play a significant role in solving the problems of mine field stress, slope instability, hydro fracturing as well as rockburst hazards, and promote the application and development of microseismic monitoring technique worldwide.

This chapter attempts to first of all give a brief introduction of microseismic monitoring technique, including the basic principle, the development history, constitution, positioning methods and waveform identification. Then engineering application of microseismic monitoring technique have been investigated through three typical hydroelectric projects such as Jinping I hydropower station, Dagangshan hydropower station and Jinping II hydropower sation in southwest of China. By analyzing the excavation-induced microseismicity, it is possible to identify and delineate the particular failure regions that underlie the seismic activity in order to perform early warning and advance support measures in rock slopes and tunnels. Meanwhile, tempo-spatial distribution of seismic source locations during construction and impoundment of dams can be used to determine the existence of micro-fractures associated with buildup, stress shadow and stress transfer in deep rockmass of rock slopes.

2. Microseismic monitoring technique

2.1 The basic principle

It is well known that rocks loaded in testing machines and rock masses that are stressed near underground excavations or inside the rock slope emit detectable acoustic or seismic signals, and seismic monitoring techniques have been developed as powerful methods for remotely determining the integrity of the rock mass in rock engineering practice (see Fig.1). This technique can be used to provide information on location, extent, and mechanism of any damage processes occurring in the rock mass. By incorporating source location and source parameter estimates, it is now possible to visualize the development of microseismic events in 3D space (Cai and Kaiser, 2005). Compared with traditional measure technology, microseismic monitoring is a remote, three-dimensional, and real-time monitoring technique. Moreover, further analysis of the scale and characteristics of rock failure can be performed according to hypocenter parameters recorded by the system. Having recorded and processed a number of seismic events within a given volume of interest over time, one can then quantify the changes in the strain and stress regimes and in the rheological properties of the rock mass deformation associated with the microseismicity (Trifu and Shumila, 2010). This in turn allows estimation of quantities like rock mass stability over space and time, and thereby prediction of slope failure precursor or rockburst of deep-buried tunnels.

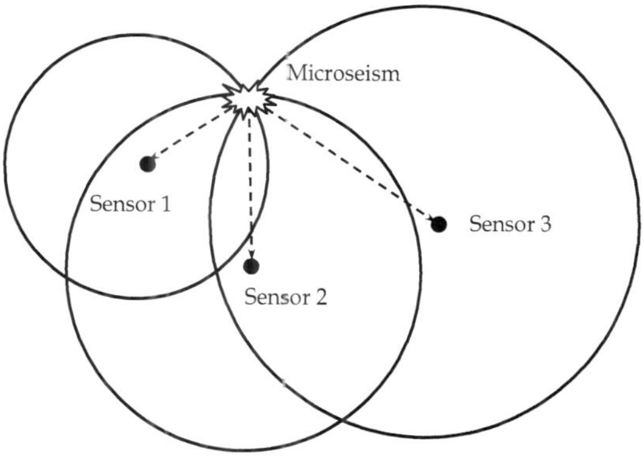

Fig. 1. Sketch diagram of the principle of microseismic monitoring.

2.2 The development history

The monitoring of microseismicity can be dated from 1908 when the first earthquake observation station was installed at Bochnum mine in Ruhr, Germany. The application of monitoring the signals of elastic wave induced by rock mass fractures in underground mines can be traced back to 1942 (Potvin and Hudyma, 2001). However, the significant achievement of seismic monitoring in understanding and investigating mine earthquakes became reality until extensive seismic sensors monitoring network was installed at Erpm mine in South Africa on 1960s (Cook, 1964). The manufacture and monitoring researches of mine seismic monitoring system have been highly paid attention to and rapidly developed when mine seismic monitoring was regards as significant approach of monitoring and prediction of mine rockburst and instability. Nowadays, the mine structures and three dimensional visualization seismic data can be integrated to provide a valid monitoring tool for investigating seismicity and excavation-induced effect in mining, petroleum, and hydropower fields (Kaiser, 2005; Lynch and Mendecki, 2001).

Currently, there are several typical microseismic monitoring systems all over the world such as ISS in South Africa, ESG in Canada, SOS in Poland and ASC in England. ISS systems were popularly used in mines and open pit slopes (Lynch et al., 2005). ESG systems have an extensive application in Canada, Australia and China (Urbancic and Trifu, 2000; Xu et al., 2011c). SOS and ASC systems are also adopted in some projects worldwide. In China, the beginning of studying microseismic monitoring was a little late, and there were few engineering applications in the early stage. Nevertheless, with decades of continuous development, there are some achievements in equipment manufactures and software designing of microseismic monitoring in recent years. Currently, the popular seismic monitoring systems above were imported to further-investigate rockburst in mines and deep buried tunnel, failure mechanism in rock mechanic and rock engineering in China. The technique develops continuously with the rapid progress of computer and data acquisition technology. In brief, it can provide a new approach to investigate spatial rupture pattern of overburden and excavating-induced stress distribution in mines, tunnels and rock slopes.

2.3 The constitution of microseismic monitoring system

It is worth noting that the microseismic monitoring systems adopted in the present study are manufactured by the Engineering Seismology Group (ESG) in Canada, although the wireless transmission system MMS-View was developed by Mechsoft Co. Ltd. in China to be supplemented into the monitoring system. The seismic monitoring network consists of Hyperion digital signal processing system, Paladin digital signal acquisition system, a number of uniaxial/triaxial acceleration transducers deployed in boreholes drilled from rock mass in rock slopes or tunnels, and the 3-D visualization system namely MMS-View based on remote wireless transmission (see Fig.2). Accelerometers are connected to the

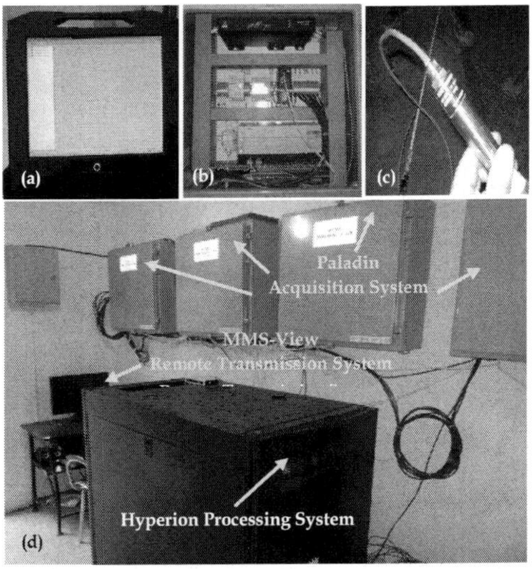

Fig. 2. Constitution of microseismic monitoring system: (a) Hyperion digital signal processing system, (b) Paladin digital signal acquisition system, (c) acceleration transducer, and (d) the center of the site monitoring system.

substations (namely Paladins acquisition system) using copper twisted-pair cables. The substations are connected in series to the central system (namely Hyperion processing system) using optical fiber cables and network cables. Paladin units rely on a pulse per second (PPS) signal originated from the Paladin Timing Source over the network. This can allow the data from each Paladin to be accurately time-stamped, ensuring that multiple units are in synchronization (Urbancic and Trifu, 2000). Finally, the seismic data recorded are transmitted from the on-site center to the office by GPRS for further analysis. The three systems are fully described in (Tang et al., 2011.; Xu et al., 2011a; Xu et al., 2011c; Xu et al., 2010). The processing system in the central station digitizes the data with a sampling frequency of 10 kHz and performs preliminary event detection when the recorded signals of the substations exceed a given threshold. using the Short Time Average vs Long Time Average algorithm (STA/LTA). The system then calculates in real time a preliminary hypocenter determination with a homogeneous velocity model and records the data on hard disk. This detailed specification can be found in (Trifu and Shumila, 2010; Xu et al., 2011b; Xu et al., 2010).

2.4 Positioning accuracy and waveforms identification

In practice, artificial fixed position blasting tests are usually adopted to calculate the mean wave velocity. Moreover, the typical waveforms such as blasting signal, micro-fracture of rock mass, tap-test signal and background noise etc. are also identified and distinguished. Herein, the blasting tests for wave configuration in Dagangshan project is taken as an example (Xu et al., 2011b).

2.4.1 Wave velocity testing

The wave velocity influences the first break time that elastic wave arrives to sensors. Therefore, wave velocity configured in the monitoring system has a great influence on seismic source location. In this respect, wave velocity of rock mass within the monitoring scope must be calibrated before testing the positioning error of the system. The mean velocity of elastic wave is first determined as 4500m/s according to wave velocity tests of rock samples on-site in Dagangshan project. Then the wave velocity of blasting tests is calculated inversely and adjusted through data analysis program. The waveform of the first blast test recorded is shown as Fig. 3, which occurred inside the drainage tunnel of the dam foundation at 1081m level on 04:25:41, May 7, 2010. The testing results show that the mean velocity of P wave and S wave are 4400m/s and 2540m/s respectively in the scope of monitoring, which coincide with testing results of wave velocities on-site. The wave velocities can be then used to locate seismic events.

After calibration of wave velocity inside rock mass in the scope of monitoring, three localization tests using artificial blasts method were performed in order to check positioning accuracy of the seismic monitoring installation at the right bank slope. The coordinates of three artificial blasting tests location and their positioning recorded by the monitoring system are shown as Table.1. The results of blasting tests show that microseismic source location error is less than 10 m in the scope of the sensor array. This validates that the accuracy of the microseismic monitoring system installed at the right bank rock slope is high. Fig.4 shows the spatial comparison between an artificial blast test location and its positioning result recorded by the system.

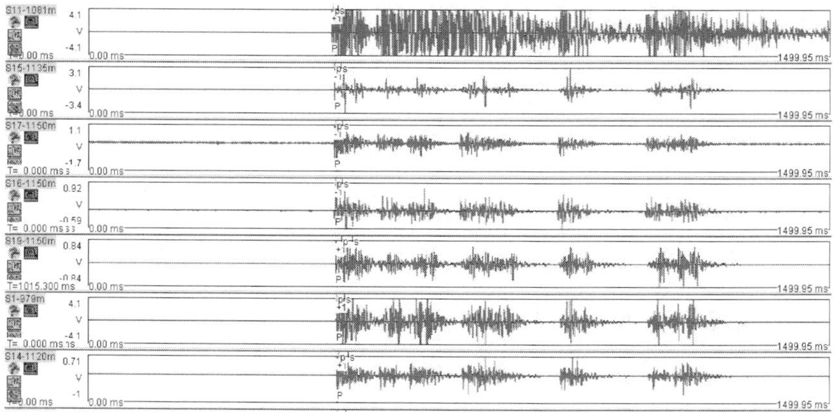

Fig. 3. Waveform of the first artificial blast recorded by the system.

No	Blast date	Blast time	Height /m	Blast test coordinates/m			Locating coordinates/m			Error/m			Absolute error/m
				X	Y	Z	X	Y	Z	X	Y	Z	
1	05-07	04:25:41	1081	520837.2	3259177.8	1082.7	520835.2	3259181.7	1074.5	2.0	3.9	8.2	9.3
2	05-07	19:08:52	1081	520829.4	3259169.8	1082.8	520823.9	3259173.5	1080.8	5.5	3.7	2.0	6.9
3	05-09	20:47:39	1081	520823.5	3259172.3	1082.8	520826.6	3259179.9	1084.3	3.0	7.6	1.5	8.3

Table 1. Comparison of coordinates of artificial blasting with those of microseismic monitoring localization (in 2010).

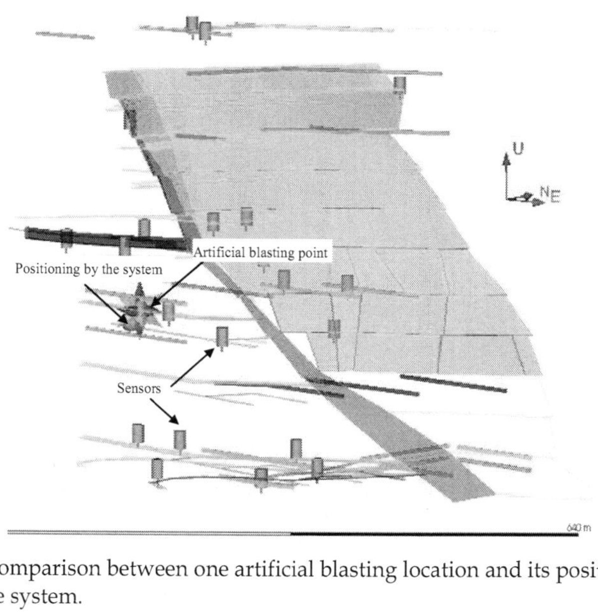

Fig. 4. Spatial comparison between one artificial blasting location and its positioning result recorded by the system.

2.4.2 Waveform identification and analysis

Through field observation and investigation, the preliminary events recorded by the microseismic monitoring system at the right bank slope are classified into three main types: rock micro-fracture events, excavation blasting events, vibration and noisy events (Xu et al., 2010). Fig.5 shows the typical waveform of microseismic event. It can be observed that such waveforms are very smooth with amplitude range from dozens of mV to hundreds of mV, magnitude distribution nearby -1.2, and energy release around about 102 Joule. The waveform of blasting tests is shown above in Fig.3. The sensor No.11 (S11-1081m), which was close to the shot point, picked up the elastic wave induced by blasting tests at first. The

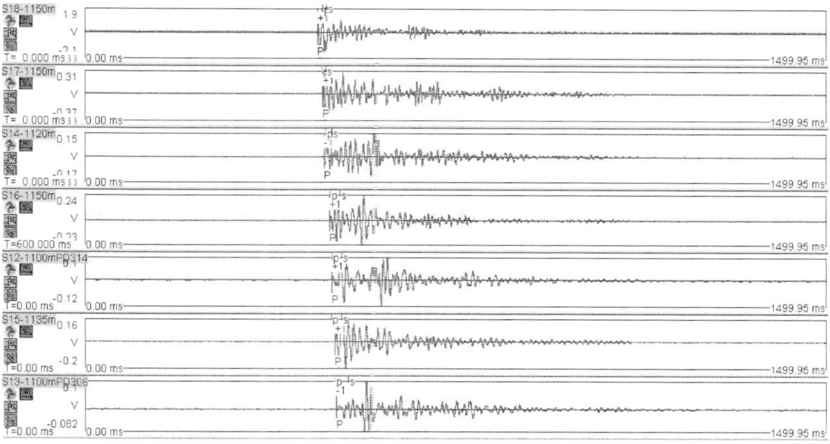

Fig. 5. Waveform of a typical microseimic event.

(a)

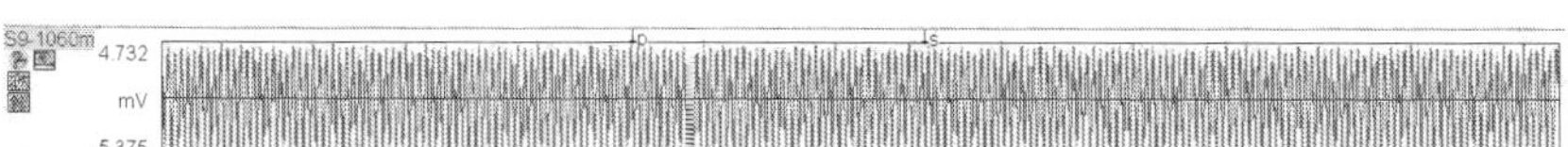

(b)

Fig. 6. Waveforms of vibration and noise (a) Waveform of tap tests, and (b) Waveform of electricity interference.

amplitude of the blasting event was about 4.1V and the moment magnitude was -0.91. Fig.6 shows the waveforms of different machinery vibration and noise. The characteristics of such waveforms are repeat shaking along time shaft and stripped. In order to verify the installation accuracy of the sensors, artificial tap tests corresponding to each sensor are performed as shown in Fig.6 (a). It can be seen that the amplitude of sensor No.5 (S5-979m) in acquisition interface is greatly higher than other sensors when knocking rock mass nearby the sensor No.5 at 979m level. The results demonstrate that the coordinates for each sensor input into the system are correct.

3. Engineering application

3.1 Stability evaluation of the left bank slope of Jinping I hydropower station

3.1.1 Project background

Jinping I hydropower station is located at the sharp bend of Jinping on the middle reach of Yalongjiang River, near Xichang, about 500 km southwest of Chengdu, Sichuan province, China. It is situated within the aslope transition zone from the Qinghai-Tibet Plateau to the Sichuan Basin. The project has a double-curvature arch dam with a maximum height of 305 m and a total installed capacity of 3300 MW, which will be the highest arch dam in the world nowadays. The geological structure of this area is complex due to the great variability in lithology and lithofacies, intensive tectonic deformation and folding, abundant fractures, and widespread metamorphic terranes. Complex geological conditions greatly affect the engineering design in the project (Qi et al., 2004; Zhong et al., 2006). As illustrated in Fig.7 (a), the dam site is located in a reach 1.5 km between Pusiluo gulley and Shoupa gulley. The natural slopes and the left slope after excavation are shown in Fig.7.

Fig. 7. Geomorphic photograph of dam site: (a) before excavation (Zhong et al., 2006); (b) the left slope after excavation.

The left bank slope of the hydropower station has lots of prominent characteristics such as large scale, complex engineering and technical conditions, high and steep natural valley slopes, higher stress levels, a strong rock unloading, inter-layer extrusion zones and deep cracks in the complex geological site. The behavior of such features plays a critical role in the stability evaluation, especially in the processes of excavation of the slopes and different tunnels. Furthermore, plenty of deep cracks and faults such as F_{42}, F_5, F_8 and lamprophyres (X) form huge latent instable blocks in these areas. The instable blocks will induce the potential instability of the slope.

3.1.2 Tempo-spatial distribution of microseismicity

Although lots of researches on microseismic monitoring have been carried out, little work was found to investigate slope stability of large-scale water conservancy projects, especially high rock slopes. Our surveys present here highlighted the continuous characteristics of microseismicity induced by continuous excavation and grouting inside the rockmass in the left bank slope. After filtering out the noisy events, a dataset of 1521 seismic events with M_w -2.5 to 0.2 occurred from June 2009 to May 2011. Fig.8 shows the temporal distribution of seismic source locations. It can be observed that the daily rate of events ranges from 1 to 14 with small bursts of activity every 3-5 days. Most of the seismic events are recorded during several swarms. According to on-site observation and analysis, such swarms are partly caused by stress redistribution of deep rock mass inside the left bank slope on the basis of continuous excavation in different tunnels or at the bottom of the rock slope, and partly caused by a secondary extension of deep rock mass fissures due to high-pressure grouting. A notable phenomenon that the frequency of the recorded microseismic events increases drastically right after excavation and grouting is obtained. In addition, there is no microseismic signal during the period from 1 December to 17 December 2009. The reason is that the seismic monitoring network broken down caused by perturbation of large machinery on-site. Seismic source locations decrease sharply after the main excavation of rock slope stopped and the pouring of dam began in October 2009 (see Fig.8).

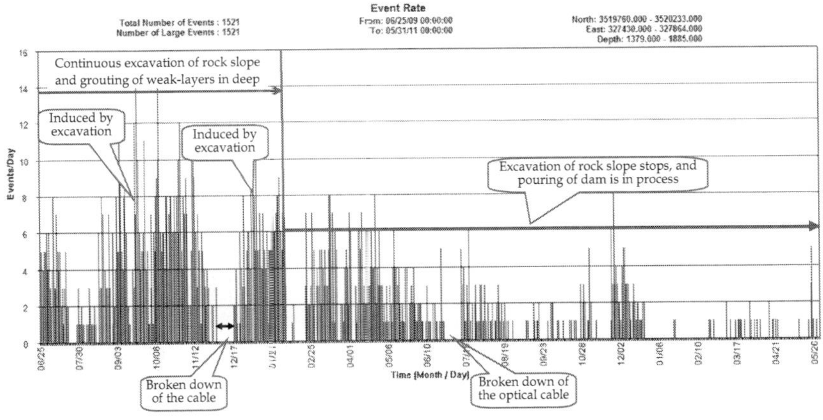

Fig. 8. Temporal distribution of seismic source locations.

The spatial distribution of the seismic source locations recorded is presented in Fig.9. The figure can highlight the distribution and migration characteristics of micro-fractures clusters in deep rock mass of the left slope. Fig.9 (a) shows that majority of microseismic events predominantly occurred along the slope of the dam spandrel, especially along the faults f_5, f_8 and X. To further emphasize the relationship between the geological structures and spatial distribution of seismic source, Fig.9 (b) displays the microseismic events in an aerial view. It can be found that seismic source clusters are concentrated at the bottom of the downstream slope nearby the dam, which implicates that the design and implementation of the sensor array meet the need of global monitoring of the interest areas at the left slope. Discarding extreme values out of the study volume, it is apparent that the regions characterized by seismic clusters show strong correlation with pre-existing geological structures, especially faults or fractures inside the rock slope (Shumila and Trifu, 2009).

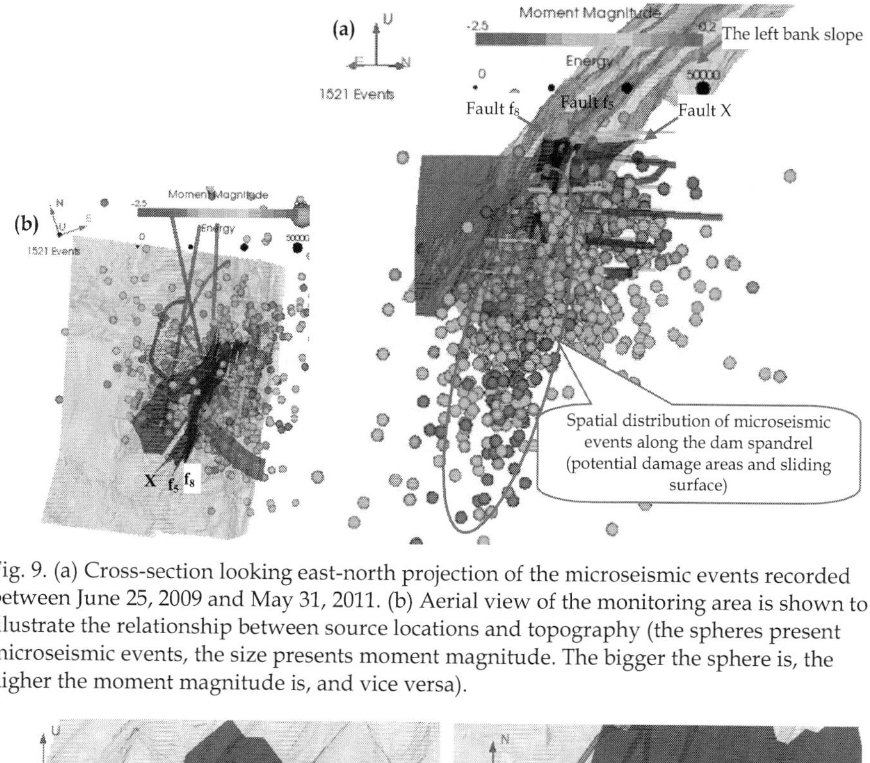

Fig. 9. (a) Cross-section looking east-north projection of the microseismic events recorded between June 25, 2009 and May 31, 2011. (b) Aerial view of the monitoring area is shown to illustrate the relationship between source locations and topography (the spheres present microseismic events, the size presents moment magnitude. The bigger the sphere is, the higher the moment magnitude is, and vice versa).

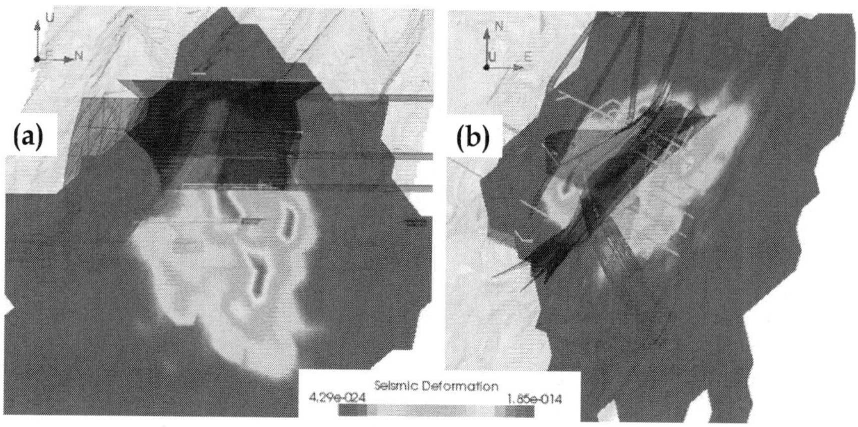

Fig. 10. Seismic deformation contours in (a) south-north vertical cross-section looking east and (b) north-easting plane at 1730m elevation looking down. Different colors represent various quantities of seismic deformation per unit volume.

Evaluating the seismic deformation is important for the understanding of stress redistribution and geomechanical processing taking place on site (Shumila and Trifu, 2009). As is shown in Fig. 10, the seismic deformation tends to be related to pre-existing geological structures such as faults f_5, f_8 and X, as well as to the excavation of tunnels inside the left slope. It can be seen from Fig.10 (b) that the maximal seismic deformation is mainly located along the faults mentioned above, and focused at the regions of elevations between 1670m and 1829m (Fig.10 (a)). Worth noting, the value of the seismic deformation is very small, which reveals that the left slope can be estimated as a stable slope currently. As expected, the seismic deformation is very prominent in the areas of pre-existing faults and man-made structures.Spatial distribution regularity of microseismic activity and seismic deformation maps can be used for two aspects: (1) support selection and tracking of rock mass condition for excavation support rehabilitation in the left slope; and (2) future routine monitoring plan and corresponding equipments rearrangement on the hazardous areas, especially after impoundment of the dam which will induce stress redistribution of deep rock mass inside the slope.

3.1.3 Comparison study between seismic monitoring and field observation

The correlation between microseismicity and excavation-induced faults and cracks are illustrated in Fig.11. The microseismic events mainly occur at the potential sliding surface, nearly 90 m away from the surface of the rock slope. It can be seen from Fig.11 (a) that microseismic events with large magnitude and high energy mainly focus at the bottom of the downstream slope. Such events are caused by the foundation pit excavation of the dam according to field observation. Conversely, microseismic events with small magnitude and low energy almost concentrate along the weak structures such as the pre-existing faults and some tunnels in the excavation phase. It is worth noting that not all geological structures are directly observed and mapped. Therefore, it is the occurrence of microseismicity that characterizes if an unknown geological structure is active. Microseismicity related to these structures is also evidence of rock mass degradation from elevated stresses as each microseismic event indicates rock fracturing (Trifu and Shumila, 2010).

Combining with on-site observation and analysis of microseism swarms, there is a direct correlation with some identified geological structure (e.g., faults f_5, f_8 and X) and newly identified geological structure. Region (I) in Fig.11 (a) shows that microseismic events are mainly located along the pre-existing faults such as f_5, f_8 and X. This can be interpreted as tensile failure subject to excavation of the dam foundation and it has an active zone along the dam abutment range from 1670 m level to 1829 m level. Region (II) in Fig.11 (a) can be interpreted as pressure-shear failure that is responsible for seismic events of large magnitude recorded at the bottom of the slope (see Fig.11 (b)) and it has an active area of about 150 m in diameter; Region (III) in Fig.11 (a) can be demonstrated that it is activated by local stress perturbations due to excavation of the corridor in the downstream slope at 1730m level (Fig.11(c)). Therefore, microseismicity in the left slope occurs due to the day to day tunnels excavation activity and grouting to the weak-layers in deep rock mass, and have typically been the result of most cracks in tunnel sidewalls and at the bottom of the slope. As such, the microseismic monitoring system implemented at the left bank slope can identify, locate and quantify microseismicity and potential damaged areas, allowing for a better understanding of the deep rock mass fracture mechanism. They then contribute to the management and mitigation of seismic hazards, for improving the rock slope stability and enhancing the capability of risk forecasting.

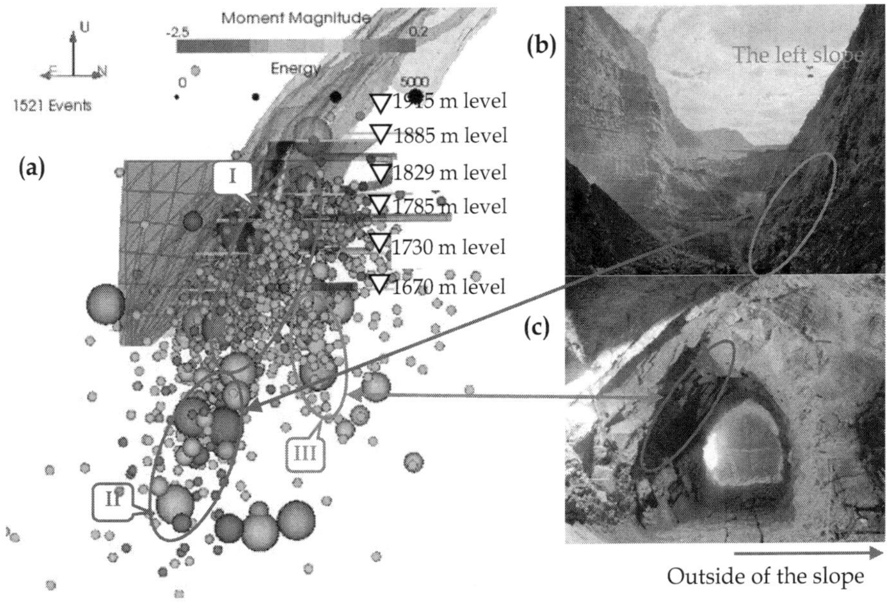

Fig. 11. Correlation between spatial distribution of microseismic events clusters and field observation: (a) Spatial distribution of source location; (b) Fractures at the foundation pit; (c) Cracks at the sidewall of the drainage tunnel on 1670 m level.

3.2 Excavation-induced microseismicity at the right bank slope of Dagangshan hydropower station

3.2.1 Project background

Dagangshan Hydropower Station is located at Dadu River in Sichuan province, southwestern China. The project is one of the large scale hydroelectric constructions which are currently exploited along the mainstream of Dadu River. It has a double-curvature dam with a maximum height of 210 m, and a total installed capacity of 2, 600 MW. Complex geological conditions greatly affect the engineering design and construction in the project. The layout of key water control and the right bank slope after excavation are presented in Fig.12 (a) and (b), respectively. The right bank slope of the project has complicated geological conditions such as faults, dykes, unloading cracks zones and joints with cracks development. In particular, faults X316-1 and f_{231} are the most significant factor that influences the stability of the rock slope. Weathering and unloading of rock mass inside the slope are also very serious. Plenty of investigations and excavations reveal that deformation failures have superficially occurred on the slope due to stress rearrangement as a result of sapping of the River. A microseismic monitoring system was adopted to further analyze the right bank slope stability in June, 2010. The spatial arrangement diagram of sensors at eleven elevations is shown in Fig.13.

Fig. 12. Geomorphic photograph of the dam site: (a) The layout of key water control, and (b) The right bank slope after excavation.

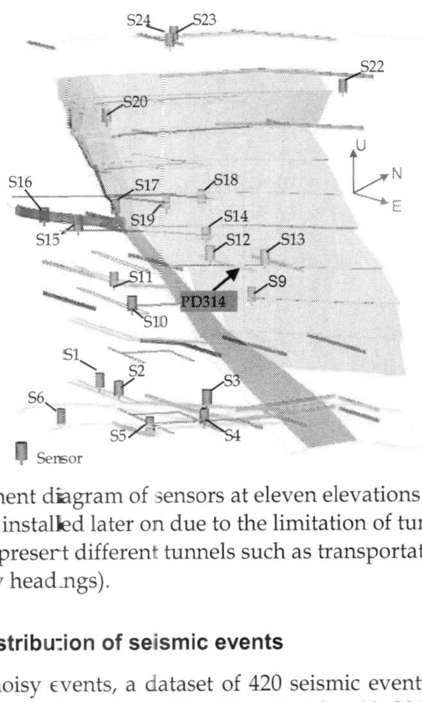

Fig. 13. Spatial arrangement diagram of sensors at eleven elevations.(Noting that three sensors No.7, 8, 21 to be installed later on due to the limitation of tunnel condition. The different color columns present different tunnels such as transportation tunnels, drainage tunnels, and exploratory headings).

3.2.2 Tempo-spatial distribution of seismic events

After filtering out the noisy events, a dataset of 420 seismic events was recorded within a given volume of interest over time from May 12 to October 20, 2010. Forty-eight percent of the data base such as blasting events, mechanical vibration and background noise events was rejected. A plot of cumulative number of seismic events recorded from the selected time shows one period of increased micro-seismic activity: July 20 to August 5, 2010. This graph is compared with the time history of excavated rock mass at the bottom of the rock slope as shown in Fig.14. It can be observed that the curve of cumulative seismic events in Fig.14 (a) goes with almost the same tendency as the curve of excavated volume of rockmass (volume blasted) in Fig.14 (b) does. This is a good correlation which illustrates a clear consistency between the volume of rock mass being excavated and the total number of seismic events. Therefore, it can be noted that excavation activities play a great role on the levels of extension fracturing within a rock slope. Seismic data recorded here confirm this link, and go further by providing routine micro-seismic monitoring to quantify the effects that excavation is having on the slope. Additionally, it can be seen from Fig.14 (a) that the daily rate of seismic event ranges from 1 to 7 with small bursts of activity per 5-7 days. The mean rate of events during the selected period is 3 per day. According to on-site investigation, swarms of micro-seismic events are caused by excavation-induced stress redistribution of deep rock mass inside the right bank slope.

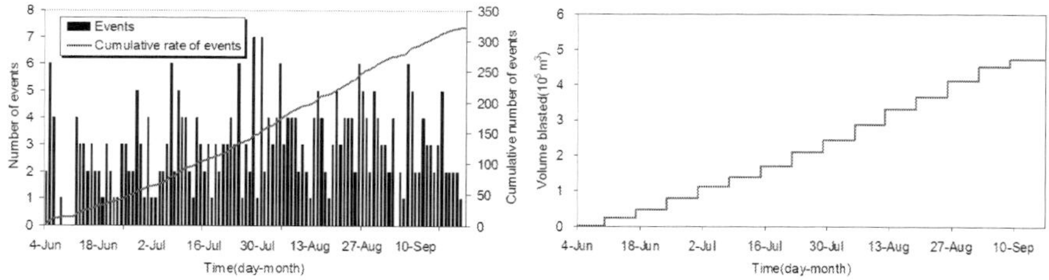

(a) (b)

Fig. 14. (a) Graph of the cumulative number of seismic events, and (b) the curve of cumulative amount of rockmass removed from the right bank slope in chronological order. The general forms of the two graphs are similar (in 2010).

The spatial distribution of the events recorded during the selected period is shown in Fig.15. Fig.15 (a) shows the absolute locations of all events including micro-seismic events, production blasts and some noises without manual processing. Fig.15 (b) shows the micro-seismic events recorded after eliminating interference events. During this period of micro-seismic monitoring, the rock slope was excavated continuously from the elevation 1070m to 980m and shearing-resistance tunnels at 1120m and 1150m elevation were also excavated, respectively. The micro-seismic event locations show that hypocenters are concentrated on these excavation zones, and this concentration is due to the excavation configuration. It can be seen from Fig.15 that hypocenters follow the working faces and they are located where the excavation is carried out, making it possible to correlate them unambiguously with the construction areas (Abdul-Wahed et al., 2006). Therefore, the regions inside the right bank slope, where micro-seismicity is active or inactive, can be identified and delineated.

In addition, Fig.15 (b) shows the majority of micro-seismic events predominantly occurred around the elevation 1180m of the upstream slope, especially focusing on the hanging wall of the fault XL316-1 (Fig.15 (c)). The top view in Fig.15 (c), looking from the upside, shows that the seismic events are being recorded at significant depths into the slope. However, there are small number of seismic events occurred at the toe of the rock slope. This may be partly attributed to residual movements (e.g., unloading crack zones, developed dikes and faults etc.) inside the rock slope at high elevations and/or the excavation of shear-resistance tunnels. On the other hand, it may be also partly due to high stress concentration at the slope toe because of slope geometry. It is known that seismic activity depends on stresses and tectonic conditions, herein the tectonic stress is probably less important in this zone and the compression stress (normal stress) is higher and helps to reduce the fracturing and the seismic emission (Xu et al., 2011a).

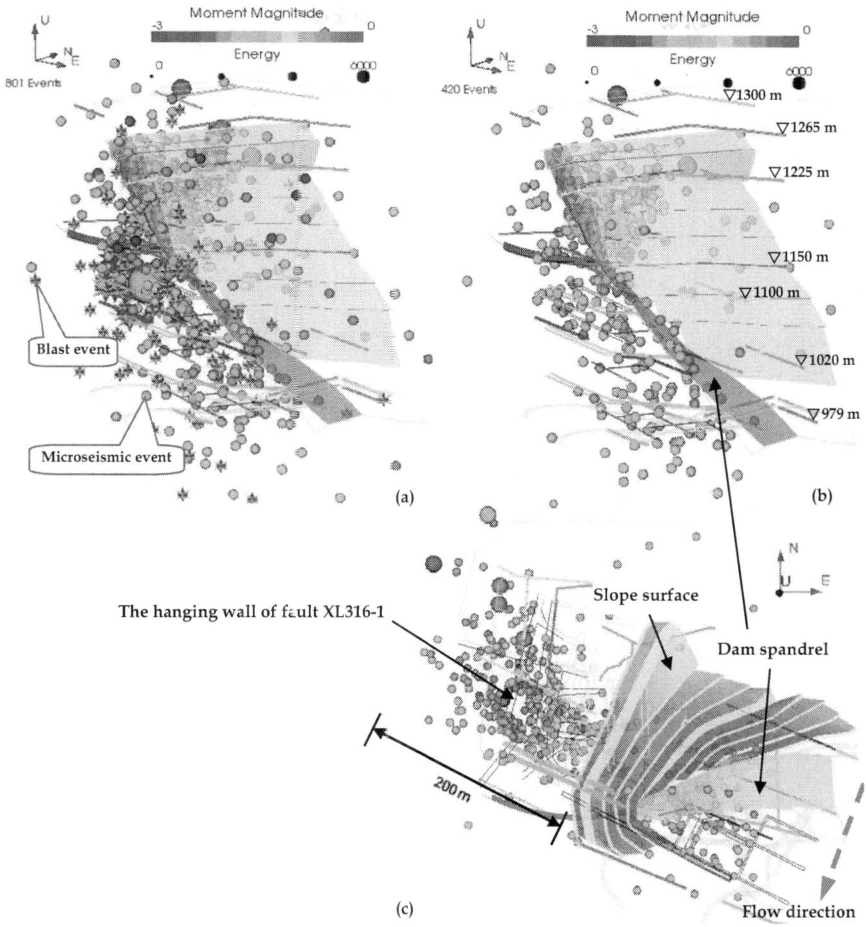

Fig. 15. Spatial distribution of events: all events without processing (a) and micro-seismic events (b), (c). Of which (c) shows top view from the upside, of seismic events recorded from May 12 to October 20, 2010. It is evident that micro-seismicity is being recorded at significant depths here (the spheres denote microseismic events, the size presents moment magnitude. The bigger the sphere is, the higher the moment magnitude is, and vice versa).

3.2.3 Excavation-induced microseismicity analysis

Fig.16 shows the area map of energy loss density induced by micro-seismicity of the rock slope since the operation of the seismic monitoring system. It can be observed that energy loss of the seismic events induced by excavation is mostly concentrated above the elevation

1135 m of the right bank slope, with an obvious spatial distribution along faults XL316-1 and f_{231}. A small volume of energy loss by micro-seismicity occurred at the bottom of the slope. However, the magnitude of the energy loss is very small, the potential impact on the rock slope stability needs continuous database of seismic source locations for in-depth investigation. With accumulation of microseismic data recorded and extension of excavation scale below 980m level at the rock slope later on, further researches on these results will be extensively studied through analysis of rock failure cases occurring in the deep rock mass. Moreover, comparison between Fig.15 and Fig.16 shows that seismic events mainly focused on the working zones. For example, seismic source locations are located precisely at, and slightly nearby the workings, meaning that they can unambiguously be correlated with the advance of the rock slope excavation. This comparison showed that most of event clusters are concentrated around the construction areas.

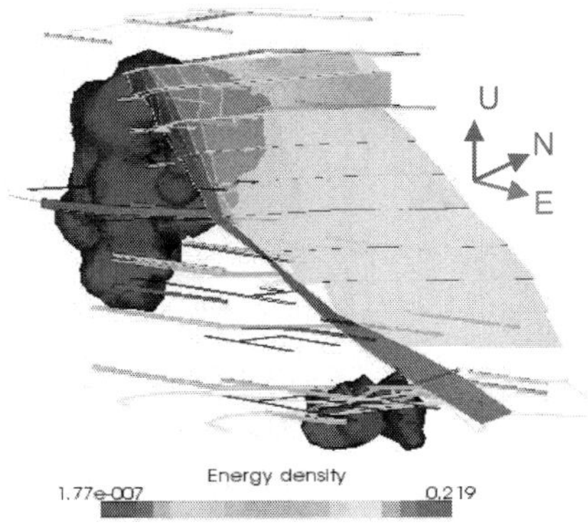

Fig. 16. Density of energy loss induced by micro-seismic events.

3.3 Rockburst prediction of TBM-excavated tunnels in Jinping II hydropower station (Tang et al., 2011)

3.3.1 Project background

Jinping II hydropower station, which is nearby Jinping I hydropower station is located on the Jinping bend of Yalong River at the junction of three counties Muli, Yanyuan, Mianning of Liangshan Yi Autonomous Prefecture, Sichuan Province, China. It takes advantage of the natural elevation drop at the Jinping bend of Yalong River, and water is diverted by a sluice dam to headrace tunnels for power generation. Jinping II Hydropower Station is an important cascade hydropower station on the main stream of Yalong River, with an installed capacity of 4800MW. The Jinping mountain is grand with multiple peaks and cut by deep valleys. The elevations of main peaks are more than 4000 m, with the maximum height of 4488 and the maximum elevation drop of over 3000 m. The project consists of 7 parallel tunnels such as the headrace tunnels No.1 to No.4, the drainage tunnel, the auxiliary tunnels A and B. Among them, headrace tunnels No.1 and No.3 are constructed by TBM tunneling and the diameter is 12.43m. The drainage tunnel is also excavated by TBM with a diameter of 7.2m. The rest tunnels are excavated by drilling and blasting method. The maximum excavated cross sections of headrace tunnels No.2 and No.4 is 13m in dimension and horseshoe-shaped. Since tunnel construction was commenced, hundreds of rockbursts in various intensities have occurred. Among them, the recent two rockbursts were strong rockburst and very strong rockburst, respectively. The region map of Jinping II hydropower station is shown in Fig.17.

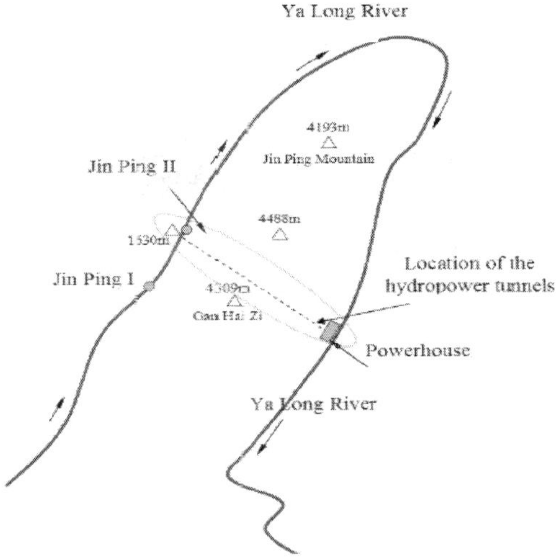

Fig. 17. Regional map of Jinping II hydropower station.

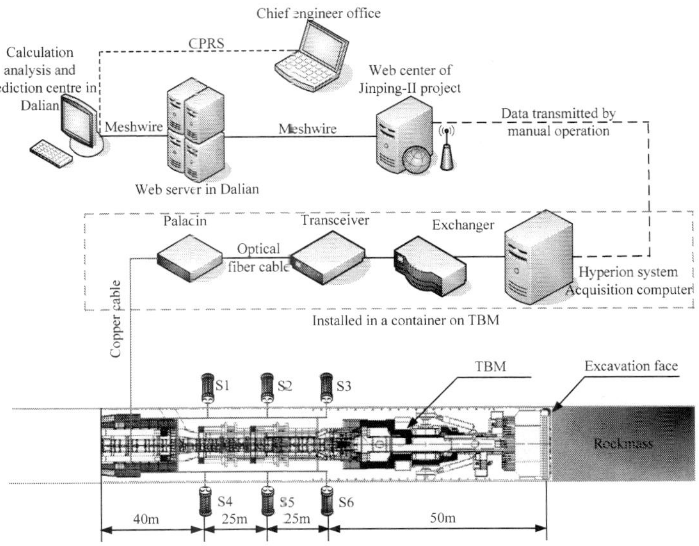

Fig. 18. The monitoring and analysis system for rockbursts during TBM tunneling for Jinping II Hydropower Station.

In September 2009, microseismic monitoring technique was employed into tunnel construction of this project. A movable microseismic monitoring system which can advance with tunneling was established. Six accelerometers were inserted into rock mass of the excavated tunnel to a depth of about 3-4 m. The sensors were placed on bottom of the boreholes. The sensors were placed 30-40 m apart, in three groups each with two sensors in 2 hour and 10 hour direction respectively. The distance between the main tunnel face and the first group of sensors was 30-40 m. The sensors at the tail end moves to the forefront as the excavation of the main tunnel advances. Data were collected continuously by the microseismic data acquisition system and transmitted to a measurement office at a distance of 5-10 km by optical fiber cable. The data were then processed and analyzed. Continuous acquisition and collective analysis of the seismic monitoring data were achieved, which provided an important platform for studies on rockburst monitoring and early warning during TBM tunneling. The microseismic monitoring system is shown in Fig.18.

3.3.2 Tempo-spatial distribution of microseismicity of the strong rockburst on November 28, 2009

The epicenter of a microseismic event may be determined from the difference between the arrival times of the initial microseismic signals at the sensors placed in the borehole. Each circle represents a relative seismic energy. Generally speaking, the everyday safety control at the site of the tunnel excavation can be based on monitoring of microseismic events, calculation of the maximum amplitudes of events measured, and the observation around the rock face. Fig.19 shows the cumulative distribution of microseismic events within 30 days before a strong rockburst in a tunnel of Jinping II Hydropower Station, where the area of concentrated microseismic events is the center of the rockburst.

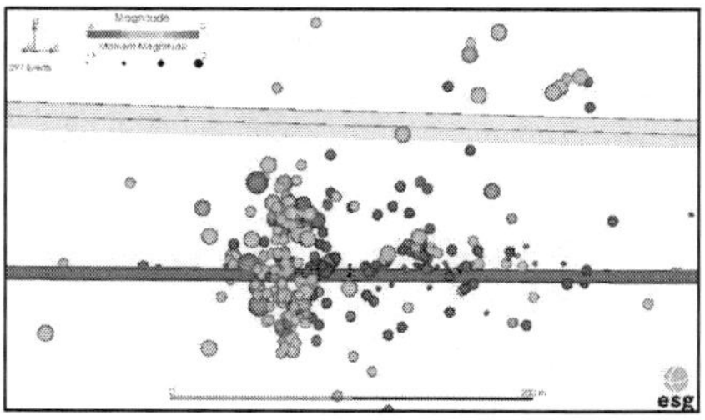

Fig. 19. The cumulative microseismic events within 30 days before the strong rockburst on November 28, 2009.

Fig.20 plots the density nephogram of microseismic events for the strong rockburst on November 28, 2009. Point B was the location of rockburst. The precursors were even more obvious for this rockburst. Anomaly was observed in the nephogram near point B 10 days before the rockburst. The nephogram core was even more evident at point B 2 days before the rockburst. Thereafter, a large-magnitude rockburst occurred, which resulted in extensive tunnel collapse (see Fig.21). Field observation showed that the crater formed by the strong rockburst was 9 m deep and indicated clears signs of structural planes. However, the microseismic location record during the rockburst process shown in Fig.22 suggested that the formation process lasted for 2 minutes, given the crater was as deep as 9 m. With the two minutes from the first microseismic event recorded at 00:42:43 to 00:44:42, about 40 microseismic events were captured by the system. Moreover, most of the seismic events were distributed along a strip, which tallied with the strike of the structural plane identified during field inspection.

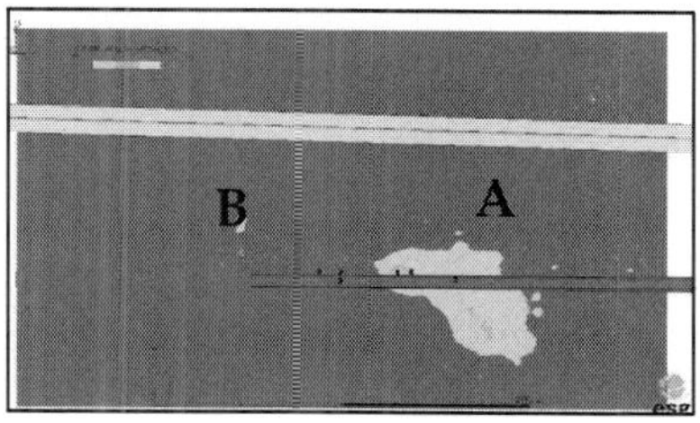

22 days before rockburst

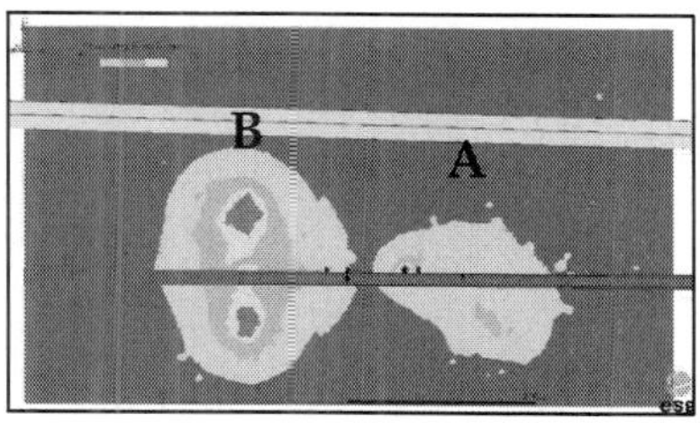

10 days before rockburst

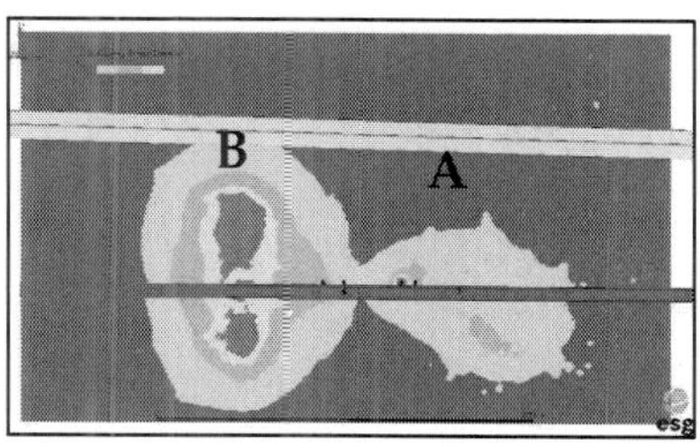

2 days before rockburst

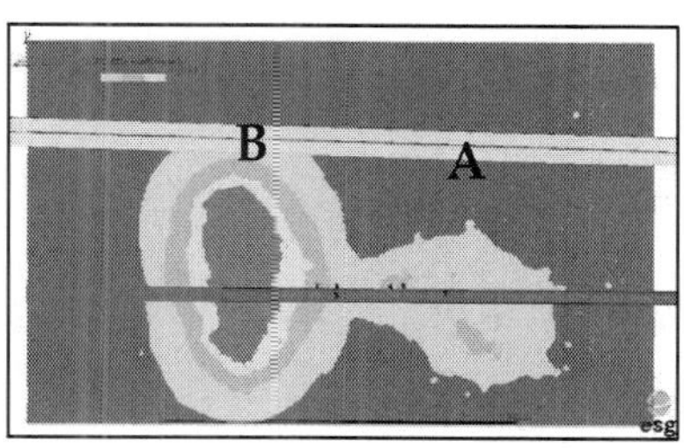

the day of rockburst

Fig. 20. Variation of density nephogram of microseismic events of the strong rockburst on November 28, 2009. (Rockburst occurred in the tunnel section below point B).

Although the prediction of time and magnitude may be very difficult at the moment, the notification of rockburst location may play an important role in mitigating or even prevent rockburst hazards. Fig.23 shows a good example of prevent a rockburst hazard using our MS-based risk-management system. As shown in the first figure in Fig.23, which is the data 8 days before the data shown in the second figure. Based on these figures, we announced a warning of a big possible rockburst in the red color area. After receiving this information, a lot of measure has been done to prevent the occurrence of the rockburst. One of the important actions is the blasting induced energy release within the potential area of rockburst, and the excavation work was successfully executed.

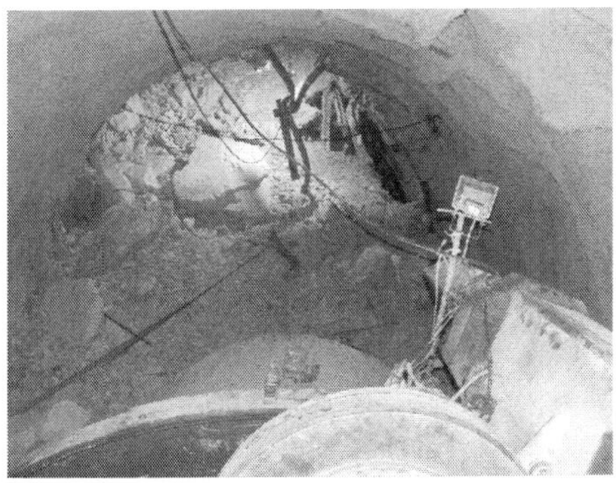

Fig. 21. The damage phenomenon of tunnel and TBM after rockburst on November 28, 2009.

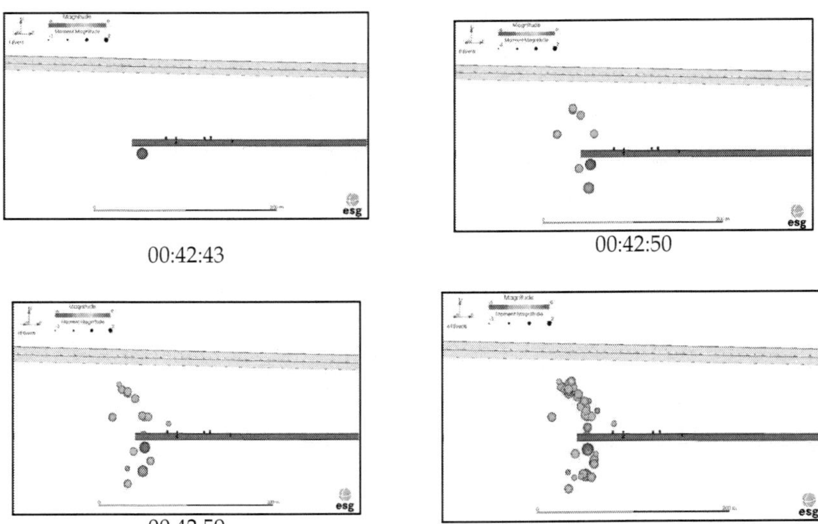

Fig. 22. Seismic source locations record during formation of rockburst crater (The formation of rockburst lasted for 2 minutes).

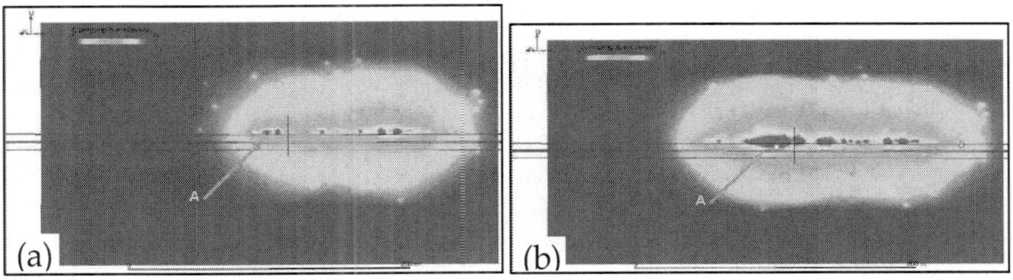

Fig. 23. Variation of density nephogram of microseismic events showing the potential location for rockburst.

4. Conclusions

This chapter focuses on understanding the microseismicity induced by continuous excavation or grouting in high rock slopes and tunnels of three large-scale hydroelectric projects such as Jinping I hydropower station, Dagangshan hydropower station, Jinping II hydropower station. The following conclusions can be drawn as follows:

1. The microseismic monitoring carried out in the left bank slope of Jinping I hydropower station highlighted stability analysis of the slope, especially the deformation of deep rock mass. The monitoring results provide invaluable information about the tempo-spatial distribution and characteristics of excavation/grouting-induced microseismicity. The regions characterized by seismic clusters show strong correlation with pre-existing geological structures, especially faults or fractures inside the rock slope. The current microseismicity occurred mainly along the slope of dam abutment where excavation was nonproductive. One proper explanation for this phenomenon is that microseismicity is caused by excavation at the dam foundation and grouting inside the slope. Thus, microseismic events could reveal the processes of initiation, development and expansion of micro-fractures inside rock slope, which could delineate the regions of potential instability. Consequently, support measures for the slope could be proposed.
2. The study of microseismic monitoring on Dagangshan project focuses on precisely determining the volume of influence at the right bank slope. The results show that microseismic monitoring can not only provide invaluable information on the tempo-spatial distribution and characteristics of excavation-induced microseismicity, but also give an indication whether a particular known geological structures is seismically active or not. Planes of weakness defined by microseismic events may indicate previously unknown geological structures. Furthermore, the current microseismicity mainly occurred around the elevation 1180 m of the upstream slope, especially concentrating on the hanging wall of the fault XL316-1. This is predominantly attributed to continuous excavation of the shearing-resistance tunnels at 1120 m and 1150 m elevation. Seismic source locations thus show a strong correlation with structure elements on-site.
3. It can be concluded that the usefulness of microseismic monitoring stems from the fact that microcrackings are located by the seismic monitoring system wherever they occur and a 3-D picture of cracks can be obtained, unlike the 1-D or 2-D data obtained from conventional monitoring. It is not so much a short-term slope failure warning

technique, but rather a system for longer-term understanding of where and when rock failures are occurring in deep rock mass of rock slopes. Therefore, microseismic activity can be regarded as the precursor of surface movement and instability failure of the rock slope.

4. The microseismicity formation recorded in Jinping II project indicates that precursory microcracking exists in prior to most rockbursts, which can be captured by the microseismic monitoring system. Rock mass failure precursors can be detected by the microseismic monitoring system for rockburst prediction tens of or more than 100 meters away. The approximate range of rockbursts can be identified and delineated by clustering of seismic source locations. Precursors usually appeared a few days before the rockburst event in terms of time. Due to the occurrence of rockburst is related to excavation progress, the exact time of rockburst can hardly be predicted although the location of rockburst may be determined in advance.

The rock mechanics and rock engineering problems are increasingly complex as the engineering scale of rock excavation at large-scale hydropower projects becomes larger and larger in China. Consequently, as a supplement and extension of traditional rock monitoring technique, high-precision microseismic monitoring technology can be regarded as an effective tool to understand rock mass deformation in complex stress conditions and the corresponding possible hazard evolution mechanism.

5. References

Abdul-Wahed, M. K., Al Heib, M.,Senfaute, G., 2006. Mining-induced seismicity: Seismic measurement using multiplet approach and numerical modeling. International Journal of Coal Geology, 66: 137-147.

Cai, M.,Kaiser, PK., 2005. Assessment of excavation damaged zone using a micromechanics model. Tunnelling and Underground Space Technology, 20: 10.

Cook, N G W, 1964. The application of seismic techniques to problems in rock mechanics. International Journal of Rock Mechanics and Mining Science and Geomechanics Abstracts, 1: 169-179.

Ge, Maochen, 2005. Efficient mine microseismic monitoring. International Journal of Coal Geology, 64: 44-56.

Hirata, A., Kameoka, Y.,Hirano, T., 2007. Safety management based on detection of possible rock bursts by AE monitoring during tunnel excavation. Rock Mechanics And Rock Engineering, 40: 563-576.

Hudyma, M R. 2008. Analysis and interpretation of clusters of seismic events in mines. Department of civil and resource engineering, University of Western Australia.

Jiang, F.X. , Yang, S.H., Cheng, Y.H., Zhang, X.M., Mao, Z.Y,Xu, F.J, 2006. A study on microseismic monitoring of rock burst in coal mine. Chinese Journal of Geophysics, 49(5): 1511-1516.

Kaiser, P.K, Vasak,P, Suorineni,F T, 2005. New dimensionals in seismic data interpretation with 3-D virtual reality visualtion for burst-prone mines. In New dimensionals in seismic data interpretation with 3-D virtual reality visualtion for burst-prone mines, eds. Y Potvin and M Hudyma, Controlling seismic risk- Proceedings of sixth international symposium on rockburst and seismicity in mines, 33-45. Nedlands: Australian Center for Geomechanics.

Lynch, R A,Mendecki, A J. 2001. High-resolution seismic monitoring in mines. In High-resolution seismic monitoring in mines, eds. G Van Aswegen, R J Durrheim and W D Ortleep, Proceedings of fifth international symposium on rockburst and seismicity in mines, 19-24. Johannesburg: The South African Institute of Mining and Metallurgy.

Lynch, R.A., Wuite, R., Smith, B.S.,Cichowicz, A. 2005. Micro-seismic monitoring of open pit slopes. In Micro-seismic monitoring of open pit slopes, eds. Potvin.Y and Hudyma.M, Proceeding of the 6th Symposium on Rockbursts and Seismicity in Mines, 581-592. Perth, Australia: ACG.

Mendecki, A J. 1997. Keynote leeture: Principle of monitoring seismic rockmass response to mining. In Keynote leeture: Principle of monitoring seismic rockmass response to mining, ed. S J Gibowiez, Proceedings of the fourth international symposium on rockbursts and seismieity in mines, 69-80. Rotterdam:A.A.Balkema.

Mendecki, A J. 2000. Data-driven understanding of seismic rockmass response to mining. In Data-driven understanding of seismic rockmass response to mining, eds. G Van Aswegen, R J , Durrheim and W D Ortleep, Proceedings of fifth international symposium on rockburst and seismieity in mines 1-9. Johannesburg: The South African Institute of Mining and Metallurgy.

Milev, A M, Spottiswoode, S M,Rorke, A J, 2001. Seismic monitoring of a simulated rock burst on a wall of an underground tunnel. Journal Of The South African Institute Of Mining And Metallurgy, 101: 8.

Potvin, Y,Hudyma, M R. 2001. Keynote address: Seismic monitoring in highly mechanized hardrock mines in Canada and Australia. In Keynote address: Seismic monitoring in highly mechanized hardrock mines in Canada and Australia, eds. G Van Aswegen, R J Durrheim and W D Ortleep, Proceedings of fifth international symposium on rockburst and seismicity in mines, 267-280. Johannesburg: The South African Institute of Mining and Metallurgy.

Shumila, V,Trifu, CI. 2009. Event mechanism analysis for seismicity induced by a controlled collapse in field II at Ocnele Mari, Romania. In Event mechanism analysis for seismicity induced by a controlled collapse in field II at Ocnele Mari, Romania., ed. CA Tang, Proceedings of the 7th international symposium on rockburst and seismicity in mines, Controlling seismic hazard and sustainable development of deep mines (RaSiM7), p.1091–1104. Dalian, China: New York: Rinton Press.

Tang, C.A., Wang, J.M.,Zhang, J.J., 2011. Preliminary engineering appliction of microseismic monitoring technique to rockburst prediction in tunneling of Jinping II project. Journal of Rock Mechanics and Geotechnical Engineering, 2: 16.

Tezuka, K.,Niitsuma, H., 2000. Stress estimated using microseismic clusters and its relationship to the fracture system of the Hijiori hot dry rock reservoir. Engineering Geology, 56: 47-62.

Trifu, CI , ,Shumila, V. , 2010. Microseismic monitoring of a controlled collapse in Field II at Ocnele Mari , Romania. Pure And Applied Geophysics, 16: 15.

Urbancic, T.I.,Trifu, C.I., 2000. Recent advances in seismic monitoring technology at Canadian mines. Journal Of Applied Geophysics, 45: 225-237.

Xu, N W, Tang, C A, Li, H,Liang, Z Z, 2011a. Excavation-induced microseismicity: microseismic monitoring and numerical simulation. Journal of Zhejiang University-Science A, In press

Xu, N W, Tang, C A, Li, H,Wu, S H, 2011b. Optimal design of microseismic monitoring networking and error analysis of seismic source location for rock slope. Open Civil Engineering Journal, 5.

Xu, N. W., Tang, C. A., Li, L. C., Zhou, Z., Sha, C., Liang, Z. Z.,Yang, J. Y., 2011c. Microseismic monitoring and stability analysis of the left bank slope in Jinping first stage hydropower station in southwestern China. International Journal Of Rock Mechanics And Mining Sciences, 48: 950-963.

Xu, N.W., Tang, C. A., Sha, C., Liang, Z. Z., Yang, J.Y,Zou, Y.Y, 2010. Microseismic monitoring system establishment and its engineering applications to left bank slope of Jinping I Hydropower Station. Chinese Journal of Rock Mechanics and Engineering, 29: 915-925.

Zhong, DH, Li, MC, Song, LG,Wang, G. , 2006. Enhanced NURRS modeling and visualization for large 3D geoengineering applications: An example from the Jinping first-level hydropower engineering project, China. Computers & Geosciences, 32: 13.

8

Limnology of Two Contrasting Hydroelectric Reservoirs (Storage and Run-of-River) in Southeast Brazil

Marcos Gomes Nogueira,
Gilmar Perbiche-Neves and Danilo A. O. Naliato
Instituto de Biociências,
UNESP – Universidade Estadual Paulista Rubião Junior s/n, São Paulo,
Brazil

1. Introduction

The present chapter is a contribution to the discussion about the limnological differences between storage and run-of-river reservoirs. We also intend to contribute to the improvement of the use of this operational concept in an ecological perspective.

Artificial reservoirs created by the construction of dams along fluvial systems tend to exhibit intermediate limnological characteristics, as compared to rivers and lakes. The tendency of a reservoir to be more similar either to a lotic or to a lentic environment is basically related to the water retention time. As shorter is the retention time the similarity with a river is higher. Conversely, the structure and functioning of a reservoir will be comparable to a lake in case of high water retention time. In both cases it is generally observed conspicuous spatial gradients, represented by riverine, intermediate and lentic zones, in the main water body of the reservoir as well as in the lateral arms, in case of dendritic systems (Thornton, 1990; Tundisi, 1990).

Vertically, the reservoirs also have a complex structure, influenced by seasonal variation in the processes of formation and breakdown of thermal stratification and depth of the thermocline (Han et al., 2000), as well as by the plant operation, especially the vertical positioning of the water intake to the turbines and total outflow (Naliato et al., 2009).

The construction of large reservoirs for hydroelectric production in Brazil was intensive after the 1960's, especially in the southeast and south hydrographic basins. Presently, it is estimated that only 40% of the main rivers of the Paraná basin, second largest one in the continent after the Amazonian, are still free of damming (Agostinho et al., 2007). The reservoirs of this basin can be classified, in terms of the operational design and engineering concept, as accumulation (storage) or run-of-river systems (Kelman et al., 1999). The accumulation systems are larger in size and volume, with high water retention time The distinctiveness in terms of physical dimensions and functioning between accumulation and run-of-river systems affects the physical and chemical limnology of these environments as well as the structure and functioning of the aquatic communities (Nogueira et al., 2008; Perbiche-Neves & Nogueira, 2010; Tundisi & Matsumura-Tundisi, 2003). Additionally, the erosive processes in the river banks downstream the dam can be intensified as a result of the different regimes of water discharges (Stevaux et al., 2009).

Studies carried out in Brazilian reservoirs indicate that the main factors influencing in the limnological conditions - seasonal and spatial patterns, are the latitude, the lake and watershed morphometry and the water retention time (e.g Naliato et al., 2009; Henry et al., 1995; Soares et al., 2009).

In the Paranapanema River, where the studied reservoirs are located, continuous limnological investigations have been carried out in the last two decades. Among the relevant results it can be mentioned the ones published by Henry (1992), Henry & Gouveia (1993), Henry & Maricatto (1996), Nogueira et al. (1999) and Nogueira et al. (2006).

Eleven reservoirs were constructed in the Paranapanema River (between 1950's and 1990's), in approximately 500 km of river stretch, in order to supply electric power (ca. 1,500 MW) for an integrated national system of distribution. Three of these reservoirs are accumulation systems, dendritic in shape, larger than 400 km^2 in area and with retention time longer than 130 days. The others are run-of-river systems with retention time not longer than to 22 days (Nogueira et al., 2006).

The main objective of the present study was to verify the distinctiveness between two kind of reservoirs, storage (Chavantes) and run-of-river (Salto Grande), despite of their geographical proximity. It was also investigated the hypothesis that the limnological conditions in the run-of-river reservoir, due to its small size and low retention time, are similar to the ones observed in the dam (lacustrine) zone of the accumulation reservoir, upstream located.

2. Material and methods

As previously mentioned, the studied reservoirs are located in the Paranapanema River, in the border of the States of São Paulo and Paraná (Brazil). Chavantes is an accumulation (storage) reservoir and Salto Grande is a run-of-river reservoir. The distance between dams is approximately 50 km and the distance between the most upstream sampling point and the most downstream one is about 105 km. There is no large cities in the region, only towns and one middle size municipality (Ourinhos municipality, 104,000 inhabitants) whose urban effluents are discharged in Salto Grande Reservoir through the tributary Pardo River. The land use is less intensive in Chavantes watershed, due to the relief limitation (acclivity). But the agriculture practices are intensive around Salto Grande Reservoir. The superficial area of Chavantes is 400 km^2, the theoretical water retention time is 418 days (annual mean value), the maximum depth is 90 m and it is considered as an oligomesotrophic system. Salto Grande is 12km^2 in size, retention time of 1.5 days (annual mean value), maximum depth of 15 m and it is considered as a eutrophic system after the Pardo River entrance. Additional information on the reservoirs can be found in Pagioro et al. (2005a), Nogueira et al. (2006), Nogueira et al. (2008), Perbiche-Neves & Nogueira (2010) and Perbiche-Neves et al. (2011). Between Chavantes and Salto Grande there is another small reservoir (Ourinhos Dam) which started to operate for electric generation, just after the beginning of the study.

The theoretical water retention time (WRT) for both reservoirs was calculated through the relation between volume and outflow. The volumes were considered as constant (normal operational cota) and for the outflow it was used the monthly mean value, calculated from continuous data registration (hourly measurements).

Samplings were performed quarterly, between October/2005 and July/2007 in Chavantes, and between November/2005 and August/2007 in Salto Grande. Six sampling stations were determined in Chavantes and 5 in Salto Grande. Figure 1 shows the sampling sites geographic position, maximum depth and trophic state classification (Carlson Index) (after Mercante & Tucci-Moura, 1999). The distribution of the sampling points was planned in order to include the main compartments in terms of the system hydrodynamics. In the lateral rivers (points VR, IR, PRM e NB), the water samples were obtained only at the subsurface (about 20 cm depth), due to the predominant small depth and turbulent condition. For the other sampling points samples were obtained in 4 different depths (Van Dorn bottle). The water samples were used for determination of total nitrogen (Mackereth et al., 1978) and total phosphorus (Strickland & Parsons, 1960), after digestion procedure (Valderrama, 1981), a chlorophyll (Talling & Driver, 1963), suspended matter (total, mineral and organic) (Cole, 1979) and turbidity (MS Tecnopon). In each sampling point a vertical profile of temperature, dissolved oxygen, pH, conductivity and redox potential was measured using a Horiba U-22 probe. The transparency (Secchi disk) and depth (Speed Tech sonar) were also measured.

Except for the tributary rivers (one sampling depth), it was calculated the data mean values and standard deviation. For statistical analyses the values were log transformed, except for pH. Pearson correlation analysis, Statistic 7.0 (Statsoft, 2006), between WRT and outflow discharges, was performed. The ANOVA test, Statistic 7.0 (Statsoft, 2006), was used to verify the spatial and temporal variations of the limnological variables.

A PCA analysis was used for ordination (Kindt & Coe, 2005) of the sampling points and sampling periods ("R Development Core Team", 2006). This analysis calculates the scores (principal components) which concentrates the variables information and reduce their dimensions.

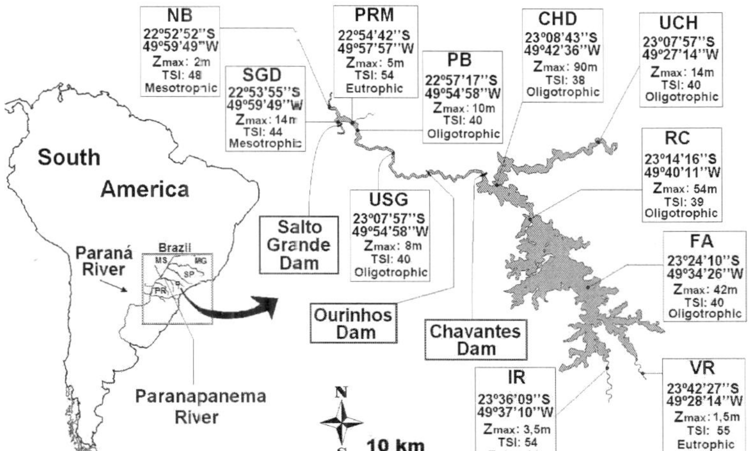

Fig. 1. Study area and the geographical positioning of the sampling stations location with information of the maximum depth (Zmax) and the trophic state index (TSI).

3. Results

During the study period the seasonal regime of rains was typical and similar for both watersheds (Chavantes and Salto Grande Reservoirs), with higher precipitation in summer (December - February) and a dry period in autumn/winter (April - September) (Figure 2). Large scale climatic phenomenon, such as "El Niño" or "La Niña", did not occur during the studied years.

The theoretical water retention in Chavantes during the study period was high (mean of 374 days) and varied between 200 days (winter - August 2006) and 500 days (autumn/winter - May to July 2007). The amplitude of variation of the water level was also high (4.5 m). It was associated with the rain seasonality and exhibited a remarkable decrease in autumn and winter 2006. A sudden elevation of the water level was seen in the beginning of summer (December/06) (Figure 3). Seasonal variation in the water level was not observed in Salto Grande. However, there was a high frequency (daily) of low amplitude (ca. 15-20cm) variation in this run-of-river reservoir (Figure 3), where the water retention time was only 1.39 days (mean value) during the study period.

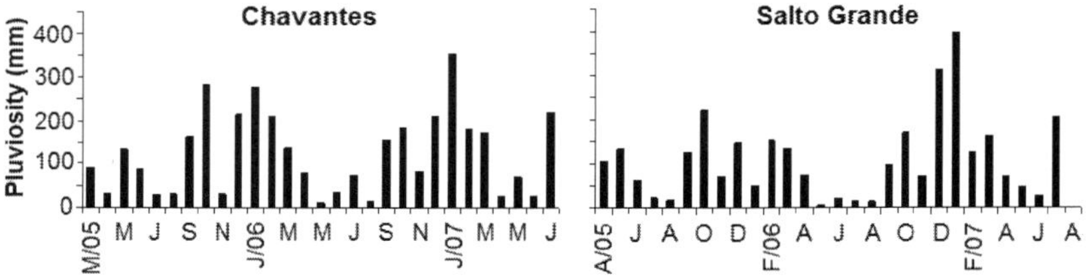

Fig. 2. Monthly rain variation in the watersheds of Chavantes and Salto Grande Reservoirs during the study period.

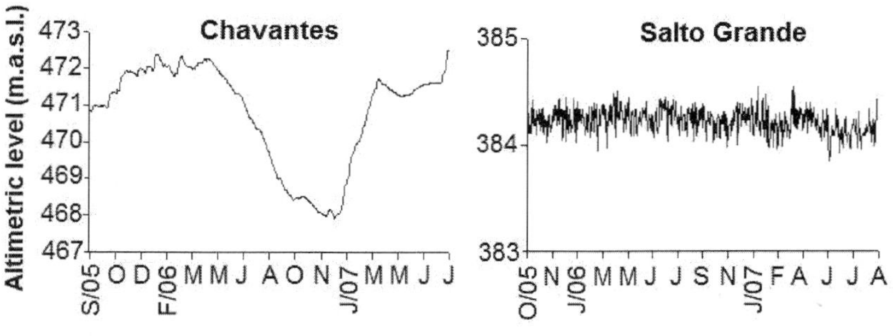

Fig. 3. Daily variation of the water level in Chavantes and Salto Grande Reservoirs during the study period.

The outflow rate for both reservoirs was inversely and significantly correlated with the water retention time, as expected (Figure 4). The water retention time was higher in the second study year (2006/2007) compared to the first (2005/2006). This inter year variation is probably determined by the needs of regulation (volume control) in the downstream reservoirs of the cascade. Despite the differences in magnitude there was a similar tendency of variation for both reservoirs, what demonstrates the strong influence of the upstream system.

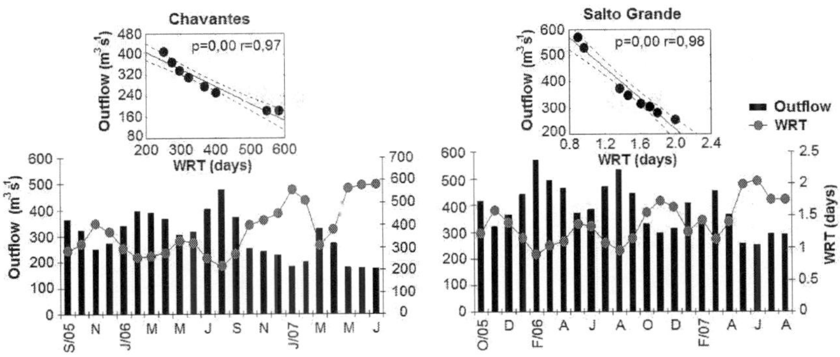

Fig. 4. Monthly variation of outflow and water retention time in Chavantes and Salto Grande Reservoirs during the study period.

The intra reservoir variation of the water transparency was distinct when considered both reservoirs (Figure 5). In the accumulation system there was a clear longitudinal increase towards the dam. Conversely, in the run-of-river reservoir the transparency decreased in the dam zone, due to the entrance of turbid waters from the tributaries (Pardo and Novo). Higher transparency in the Salto Grande upstream sampling station is directly correlated to the lacustrine water releases from Chavantes.

Seasonally the transparency increased in the autumn/winter for both reservoirs (e.g. July in Chavantes) due to the decrease in precipitation and, as a consequence, in the entrance of suspended solids from the adjacent soils and tributaries.

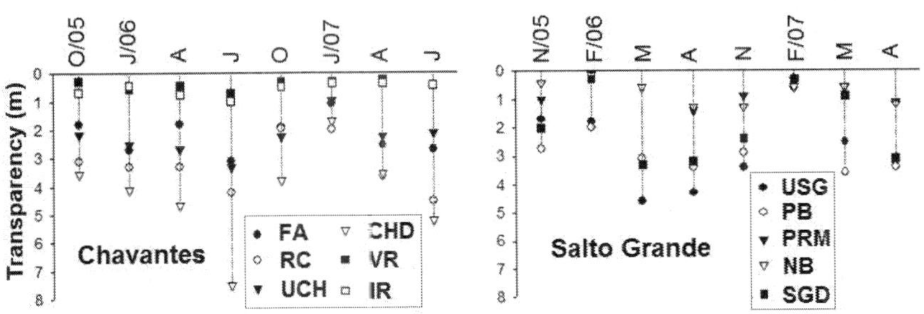

Fig. 5. Water transparency variation in Chavantes and Salto Grande Reservoirs during the study period. For abbreviations see Figure 1.

The vertical profiles data showed the existence of well-structured thermal stratification in deeper sampling stations of Chavantes Reservoir during summer. At the sampling stations CHD, RC and FA differences up to 7°C were observed along the water column; with remarkable decreases not only of the temperature, but also of pH and dissolved oxygen bellow 30 m. Homogeneous physical and chemical conditions prevailed in all sampling stations during winter.

In Salto Grande there was no thermal stratification. Only a weak tendency of stratification, probably a daily phenomenon, was seen in the dam zone (SGD) and, episodically, at NB.

The contrasting conditions in terms of the vertical structure of the water column is shown in Figure 6, which exhibits detailed profiles of temperature and oxygen in the dam zones of Chavantes and Salto Grande during summer and winter.

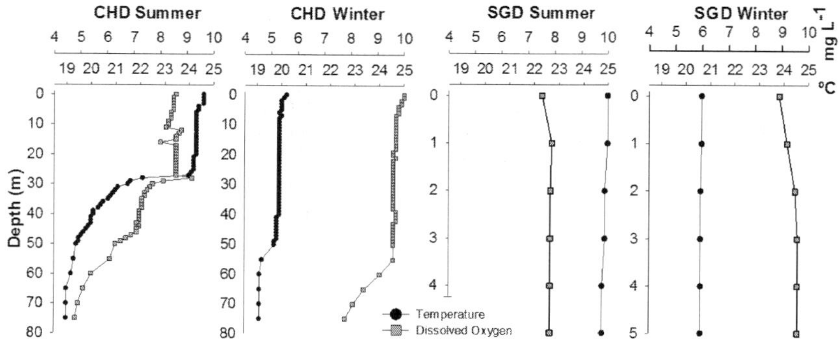

Fig. 6. Vertical profiles of temperature and dissolved oxygen in the dam zones of Chavantes (CHD) and Salto Grande (SGD) Reservoirs.

The mean values and standard deviation of the variables temperature, pH, dissolved oxygen, electric conductivity, turbidity, suspended solids, total phosphorus, total nitrogen an chlorophyll are shown in Tables 1 and 2 for Chavantes and Table 3 for Salto Grande.

In case of Chavantes, the influence of stratified conditions, previously mentioned, is indicated by high standard deviations for temperature, dissolved oxygen and pH.

Clear spatial and temporal differences were observed. Comparatively higher values of temperature were measured in lotic environments during summer and lower in the winter.

The pH variation was higher in the lentic compartments of the accumulation reservoir, due to the influence of stratification and higher depths. Acid values were detected in the deeper layers. Similar pattern occurred for the oxygen.

The distribution of the conductivity values was not influenced by stratification processes. For Chavantes Reservoir higher values, about 100 μS cm^{-1}, were associated to the tributary Itararé (IR) and lower values, about 30 μS cm^{-1}, to the tributary Verde (VR), indicating distinct geological composition of these sub basins. For Salto Grande Reservoir the variability of conductivity was also determined by the tributaries entrance. It was higher in Pardo River Mouth (PRM), about 120 μS cm^{-1}, and lower in Novo River Bay (NB), about 45 μS cm^{-1}.

Higher mean values of turbidity were detected in the tributaries mouths for both reservoirs. The highest value was verified in the Verde River (VR), 810 NTU in April/2007. The same tendency of variation was seen for suspended solids. A diminution of turbidity and suspended solids occurred towards the dam of Chavantes (CHD). But this tendency was not observed for Salto Grande due to the Pardo River entrance, as already mentioned in the description of the transparency variation. Both variables were also influenced by stratification, with increments in deeper layers of Chavantes sampling stations during warmer periods.

		Temp. (° C) M±D.P	pH M±D.P	O.D. (mg L^{-1}) M±D.P	TP(µg L-1) M±D.P	TN (µg L-1) M±D.P	Chlor. (µg L-1) M±D.P
UCH	Oct/05	23.23±0.20	8.84±0.39	7.98±0.08	22.13±1.70	403.08±59.30	1.87±0.64
	Jan/06	27.60±0.69	6.73±1.02	6.63±0.62	38.97±4.28	409.26±16.75	1.87±0.72
	Apr/06	23.81±0.04	8.30±0.00	6.64±0.17	11.58±0.58	122.90±20.70	1.16±0.09
	Jul/06	19.04±0.17	10.27±0.16	6.71±0.11	11.71±2.77	393.79±45.64	1.12±0.20
	Oct/06	22.58±0.26	6.72±0.04	9.95±0.12	16.82±0.94	352.10±6.61	1.66±0.59
	Jan/07	25.90±0.02	6.15±0.02	8.36±0.06	12.98±1.98	289.15±76.47	1.15±0.33
	Apr/07	25.78±0.09	4.82±0.01	6.96±0.05	9.15±0.99	266.98±34.78	1.15±0.24
	Jul/07	19.24±0.12	6.30±0.02	9.61±0.07	6.11±0.41	276.87±30.68	1.50±0.56
FA*	Oct/05	20.98±1.12	9.28±0.32	7.87±0.46	40.02±0.54	443.03±93.39	2.47±1.24
	Jan/06	24.43±2.33	5.31±1.97	6.44±0.80	36.95±3.30	320.85±11.27	1.72±1.57
	Apr/06	22.88±1.00	7.39±0.97	6.80±0.30	11.38±1.58	129.09±69.96	1.03±0.08
	Jul/06	19.49±0.08	9.49±0.08	6.93±0.36	12.57±1.79	466.48±54.48	1.15±0.32
	Oct/06	20.53±1.36	10.32±0.32	6.82±0.19	15.74±7.88	422.08±30.83	1.89±0.68
	Jan/07	23.62±1.92	6.87±1.27	5.82±0.28	7.99±0.99	338.70±29.26	1.35±0.33
	Apr/07	24.16±2.02	4.71±2.53	4.05±0.65	8.18±1.68	227.70±13.50	1.35±0.80
	Jul/07	19.92±0.02	8.71±0.04	6.07±0.04	4.88±0.67	320.33±64.43	1.09±0.57
RC*	Oct/05	21.85±1.79	8.88±0.63	7.61±0.69	28.24±2.86	412.61±89.31	2.55±0.62
	Jan/06	26.46±2.46	5.76±1.72	7.00±0.82	31.55±6.60	244.30±69.06	3.00±0.75
	Apr/06	23.39±1.56	7.40±1.07	6.99±0.25	9.66±1.49	70.06±51.33	1.50±0.57
	Jul/06	19.93±0.29	9.22±0.22	6.67±0.19	10.52±2.86	382.19±62.09	1.19±0.51
	Oct/06	21.63±1.84	11.28±0.50	6.98±0.34	15.21±1.35	366.63±12.06	1.79±0.39
	Jan/07	22.82±3.25	6.66±1.41	6.30±0.57	10.18±2.31	305.93±51.29	1.89±0.26
	Apr/07	23.97±2.64	5.00±1.97	5.88±0.50	6.78±1.72	249.71±57.26	1.89±0.80
	Jul/07	20.15±0.11	7.30±0.43	5.98±0.05	6.82±1.05	234.06±21.36	1.08±0.85
CHD*	Oct/05	21.85±1.42	8.97±0.47	7.75±0.30	36.42±3.90	457.71±140.87	2.89±0.92
	Jan/06	24.80±2.65	5.68±1.74	6.43±0.67	36.06±2.31	317.76±65.35	3.46±1.34
	Apr/06	22.53±2.04	7.72±1.05	7.22±0.46	8.89±1.03	58.20±7.38	1.67±0.35
	Jul/06	19.89±0.08	9.58±0.37	6.79±0.37	14.00±6.05	321.14±28.20	0.99±0.42
	Oct/06	20.48±1.60	10.65±0.37	6.37±0.29	13.29±2.45	357.43±16.76	1.13±0.46
	Jan/07	22.57±3.06	6.73±1.02	6.33±0.28	11.89±1.55	300.22±67.52	1.18±0.22
	Apr/07	22.85±2.94	4.91±1.81	5.86±0.40	9.27±1.44	218.33±55.66	1.18±0.97
	Jul/07	21.85±1.42	5.91±0.01	8.30±0.00	6.40±0.40	168.33±22.46	1.67±0.42
VR**	Oct/05	24.80	7.26	8.00	35.42	533.75	3.37
	Jan/06	27.80	7.30	5.27	51.88	468.02	9.83
	Apr/06	21.40	6.90	8.90	23.93	120.83	2.64
	Jul/06	16.70	6.47	11.30	22.21	523.70	1.24
	Oct/06	25.25	6.43	8.90	50.85	633.50	1.30
	Jan/07	26.86	6.63	6.80	27.99	642.95	6.20
	Apr/07	23.55	5.35	6.40	69.98	679.80	5.16
	Jul/07	16.22	5.65	10.30	34.86	574.80	2.69
IR**	Oct/05	24.00	7.40	8.16	42.03	769.59	1.57
	Jan/06	29.20	7.79	5.41	50.26	577.82	4.26
	Apr/06	21.30	7.10	8.70	19.35	110.01	3.30
	Jul/06	17.07	6.36	11.60	24.88	676.80	2.86
	Oct/06	26.00	6.59	8.70	47.43	675.90	2.80
	Jan/07	25.93	5.80	9.20	39.76	825.80	4.50
	Apr/07	24.76	6.00	6.30	27.97	420.00	5.84
	Jul/07	16.65	5.92	9.80	64.07	511.65	6.62

Table 1. Water column mean values and standard deviation for temperature, pH, dissolved oxygen, total phosphorus, total nitrogen and chlorophyll at the sampling stations of Chavantes Reservoir during the study period. * Stratified in summer. ** Only surface samples.

The nutrient phosphorus and nitrogen exhibited a remarkable spatial and temporal variation in the accumulation reservoir. Higher concentrations were measured in the zones under influence of tributaries, as well as a decreasing tendency in direction to the dam. For the run-of-river reservoir the variation was not so regular.

Higher values of chlorophyll were generally observed in the lotic compartments of both reservoirs, especially in case of Salto Grande. In this reservoir values varied between 0,48 e 5,1 μg L^{-1}. Seasonally higher concentrations were measured at PB, PRM, NB and SGD in February/2006, with the maximum in NB.

		Cond.(μS cm^{-1}) M±D.P	Turb. (NTU) M±D.P	T.S. (mg L^{-1}) M±D.P		Cond.(μS cm^{-1}) M±D.P	Turb. (NTU) M±D.P	T.S. (mg L^{-1}) M±D.P
UCH	Oct/05	56.00±0.00	4.74±0.27	2.15±0.84	CHD*	60.00±3.74	2.86±0.19	1.13±0.15
	Jan/06	44.00±1.41	3.27±1.05	2.13±0.91		47.75±5.56	1.85±0.60	0.85±0.41
	Apr/06	80.00±0.00	1.38±1.51	1.63±0.23		90.18±1.34	8.73±1.30	1.38±0.22
	Jul/06	60.67±2.58	3.60±0.98	3.53±1.66		67.32±4.47	9.69±18.01	1.02±0.73
	Oct/06	80.00±0.00	7.26±2.22	2.47±1.05		90.00±0.00	3.94±0.96	1.11±0.29
	Jan/07	100.00±0.00	13.40±0.61	5.82±4.25		104.63±5.03	8.85±3.14	2.51±0.56
	Apr/07	64.06±0.25	13.88±7.99	2.37±0.60		63.77±2.50	3.42±1.91	1.88±1.33
	Jul/07	62.00±0.00	5.10±0.05	3.63±1.01		84.18±0.87	1.09±0.28	0.60±0.02
FA*	Oct/05	55.00±1.63	7.65±1.04	2.82±0.11	VR**	42.00	36.30	27.80
	Jan/06	49.00±2.71	8.69±8.87	2.61±0.37		33.00	70.20	81.03
	Apr/06	90.91±2.92	10.33±6.17	2.30±0.30		70.00	42.00	28.74
	Jul/06	70.00±0.00	1.47±0.78	3.79±0.95		50.00	34.00	14.10
	Oct/06	90.00±0.00	7.73±1.40	2.30±0.15		80.00	173.00	103.35
	Jan/07	110.00±0.00	11.95±0.62	3.16±0.26		90.00	63.90	65.13
	Apr/07	69.73±2.78	7.87±1.68	1.28±0.72		47.00	810.00	302.43
	Jul/07	75.75±0.52	4.24±1.77	1.34±0.22		39.00	48.80	40.76
RC*	Oct/05	53.75±4.35	5.21±0.85	3.70±2.10	IR**	67.00	69.20	107.30
	Jan/06	39.75±2.99	8.06±2.65	6.22±6.07		51.00	45.60	57.26
	Apr/06	93.24±4.75	9.30±1.43	1.34±0.05		120.00	19.00	10.94
	Jul/06	70.00±0.00	1.09±0.36	3.23±1.32		100.00	29.00	13.69
	Oct/06	90.00±0.00	5.09±0.78	1.53±0.16		120.00	44.00	45.17
	Jan/07	110.00±0.00	9.28±1.77	2.64±0.15		100.00	76.30	81.23
	Apr/07	70.47±4.25	2.96±0.71	1.75±0.39		85.00	45.70	45.45
	Jul/07	70.08±0.36	2.54±0.70	0.84±0.23		66.00	55.70	35.14

Table 2. Water column mean values and standard deviation for conductivity, turbidity and suspended solids at the sampling stations of Chavantes Reservoir during the study period. * Stratified in summer. ** Only surface samples.

The ANOVA test results evidenced the spatial and temporal differences for the limnological variables. For most variables the differences were significant when compared the sampling stations and the sampling periods for both reservoirs (Table 4).

A high percentage of data variance was explained by the PCA analysis (linear correlations), 88% for Chavantes and 81% for Salto Grande (Figure 7). The results evidence the river-reservoir spatial compartmentalization.

Transparency was associated to deeper sites in Chavantes (CHD and RC) - first component (positive side). Conversely, these variables (depth and transparency) were inversely correlated with suspended solids, turbidity, total phosphorus and total nitrogen, which were associated to the negative side of the first component and to the sampling stations VR and IR.

		Temp. (° C) M±D.P.	pH M±D.P	O.D. (mg L) M±D.P	Cond. (µS cm^{-1}) M±D.P	Turb. (NTU) M±D.P	T.S. (mg L^{-1}) M±D.P	TP (µg L^{-1}) M±D.P	TN (µg L^{-1}) M±D.P	Chlor. (µg L^{-1}) M±D.P
USG	Nov/05	24.60±0.00	7.88±0.01	7.66±0.02	77.60±0.55	5.99±0.23	4.51±0.44	27.08±0.81	241.98±56.13	1.99±0.97
	Feb/06	25.70±0.00	6.24±0.11	7.98±0.04	74.00±5.48	7.67±3.79	4.87±1.32	15.25±2.10	211.82±27.06	1.52±0.08
	May/06	21.10±0.00	7.24±0.12	9.20±0.00	60.00±0.00	0.33±0.58	1.10±0.26	14.19±0.00	430.14±0.00	1.07±0.00
	Aug/06	20.37±0.00	6.88±0.04	9.24±0.05	90.00±0.00	2.22±0.72	1.55±0.31	9.87±1.41	269.03±13.92	0.79±0.15
	Nov/06	26.25±0.01	6.71±0.01	10.05±0.05	90.00±0.00	2.50±0.45	1.38±0.30	6.36±0.43	355.37±27.08	2.09±0.41
	Feb/07	27.19±0.01	6.79±0.01	8.08±0.08	100.00±0.00	4.65±0.00	12.72±1.11	17.73±3.06	294.20±51.37	2.40±0.12
	May/07	22.30±0.00	6.98±0.03	9.58±0.04	65.83±0.41	2.91±0.39	3.07±0.17	11.74±1.39	242.23±33.64	1.40±0.17
	Aug/07	19.62±0.04	5.77±0.01	9.40±0.00	56.00±0.00	3.09±0.34	2.25±0.17	35.20±45.86	270.10±5.38	1.21±0.37
PB	Nov/05	24.54±0.05	7.60±0.05	7.58±0.11	47.20±1.64	5.50±0.62	2.51±0.57	28.26±1.46	318.02±75.11	1.12±0.16
	Feb/06	26.80±0.00	6.44±0.05	8.00±0.00	70.00±0.00	4.60±0.89	16.22±16.09	17.77±2.75	238.37±11.70	2.21±1.27
	May/06	21.40±0.00	7.10±0.18	9.10±0.00	60.00±0.00	1.40±0.89	1.57±0.49	14.88±0.31	421.64±4.38	1.24±0.20
	Aug/06	20.69±0.01	6.94±0.04	6.36±0.23	90.00±0.00	6.00±0.00	1.20±0.00	12.68±0.23	248.37±30.91	0.78±0.44
	Nov/06	25.83±0.05	6.89±0.08	8.40±0.00	90.00±0.00	3.25±0.85	1.70±0.29	7.72±1.48	333.52±30.85	1.59±0.18
	Feb/07	26.16±0.02	6.30±0.03	7.55±0.05	100.00±0.00	42.43±3.09	18.00±1.59	25.08±1.62	355.93±35.80	2.02±0.30
	May/07	23.10±0.00	6.93±0.03	9.38±0.04	69.67±0.52	1.60±0.16	1.47±0.37	9.82±0.20	892.41±1122.30	1.03±0.28
	Aug/07	20.22±0.04	5.83±0.02	9.32±0.04	61.17±0.41	4.01±0.52	2.04±0.73	34.62±46.93	279.07±9.58	0.67±0.06
PRM	Nov/05	25.65±0.06	7.94±0.05	7.48±0.35	58.40±0.55	12.47±1.85	9.27±4.31	65.03±13.58	401.92±23.51	3.82±0.16
	Feb/06	24.70±0.00	6.56±0.51	8.34±0.09	80.00±0.00	356.67±5.77	83.97±55.96	40.83±4.22	357.07±41.01	5.10±2.71
	May/06	17.95±0.05	7.04±0.05	9.80±0.00	70.00±0.00	26.33±3.51	13.30±2.42	34.68±1.71	481.17±0.00	1.69±0.23
	Aug/06	21.13±0.01	7.11±0.02	8.50±0.35	120.00±0.00	12.39±0.00	2.85±0.40	48.04±2.36	455.67±16.84	0.83±0.06
	Nov/06	27.30±0.33	6.79±0.04	7.16±0.09	120.00±0.00	12.80±0.80	5.05±0.84	31.19±0.56	420.52±51.46	1.07±0.10
	Feb/07	24.33±0.02	6.57±0.02	8.20±0.07	114.00±5.48	44.37±6.68	20.78±5.59	28.18±3.06	429.77±71.24	1.54±0.03
	May/07	20.80±0.00	6.76±0.05	9.50±0.00	82.00±0.00	9.56±2.37	16.67±1.20	37.48±2.28	514.03±100.14	1.59±0.42
	Aug/07	19.08±0.13	5.95±0.04	9.44±0.21	72.80±1.30	18.07±4.13	8.38±4.14	25.18±1.73	342.77±259.26	0.78±0.44
NB	Nov/05	24.80±0.00	7.43±0.00	7.54±0.00	68.00±0.00	15.00±0.00	22.03±2.25	52.84±0.00	389.93±0.00	4.16±1.43
	Feb/06	24.70±0.00	5.87±0.21	7.67±0.57	50.00±0.00	130.00±0.00	47.75±1.73	34.29±0.00	350.49±0.00	4.36±0.60
	May/06	17.96±0.04	6.74±0.16	10.27±0.15	40.00±0.00	23.67±4.04	4.53±0.06	29.27±0.00	211.31±0.00	1.35±0.00
	Aug/06	20.80±0.00	6.71±0.01	7.57±0.06	80.00±0.00	11.50±0.00	2.86±0.00	28.29±0.86	239.75±29.77	1.01±0.16
	Nov/06	26.15±0.38	6.03±0.01	7.40±0.00	80.00±0.00	11.50±0.71	4.33±0.21	12.94±1.58	278.70±13.72	3.60±0.32
	Feb/07	23.55±0.01	6.47±0.01	8.25±0.07	90.00±0.00	58.00±36.77	13.66±0.11	15.83±0.81	330.00±3.39	1.01±0.32
	May/07	20.84±0.00	6.39±0.02	9.45±0.07	53.00±0.00	6.86±0.00	14.01±0.39	17.35±2.64	286.75±11.10	3.93±1.11
	Aug/07	18.74±0.02	5.49±0.08	9.30±0.14	45.50±0.71	10.30±0.14	4.53±0.06	10.63±1.17	380.50±51.19	1.91±0.16
SGD	Nov/05	25.52±0.66	7.83±0.05	7.84±0.03	49.60±0.55	5.93±1.22	2.88±1.23	48.65±14.68	295.59±155.43	1.09±0.57
	Feb/06	26.00±0.00	6.28±0.04	8.80±0.00	70.00±0.00	76.50±4.51	22.86±2.41	29.77±1.14	303.71±32.26	3.76±1.19
	May/06	20.44±0.15	6.84±0.15	9.26±0.05	60.00±0.00	2.50±1.73	1.10±0.26	14.45±0.62	304.61±23.59	0.82±0.06
	Aug/06	20.61±0.02	6.92±0.03	9.50±0.17	90.00±0.00	3.37±1.59	1.63±0.10	13.08±0.62	297.87±10.46	0.61±0.29
	Nov/06	26.10±0.34	6.79±0.10	8.24±0.15	97.50±4.63	6.47±0.59	1.66±0.23	19.14±4.02	456.18±17.95	1.31±0.23
	Feb/07	25.53±0.43	6.57±0.02	7.65±0.05	110.00±0.00	54.83±7.00	14.76±2.31	27.80±1.87	396.57±96.56	1.63±0.17
	May/07	22.45±0.08	6.76±0.01	9.30±0.00	72.00±0.93	7.19±1.62	4.89±0.63	84.75±110.04	315.03±15.20	0.99±0.26
	Aug/07	20.34±0.24	5.95±0.06	9.38±0.10	62.13±1.25	4.75±1.25	1.38±0.16	10.41±0.34	318.13±9.41	0.48±0.03

Table 3. Water column mean values and standard deviation for temperature, pH, dissolved oxygen, conductivity, turbidity, suspended solids, total phosphorus, total nitrogen and chlorophyll at the sampling stations of Salto Grande Reservoir during the study period.

Total nitrogen and total phosphorus were also associated to periods of intense precipitation (second component).

In Salto Grande the variables suspended solids, turbidity, total phosphorus and chlorophyll were correlated (first component positive side) with the sampling sites PRM and NB (eventually SGD) during most sampling periods. Inversely, the variables transparency and depth were associated to PB and USG (eventually SGD).

		Chavantes		Salto Grande	
		F	p	F	p
Temperature	Site	0.79	0.56	3.92	**0.01**
	Month	16.11	**0.00**	55.14	**0.00**
pH	Site	1.52	0.20	11.00	**0.00**
	Month	19.95	**0.00**	68.42	**0.00**
Dissolved oxygen	Site	3.89	**0.00**	1.42	0.25
	Month	4.08	**0.00**	6.97	**0.00**
Conductivity	Site	18.29	**0.00**	10.73	**0.00**
	Month	45.11	**0.00**	25.70	**0.00**
Turbidity	Site	29.82	**0.00**	1.73	0.16
	Month	1.96	0.08	2.85	**0.02**
Total N	Site	20.48	**0.00**	2.72	**0.04**
	Month	42.38	**0.00**	1.25	0.30
Total P	Site	15.10	**0.00**	3.89	**0.01**
	Month	8.41	**0.00**	1.48	0.21
Suspended solids	Site	57.44	**0.00**	3.49	**0.01**
	Month	1.70	0.13	5.50	**0.00**
Chlorophyll	Site	9.03	**0.00**	4.17	**0.00**
	Month	2.74	**0.02**	5.20	**0.00**
Transparency	Site	120.13	**0.00**	17.42	**0.00**
	Month	10.84	**0.00**	7.32	**0.00**

Table 4. Results of the ANOVA two-way (F and p) for the limnological variables tested for sampling points and sampling periods. Significant differences ($p<0.05$) in bold. For spatial comparison freedom degree is 5 for Chavantes and 4 for Salto Grande. For temporal comparison freedom degree is 7 for both reservoirs.

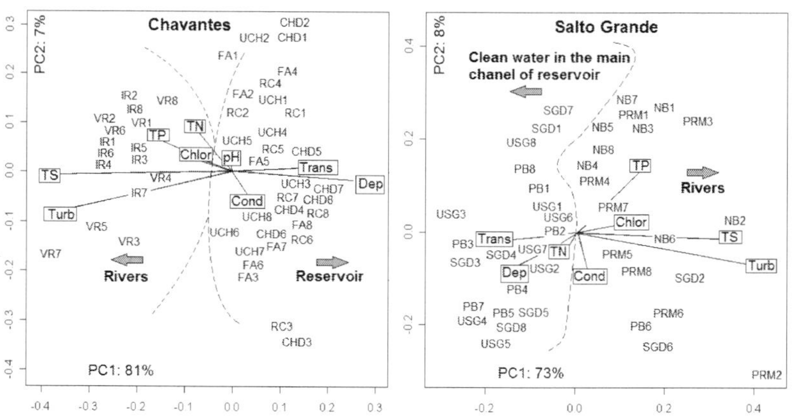

Fig. 7. Graphic results of the PCA analysis (components 1 and 2) for ordination of the limnological variables considering the sampling sites and sampling periods for Chavantes and Salto Grande Reservoirs.

4. Discussion

The rainfall pattern was similar for both reservoirs, with higher values in summer (December-February) and lower in autumn/winter (April to September). This seasonal regime had already been identified as an important factor influencing the limnological functioning of the Paranapanema River reservoirs (Nogueira et al., 2002b, 2006; Jorcin & Nogueira, 2005; Pagioro et al., 2005 b; Henry et al., 2006).

Higher values of turbidity, suspended solids, nutrients and conductivity were observed in the tributary rivers mouths for both reservoirs. However, only a minor influence of these rivers entrance was observed in the main water body of the accumulation reservoir, Chavantes, due to its large size and high water retention time. Conversely, for the run-of-river reservoir the influence of the tributaries was prominent, with major changes of the water quality downstream the rivers entrance, towards to the dam zone. This process is particularly evident in the rainy period (summer).

Our results corroborate previous studies carried out in Chavantes and Salto Grande by Pagioro et al. (2005 a) and Nogueira et al. (2006). These authors considered Chavantes as an oligotrophic system, eventually mesotrophic in some periods of the year, and Salto Grande a meso/eutrophic system, due to the introduction of large nutrient loads from the lateral tributaries.

In Chavantes, different from Salto Grande, the longitudinal zones (*sensu* Thornton et al., 1990) were clearly observed. This spatial organization is a consequence of the high water retention time, large size and complex morphometry, which determine the occurrence of distinct compartments.

The water retention time is key variable for reservoir ecology, interfering on physical, chemical and biological characteristics and it depends on the interactions among distinct factors, such as precipitation, outflow, evaporation an infiltration (Tundisi, 1990; Nogueira et al., 1999; Nogueira 2000; Nogueira et al., 2006). The vertical structure of the water column in Chavantes and Salto Grande is directly associated to the water retention time features of each reservoir.

Except for the winter season, the deeper sampling points located in Chavantes Reservoir (CHD, RC and FA) exhibited a well-defined thermal and chemical stratification. The temperature profiles mean values varied up to 11 °C (from 18 to 29°C). The oxygen and pH were much lower in deep layer, indicating reduced chemical conditions. Thomaz et al. (1997) also found stratification only in the dam (lacustrine) zone of Segredo Reservoir in the Iguaçu River (Paraná State, Brazil).

Pagioro et al. (2005b) and Nogueira et al. (2006) had already observed stratification in the water column of larger reservoirs of the Paranapanema basin during spring and summer and complete mixture in winter. These authors pointed out to the occurrence of low oxygen concentrations in the bottom layers of Chavantes. Henry & Nogueira (1999) also verified seasonal thermocline in Jurumirim Reservoir, the first in the Paranapanema reservoir cascade. Other similar patterns in reservoirs of southeast and central Brazil have been registered. De Filippo et al. (1999) observed low oxygen concentrations, and even anoxic condition, bellow 20 m depth in Serra da Mesa Reservoir (Goiás State, Brazil), right after its filling up and still under strong influence of decomposition of organic loads. Clinograde patterns were measured by Pinto-Coelho et al. (2006) in São Simão Reservoir, located in Parnaíba River (border of Goiás and Minas Gerais states, Brazil).

Continuous vertical mixing and absence of thermal and chemical stratification in Salto Grande are associated to the reservoir semi-lotic conditions (very low water retention time) and low depth. This condition was also observed by Pagioro et al. (2005b) in run-of-river reservoirs of Paraná State.

When compared the maximum temperature of both reservoirs it is observed low values in Salto Grande. The low heat retention capacity in this reservoir is also associated to its morphometry and functioning (run-of-river system). As a small size reservoir, Salto Grande is also more influenced by the tributary entrances, which exhibit colder waters in some periods of the year, especially in the winter.

The contrast in terms of retention time and morphometry among Chavantes and Salto Grande also explain the distinct magnitude of the tributaries influence. The modifications due to the tributaries entrance in the run-of-river is higher, especially bellow the Pardo River Mouth (PRM). There is no sufficient time/distance for sedimentation of the introduced loads. The simple morphometry do not favor the depositional process as well. The opposite situation occurs in Chavantes, due to its dendritic shape and long main axis, which favor the particles settlement and a longitudinal increase in transparency.

In the sampling station CHD, the deepest one in Chavantes, it was observed high transparency and low values of turbidity and suspended solids. This characteristic corroborates the predictable modifications towards the dam zone due to the reduction of water velocity and increase in depth (Thornton et al., 1990; Henry & Maricatto, 1996; Thomaz et al., 1997; Nogueira et al., 1999; Pinto-Coelho et al., 2006).

The sampling points located in the transitional (river-reservoir) zone of Chavantes, exhibited some common characteristics, such as higher values of turbidity and total nitrogen and lower transparency, when compared to the lacustrine zone. However, in relation to dissolved oxygen there was contrasting tendencies. Higher concentrations were observed at UCH, probably due to the predominant high water velocity and relatively low depth. Conversely, at FA the values were low, associated to the water column stratification and reduced concentrations in the bottom. Besides the input of sediments from agriculture activities, in this region there is also the influence of erosive process of the river banks associated to intensive wind action. The sediments are deposited in deep layers, bellow the mixture zone (ca. 20m), where is verified an increase in turbidity.

In Salto Grande the sampling points located in the reservoir main axis exhibited similarity for some limnological variables. Nevertheless, the effect of the Pardo River entrance in the intermediate region of the reservoir is remarkable, with the increase of nutrients, turbidity and diminution of transparency. It is clearly observed the distinctiveness of the water masses, the brownish one coming from Pardo River and the greenish and much more transparent one from the Paranapanema River (lacustrine zone of Chavantes). Sampaio et al. (2002), Britto (2003) and Nogueira et al. (2006) have already mentioned the effects of Pardo River on the increase of turbidity in Salto Grande.

Comparisons between different kinds of reservoirs show the occurrence of higher concentrations of turbidity and suspended solids in run-of-river systems, which are more similar to rivers in terms of hydrodynamics (Pagioro et al., 2005 b; Nogueira et al., 2006).

The Novo River (NB), despite its reduced flow, is another important tributary contributing to the introduction of nutrients and suspended solids (and decrease of transparency) for the dam zone of Salto Grande (SGD). In NB there are extensive macrophytes banks and also episodic algal blooms, indicate by chlorophyll peaks, due to high nutrient concentration and reduced flow.

The transportation and retention of suspended particles, from river inputs, are considered as the main processes determining the reservoirs compartmentalization in terms of their limnology features and organization of biotic communities (Thornton et al., 1990).

The sampled lotic stretches exhibited mean values for the limnolgical variables distinct from the ones calculated for the reservoirs (more lacustrine) sampling sites, as evidenced by ANOVA tests.

The discrimination among the riverine sampling points, VR, IR, USG e PB; the tributaries mouth zones, PRM e NB; the river-reservoir transitional compartments, UCH, FA e SGD and the lentic environments, RC e CHD, was possible through the obtained results. For the riverine compartments there was two different conditions; rivers without upstream dams (Verde, Itararé, Pardo and Novo) and stretches under dams influence (USG and PB). In the first case there is a direct response of the watershed to precipitation events and the predominance of lower temperature, especially in autumn and winter season, higher dissolved oxygen, nutrient concentration, turbidity and suspended solids and lower transparency. Some of these characteristics are amplified by the intensive land use for agriculture (Brigante et al., 2003; Nogueira et al., 2006; Pinto-Coelho et al., 2006).

Feitosa et al. (2006) verified that the environmental degradation is more intensive in the lateral tributaries compared to the Paranapanema River main channel and reservoirs. In our study, except for the oxygen values, this tendency was also verified.

A strong influence of Chavantes Reservoir on the upstream sampling points of Salto Grande Reservoir (upstream the Pardo River entrance) was evident for some periods. The storage reservoir releases transparent waters with low concentrations of suspended solids, because most sediments are retained in upstream compartments due to its large size, complex morphometry and high water retention time, as previously mentioned. Similar structure and functioning was already described for Jurumirim Reservoir, the first in the cascade and about 50 km upstream the transitional zone of Chavantes (Henry & Maricatto, 1996; Nogueira et al., 1999).

In January/2007 it was observed high conductivity values in Chavantes Reservoir, including the sampling points without the influence of the tributaries, such as RC and CHD. This fact can be related to three causes. The first would be a consequence of the superficial water intake to the turbines, promoting a water mass displacement, including the ones coming from the tributaries, towards the dam zone. The second cause would be related to the sudden water level increase, which inundates exposed littoral areas during the dry season where the vegetation grew and its decomposition would change the physical and chemical characteristics of water. Finally, the high frequency and volumes of rains in this period would promote a significant runoff of superficial adjacent soils, resulting in the introduction of suspended and dissolved material.

In relation to the conductivity it is also important to note the relatively low values of the Verde River (VR), especially if compared to the ones measured in the Itararé River (IR).

Despite their proximity, the considered rivers must have particular geological substrate responsible for the observed differences. Another hypothesis is that the land use (urban and agriculture activities) in Itararé basin would be more intensive. In fact in this basin it is observed intensive commercial forestation with *Pinus* sp. and *Eucaliptus* sp., for instance.

Despite of the relative higher loads of nutrients in Salto Grande, associated to intensive human activities in the tributaries watersheds and around the reservoir, the chlorophyll concentrations in this reservoir was low. Certainly this fact is associated to the very low water retention time. Probably an increase in the WRT would result in higher phytoplankton density and biomass, as commonly observed in other tropical reservoirs exposed to human induced eutrophication. Some examples are provided by the literature such as in Billings and Guarapiranga Reservoirs (Sendacz & Kubo, 1999) and Garças Lake (reservoir) (Sant'Anna et al., 1997; Bicudo et al., 1999) in São Paulo metropolitan region; Paranoá Reservoir in Brasília metropolitan region (Branco & Cavalcanti, 1999) and Iraí Reservoir in Curitiba metropolitan region (Bollmann et al., 2005; Pagioro et al., 2005a). According to Abe et al. (2006), in Brazil the highest concentration of nutrients potentially important for eutrophication (nitrogen and phosphorus) are mainly associated to urban development.

5. Conclusion

In order to represent the main results of this study it is presented a schematic drawing (Figure 8) of the studied reservoirs, synthesizing the main limnological differences. The storage reservoir showed intense thermal and chemical stratifications in summer and isothermal and isochemical profiles in winter, what is directed associated to the physical characteristics and operational functioning. The run-of-river reservoir exhibited only daily (ephemerons) thermal stratification tendency near the dam during summer, what was also a consequence of its physical characteristics and operational procedures.

The influence of the accumulation reservoir on the run-of-river reservoir, downstream located, seems to be less intense in summer and more evident in winter (dry period).

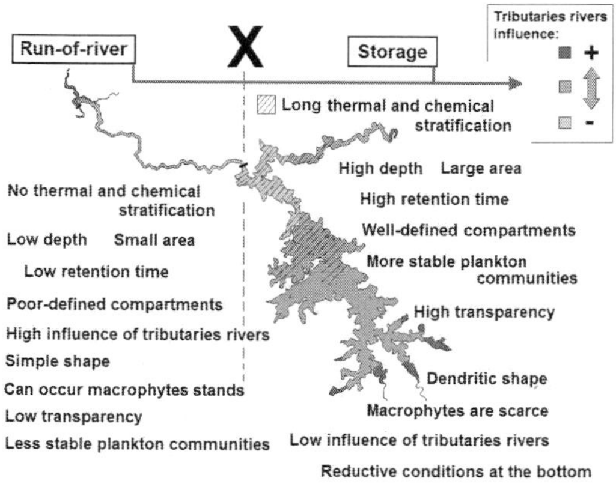

Fig. 8. Schematic drawing synthesizing the main limnological differences between Chavantes (storage) and Salto Grande (run-of-river) Reservoirs.

The distinctiveness between both reservoirs, despite their geographical proximity, was verified. Nevertheless, the second hypothesis was not confirmed as the influence of the lacustrine zone of upstream reservoir was limited to certain periods of the year. The tributaries had a strong influence on the run-of-river reservoir, modifying the turbidity and nutrient concentrations after their mouths, especially in the rainy season.

6. Acknowledgements

The authors are grateful to Fapesp for financial support (process 2005-02811-0) and for the scholarship conceived for the second author (process 2005/03311-0); to Silvia M. C. Casanova and Fabiana Akemi Kudo for collaboration during the field work and laboratory analyses; to Duke Energy - Geração Paranapanema for financial and logistic support and for the hydrological data and to Casa de Agricultura de Fartura (SP) and Instituto de Águas do Paraná (PR) for the precipitation data.

7. References

Abe, D.S.; Tundisi, J.G.; Matsumura-Tundisi, T.; Tundisi, J.E.M.; Sidagis Galli, C.; Teixeira-Silva, V.; Afonso, G.F.; Von Haehling, P.H.A.; Moss, G. & Moss, M. (2006). Monitoramento da qualidade ecológica das águas interiores superficiais e do potencial trófico em escala continental no Brasil com uso de hidroavião, In: *Eutrofização na América do Sul: causas, conseqüências e tecnologia de gerenciamento e controle. São Carlos: Instituto Internacional de Ecologia*, J.G. Tundisi, T. Matsumura-Tundisi, & C.S. Galli. (Eds.), 225-239, DMD Propaganda, São Carlos, Brazil

Agostinho, A.A.; Gomes, L.C. & Pelicice, F.M. (2007). *Ecologia e manejo de recursos pesqueiros em reservatórios do Brasil.* Eduem, Maringá, Brazil

Bicudo, C.E.M.; Ramírez, R.J.J.; Tucci, A. & Bicudo, D.C. (1999). Dinâmica de populações fitoplanctônicas em ambiente eutrofizado: o lago das Garças, São Paulo, In: *Ecologia de Reservatórios: estrutura, função e aspectos sociais,* R. Henry, (Ed.), 451-507, Fapesp/Fundibio, Botucatu, Brazil

Bollmann, H.A. & Andreoli, O.R. (2005). Água no sistema urbano, In: *Gestão Integrada de Manaciais de Abastecimento Eutrofizados,* C.V. Andreoli & C. Carneiro, (Eds.), 83-119, Capital, Curitiba, Brazil

Branco, C.W.C. & Cavalcanti, C.G.B. (1999). A ecologia das comunidades planctônicas no lago Paranoá, In: *Ecologia de Reservatórios: Estrutura, Função e Aspectos Sociais,* R. Henry, (Ed.), 575-595, Fapesp/Fundibio, Botucatu, Brazil

Brigante, J.; Espíndola, E.L.G.; Povinelli, J. & Nogueira, A.M. (2003). Caracterização física, química e biológica da água do rio Mogi-Guaçu, In: *Limnologia fluvial,* J. Brigante, & E.L.G. Espíndola, (Eds.), 55-76, Rima, São Carlos, Brazil

Britto, Y.C.T. (2003). *Associações de Cladocera (Crustacea Branchiopoda) do sistema de reservatórios em cascata do rio Paranapanema (SP-PR).* Dissertação de mestrado, UNESP, Botucatu, Brazil

Cole, G.A. (1979) *Textbook of limnology* (2nd.), The C.V. Mosby Company, Saint Louis

De Fellipo, R.; Gomes, E.L.; Lenz-César, J.; Soares, C.B.P. & Menezes, C.F.S. (1999). As alterações na qualidade de água durante o enchimento do reservatório da UHE Serra da Mesa (GO), In: *Ecologia de reservatórios: estrutura, função e aspectos sociais,* R. Henry, (Ed.), 324-345, Fapesp/Fundibio, Botucatu, Brazil

Feitosa, M.F.; Nogueira, M.G. & Vianna, N.C. Transporte de Nutrientes e Sedimentos no Rio Paranapanema (SP/PR) e seus Principais Tributários nas Estações Seca e Chuvosa. In: *Ecologia de reservatórios: Impactos potenciais, ações de manejo e sistemas em cascata*, M.G. Nogueira, R. Henry & A. Jorcin, (Eds.), 435-459, Rima, São Carlos, Brazil.

Han, B.; Armengol, J.; Garcia, J. C.; Comerna, M R.; Dolz. J. & Straskraba, M. (2000). The thermal structure of Sal Reservoir (NE: Spain): a simulation approach. *Ecological Modelling*, Vol.125, pp. 109-122

Henry, R. & Gouveia, L. (1993). O fluxo de nutrientes e seston em cursos de água do alto Paranapanema (São Paulo) - sua relação com usos do solo e morfologia das bacias de drenagem. *Na. Acad. Bras. Ciênc*, Vol.65, pp. 439-51

Henry, R.; Carvalho, E.D.; Nogueira, M.G.; Pompeo, M.L.M.; Moschini-carlos, V.; Santos, C.M. dos; Luciano, S. de C. & Fujihara C.Y. (1995), The Jurumirim Reservoir, In: *Mid. Congress Excursions, XXVI SIL Congress*, R. Henry & P.A.C. Senna (Eds.), 13-33, São Paulo, Brazil

Henry, R. & Maricatto, F.E. (1996). Sedimentation rates of tripton in Jurumirim Reservoir (São Paulo, Brasil). *Limnologica*, Vol.25, pp. 15-25.

Henry, R. & Nogueira, M.G. (1999). A Represa de Jurumirim (São Paulo): Primeira síntese sobre o conhecimento limnológico e uma proposta preliminar de manejo ambiental, In: *Ecologia de reservatórios: estrutura, função e aspectos sociais*, R. Henry, (Ed.), 651-685, Fapesp/Fundibio, Botucatu, Brazil

Henry, R. (1992). The oxygen deficit in Jurumirim Reservoir (Paranapanema River, São Paulo, Brazil*). Jpn. J. Limnol*, Vol.53, pp. 379-84

Henry, R.; Nogueira, M.G.; Pompeo, M.L.L. & Moschini-Carlos, V. (2006). Annual and short-term variability in primary productivity by phytoplankton and correlated abiotic factors in the Jurumirim Reservoir (São Paulo, Brazil). *Braz. J. Biol.*, Vol.66, pp. 239-261

Jorcin, A. & Nogueira, M.G. (2005) Temporal and spatial patterns along the cascade of reservoirs in the Paranapanema River (SE Brazil) based on the characteristics of sediment and sediment-water interface. *Lakes & Reserv. Res. Manage.*, Vol.10, pp. 1-12

Kelman, J.; Pereira, M.V.; Araripe-Neto, T.A. & Sales P.R.H. (1999). Hidroeletrecidade, In: *Águas doces no Brasil: Capital ecológico, uso e conservação*, A.C. Rebouças, B. Braga & Tundisi J.G, (Eds.), 371-418, Editora Escrituras, São Paulo, Brazil

Kindt, R. & Coe, R. (2005). *Tree diversity analysis*. World Agroforestry Centre, Kenya

Marckereth, F.I.H.; Heron J. & Talling J.F. (1978) *Water analysis: some revised methods for limnologists*. Freshwater Biological Association, London

Mercante, C.T.J. & Tucci-Moura, A. (1999). Comparação entre os índices de Carlson e de Carlson modificado aplicados a dois ambientes aquáticos subtropicais, São Paulo, SP. *Acta Limnologica Brasiliensia*, Vol.11, No.1, pp. 1-14

Naliato, D.A.O.; Nogueira, M.G. & Perbiche-Neves, G. (2009). Discharge pulses of hydroelectric dams and their effects in the downstream limnological conditions: a case study in a large tropical river (SE Brazil). *Lakes & Reser.: Res. and Manag.*, Vol.14, pp. 301-314

Nogueira M.G.; Henry, R. & Maricatto, F.E. (1999). Spatial and temporal heterogeneity in the Jurumirim Reservoir, São Paulo, Brazil. *Lakes & Reser.: Res. and Manag.*, Vol.4, pp. 107-120

Nogueira, M.G.; Jorcin A.; Vianna, N.C. & Britto Y.C. (2006). Reservatórios em cascata e os efeitos na limnologia e organização das comunidades bióticas (fitoplâncton, zooplâncton e zoobentos): Um estudo de caso no rio Paranapanema (SP/PR). In: *Ecologia de reservatórios: Impactos potenciais, ações de manejo e sistemas em cascata*, M.G. Nogueira, R. Henry & A. Jorcin, (Eds.), 83-125, Rima, São Carlos, Brazil

Nogueira, M.G.; Reis-Oliveira, P.C. & Britto Y.T. (2008) Zooplankton assemblages (Copepoda and Cladocera) in a cascade of reservoirs of a large tropical river (SE Brazil). *Limnetica*, Vol.27, pp. 151-170

Nogueira M.G. (2000). Phytoplankton composition, dominance and abundance as indicators of enviromental compartmentalization in Jurumirim Reservoir (Paranapanema River), São Paulo, Brazil. *Hydrobiol.*, Vol.431, pp. 115-128

Nogueira, M.G.; Jorcin, A.; Vianna, N.C. & Britto, Y.C.T. (2002a). Uma avaliação dos processos de eutrofização nos reservatórios em cascata do Rio Paranapanema (SP-PR), Brasil, In: *El Água en Iberoamerica, de la limnologia a a gestión en Sudamerica, Argentina*, A. Cirelli & G. Marquisa (Eds.), 91-106, Cyted XVII, Argentina

Nogueira, M.G.; Jorcin, A.; Vianna, N.C. & Britto, Y.C.T. (2002b). A two- year study on the limnology of a cascade reservoir system in a large tropical river in Southeast Brazil. *4th International conference on reservoir limnology and water quality*, Èeské Bud˜jovice, August, 2002

Pagioro, T.A.; Thomaz, S.M. & Roberto, M.C. (2005a). Caracterização limnológica abiótica dos reservatórios. In: *Biocenoses em reservatórios: padrões espaciais e temporais,* L. Rodrigues, S.M. Thomaz, A.A. Agostinho & L.C. Gomes (Eds.), 17-37, Rima, São Carlos, Brazil

Pagioro, T.A.; Velho, L.F.M.; Lansac-Tôha, F.A.; Pereira, D.G. & Nakamura, A.K.S. (2005b). Influência do grau de trofia sobre os padrões de abundância de bactérias e protozoários planctônicos em reservatórios do Estado do Paraná. In: *Biocenoses em reservatórios: padrões espaciais e temporais,* L. Rodrigues, S.M. Thomaz, A.A. Agostinho & L.C. Gomes (Eds.), 47-56, Rima, São Carlos, Brazil

Perbiche-Neves, G. & Nogueira, M.G. (2010). Multidimensional effects on cladoceran (Crustacea, Anomopoda) assemblages in two cascade reservoirs (SE - Brazil). *Lakes and Res.: Research and Manag.*, Vol.15, pp. 151-164

Perbiche-Neves, G.; Ferreira, R.A.R. & Nogueira, M.G. (2011). Phytoplankton structure in two contrasting cascade reservoirs (Paranapanema River, Southeast Brazil). *Biologia (Bratislava) Section Botany,* Vol. 66 (6), pp. 967-976

Pinto-Coelho, R.M.; Azevedo, L.M.; Rizzi, P.E.V.; Bezerra-Neto, J.F. & Rolla, M.E. (2006). Origens e efeitos do aporte externo de nutrientes em um reservatório tropical de grande porte: Reservatório de São Simão (MG/GO), In: *Ecologia de reservatórios: Impactos potenciais, ações de manejo e sistemas em cascata*, M.G. Nogueira, R. Henry & A. Jorcin, (Eds), 127-164, Rima, São Carlos, Brazil

R Development Core Team (2009). *A language and environment for statistical computing.* Vienna, Austria, R Foundation for Statistical Computing ISBN 3-900051-07-0, Date access 15.04.2009, available from: <http://www.R-project.org>.

Sampaio, E.V.; Rocha, O.; Matsumura-Tundisi, T. & Tundisi, J.G. (2002). Composition and Abundance of Zooplankton in the Limnetic of Seven Reservoir of the Paranapanema River, Brazil. *Braz. J. Biol.*, Vol.62, No.3, pp. 525-545

Sant'Anna, C.L.; Sormus, L. Tucci, A. & Azevedo, M.T.P. (1997). Variação sazonal do fitoplâncton do lago das Garças, São Paulo, SP, Brasil. *Hoehnea*, Vol.24. pp. 67-86

Sendacz, S. & Kubo, E. (1999). Zooplâncton de reservatórios do alto Tietê, Estado de São Paulo, In: *Ecologia de reservatórios: estrutura, função e aspectos sociais*, R. Henry, (Ed.), 511-529, Fapesp/Fundibio, Botucatu, Brazil

Soares, M.C.S.; Marinho, M.M.; Huszar, V.L.M.; Branco, C.W.C. & Azevedo, S.M.F.O. (2008) The effects of water retention time and watershed features on the limnology of two tropical reservoirs in Brazil. *Lakes & Reserv.: Res. and Manag.*, Vol.13, pp. 257–269

StatSoft, Inc. (2006) *STATISTICA (data analysis software system), version 6.0* Available from: <www.statsoft.com>

Stevaux, J.C.; Martins, D.P. & Meurer, M. (2009). Changes in regulated tropical rivers: The Paraná River downstream Porto Primavera dam, Brazil. *Geomorphology*, Vol.113, pp. 230-238

Strickland, J.D. & Parsons, T.R. (1960). A manual of sea water analysis. *Bull. Fish. Res. Board of Can.*, Vol.125, pp. 1-185

Talling, J.F. & Driver D. (1963). Some problems in the estimation of chlorophyll a in phytoplankton. In: *Proceedings, Conference of primary productivity measurements in marine and freshwater*, 142-146, Atomic Energy Commision, Hawaii, USA

Thomaz, S.M.; Bini, L.M. & Alberti, S.M. (1997). Limnologia do reservatório de Segredo: padrões de variação espacial e temporal, In: *Reservatório de Segredo: bases ecológicas para o manejo*, A.A. Agostinho & L.C. Gomes, (Eds.), 19-37, Eduem, Maringá, Brazil

Thornton, W.K. (1990). Perspectives on reservoir limnology, In: *Reservoir Limnology: ecological perspectives*, K.W. Thornton, B.L. Kimmel & E.F. Payne (Eds.), 1-13, John Wiley & Sons Inc., New York

Tundisi, J.G. (1990). Key factors of reservoir functioning and geographical aspects of reservoir limnology - chairman's overview. *Arch. Hydrobiol. Beih. Ergebn.*, Vol.33, pp. 654-646

Tundisi, J.G. & Matsumura-Tundisi, T. (2003) Integration of research and management in optimizing multiple uses of reservoirs: the experience in South America and Brazilian case studies. *Hydrobiologia*, Vol.500, pp. 231–42

Valderrama, J.G. (1981). The simultaneous analysis of total nitrogen and phosphorus in natural waters. *Mar. Chem.*, Vol.10, pp. 109-122

9

Reservoir Operation Applied to Hydropower Systems

João Luiz Boccia Brandão
University of São Paulo, Brazil

1. Introduction

Currently, studies on hydropower systems operations primarily focus on multiple uses of water, i.e., water resources exploitation and control systems to satisfy human needs and demands connected to economic and social activities. Focuses include power generation, urban and industrial water supply, irrigation, navigation, leisure and recreation-related uses, flood control and water pollution control.

Conflicts often occur because the resources available cannot meet the demands of all the users in a given system. Therefore, it is essential to objectively evaluate the system potential and the best form of operation.

This chapter will describe how to analyze operational research techniques, such as non-linear programming, which can be applied to the operation of hydropower reservoir systems with multiple uses and be used to evaluate their performance.

In this chapter, most of the analyses and evaluations were obtained from case studies, based on the system of reservoirs in the São Francisco River basin in Brazil. In this basin, there are current water-use conflicts due to an increase in water demand for irrigation and a need to possibly transfer water from the São Francisco River to the semi-arid region in the northern part of northeast Brazil. There are also problems with water pollution and environmental conservation problems in certain stretches of this river.

2. Literature review

Over the last three decades, articles involving the optimization of reservoir systems have increasingly been published. An important review of the state of the art on the subject was written by Yeh (1985), who discussed various types of models for reservoir systems. However, he placed the greatest emphasis on optimization models, such as linear programming (LP) and dynamic programming (DP) and its variations (stochastic DP, incremental DP with successive approximations, DP with probabilistic restrictions and progressive optimality). According to Yeh, LP offers the following advantages: a) allows solutions of problems with large dimensions; b) there are widely accepted computational packages in the market, e.g., Simplex; and c) reaches the optimal global point.

However, according to Yeh, DP is more adaptable for nonlinear problems, sequential decision-making and stochastic aspects characteristic of the operation of reservoirs.

However, problems of dimensionality, when the number of variables grows exceptionally large with the number of reservoirs, make it very difficult to apply DP to large systems. In such cases, techniques, such as successive approximation, incremental and differential DP, etc., can be used.

In 1985, Yeh noted problems with nonlinear programming (NLP) models that today have been overcome. The problems were based on low memory capacities and low processing speeds of the computers used at the time.

In the last few decades, efficient mathematical algorithms have been developed to solve linear and nonlinear optimization problems. The development of efficient optimization routines, along with the rapid development of informatics, which resulted in portable computers with high processing and data storage capabilities, enabled the development of models that are more efficient and easier to process.

A program that incorporates such algorithms is the MINOS (Modular In-core Nonlinear Optimization System) package developed by Stanford University's System Optimization Laboratory (Murtagh & Saunders, 1995). An important application of this program was reported by Tejada-Guilbert et al. (1990), who used MINOS for NLP to optimize the operation of the California Central Valley Project. The package was used to maximize the economic value of the energy generated each month. This work presented a very interesting discussion about the optimization of nonlinear systems and the applicability of MINOS.

An example of stochastic DP, Barros (1989) analyzed the operational problem of reservoirs with an implicit stochastic focus, where the randomness of the process was considered, beginning from the generation of a synthetic series based on the Monte Carlo method.

In an article authored by Kelman et al. (1990), a sampling stochastic dynamic programming technique was used to model the complex structure of the spatial and temporal correlation of flows into reservoirs using a large number of samples in a temporal flow series.

Braga et al. (1991) presented an application of stochastic DP with an explicit focus, using the one-at-a-time technique, which is similar to the stratagem of successive approximations to attenuate the "curse of dimensionality".

An application to compare deterministic and stochastic optimization was presented by Lund & Ferreira (1996). The methodology was applied to a system of six reservoirs in the Missouri River (USA). The results involved issues of applicability and limitations in the use of deterministic optimization for large systems.

With respect to the operation of systems with conflicting uses, Ponnambalam & Adams (1996) used stochastic optimization to define rule-curves for a system of reservoirs used for electric energy production and irrigation in India. According to the authors, the results obtained from the application of the optimized operational rules to a simulation model indicated a gain in the system's performance compared with real operational data.

In the context of the problem of defining operational rules for reservoir systems, Oliveira & Loucks (1997) used genetic search algorithms and presented a methodology to generate a set of operational policies, which were tested on a simulation model. The policies that resulted in the best performance were selected and used to define new policies, which were again tested. The process evolved until the performances ceased to improve. The algorithm was applied to an electrical energy production and supply system with promising results.

Francato & Barbosa (1997) analyzed several factors that could potentially influenced the results from hydroelectric system optimization models. Their study focused on aspects relating to the type of objective function and the topology of the system. Their analyses were performed based on the models of Emborcação and Itumbiara reservoirs located in the Paranaíba River in Brazil.

Labadie (1998) authored a critical review of the principal optimization models, emphasizing implicit and explicit stochastic optimization to treat the randomness of the processes involved in the operation of reservoirs. It is important to note the author's concern in placing the operational problem as part of a decision-making support system to ensure the effective implementation of policies originating from research and development groups.

A reference integrated LP and DP models for the operation of reservoir systems was presented by Braga et al. (1998), where the authors developed the SISCOM (Sistema Computadorizado de Apoio ao Planejamento e Operação de Sistemas Hidroeléticos) model to optimize the operation of Brazil's hydro energy system.

Philbrick Jr. & Kitanidis (1999) also analyzed the problem concerning the operation of reservoirs, comparing results produced by deterministic optimization and by stochastic optimization. The authors concluded that the deterministic approach tended to produce pseudo-optimal results that may underestimate the benefits associated with the systems.

An article by Peng & Buras (2000) presented another MINOS application to optimize the operation of reservoirs, emphasizing that the evolution of computers and operational research algorithms have expanded the use of packages for solving large LP and NLP problems. The authors developed a model of reservoir systems with multiple objectives, using the implicit method to consider the stochastic nature of the inflows.

Lopes (2001) discussed NLP applications in the operation of hydroelectric plant systems, obtaining rules of operation according to the system's topology. For parallel configurations, the author suggested that reservoirs with a lower head loss per unit volume (drop reduction factor) should be emptied first. For systems in series, the reservoirs should be emptied in an upstream to downstream sequence, except when the differences between the head reduction factors indicate the contrary. This work also emphasized the need to consider nonlinear treatment when the reservoirs are used to generate hydroelectricity to obtain correct productiveness values (magnitude expressed in $MW/m^3.s^{-1}$) as a function of the head.

Barros et al. (2003) also used MINOS in a SISOPT model to deterministically optimize the operation of large electrical energy generating systems. The model optimized the operation using the LP technique and successive linear programming (SPL) and allowed the user to define objective functions, such as the minimization of spillage and minimization of quadratic deviations in relation to a rule-curve, among others.

Motivated by the current conjuncture that imposes the intensification and improvement of the performance of existing systems, Labadie (2004) authored a new review on the state-of-the-art on operation of reservoir systems. In the article, the author discusses stochastic optimization methods for both implicit and explicit schemes, involving DP (and its derivations), linear and nonlinear programming, network-flow models and multi-objective optimization models. He also described heuristic programming methods, such as neural network models and fuzzy mathematics techniques. It is interesting to note the author's comments concerning the present difficulties concerning how to deal realistically with the problems relating to hydrological uncertainties.

Lopes (2007) developed optimization models based on the equivalent reservoir approach applied to the Brazilian electrical system using the commercial nonlinear programming packages MINOS, SNOPT (Sequential Non-linear Optimiser) and CONOPT (Constrained Optimization) with the support of the optimization language GAMS (General Algebraic Modelling System), where the spreadsheet optimizer SOLVER was also used.

The HIDRO model, proposed by Zambon (2008), evolved from the SISOPT model. However, the LP and SLP were no longer used. The objective function sought to minimize the supplement of power generated by thermal power plants to meet energy demand. The objective function proposed by Zambon minimized the square of the difference between the energy demand and energy production. It is therefore a model to be solved by NLP. The decision variables include the water discharge and spillage from each reservoir. To generate a complete and feasible initial solution for the NLP, it was developed a simulator that had three options: operation as run-of-river power plants, operation considering the maximum flow through turbines and flow through turbines and spillage defined by the user.

In the same study, Zambon (2008) proposed the TERM model, which aggregated the thermal power plants and the energy demands into subsystems. The objective function proposed by Zambon minimized the total cost of thermal generation, exchanges and deficits in each time interval. The input data for each subsystem included the demand forecast, the generation by small hydro and nuclear plants, the resulting HIDRO model generation, the imports and exports of energy, the inflexible thermal generation limits, the limits of exchange between the subsystems and the cost of thermal generation for each subsystem. The decision variables included the additional thermal generation and exchanges in each time interval for each subsystem.

Zambon (2008) also proposed the implementation of the model, HIDROTERM, which used an iterative process between the TERM and HIDRO models or by a unified formulation. In the model, one could choose the minimization of the quadratic thermal complementation used in the HIDRO model or the minimization of the sum of the cost of thermal generation, exchanges and the deficit used by the TERM model as the objective function.

From the most recent works, it can be concluded that many studies and applications have used packages that solve linear and nonlinear programming problems, which is associated with the development of programs and computers that provide increasingly fast solutions to increasingly complex problems. Moreover, many of these packages can be used in conjunction with graphic interfaces that allow for a high degree of generalization of the problems and their use in modeling a variety of systems.

3. Methodology

The methodology described in this chapter is intended to improve the modeling of multiple-use reservoir systems, where alternative ways to solve the problem are explored. Currently, the model of these systems, as used by ONS, the Brazilian agency in charge of the power generation systems operation, considers other water uses as constraints. Another aspect to be considered is the issue of hydrological risk associated with natural flows into the reservoirs.

The methodology used is general because it can be applied to any type of water resource reservoir system.

The optimization problem of a multiple-use reservoir system may be formulated as follows:

$$Maximmize\ or\ Minimize\ OF = \sum_{i=1}^{m}\sum_{t=1}^{n} R_{i,t} \tag{1}$$

where OF is the objective function; $R_{i,t}$ is a function that measures return and/or performance associated with reservoir i at an interval t; i = 1, 2,...,m (m = number of reservoirs in the system) t = 1, 2,...,n (n = number of time intervals). The equation is subject to

$$VF_{i,t} = VF_{i,t-1} + \left[QA_{i,t} - QD_{i,t} - QC_{i,t}\right].K - EV_{i,t} \tag{2}$$

$$Vminimum_i < VF_{i,t} < Vmaximum_i \tag{3}$$

$$QDminimum_i < QD_{i,t} < QDmaximum_i \tag{4}$$

$$QCminimum_i < QC_{i,t} < QCmaximum_i \tag{5}$$

$$QD_{i,t} > 0 \tag{6}$$

where $QD_{i,t}$ = outflow from reservoir i throughout time interval t (decision variable) in m^3/s; $QC_{i,t}$ = flow of consumptive use of reservoir i throughout time interval t (this may be a decision variable or only a restriction, depending on the type of objective function) in m^3/s; $QA_{i,t}$ = inflow to reservoir i throughout time interval t (including the flow in the intermediate drainage area between reservoir i and the reservoirs that lie immediately upstream plus the sum of the outflows from these reservoirs) in m^3/s; K = a constant to convert flows from m^3/s to monthly volumes in m^3, or multiples of this unit; $VF_{i,t}$ = volume of reservoir i at the end of interval t (state variable) in m^3, or multiples of this unit, $EV_{i,t}$ = volume evaporated from reservoir i during interval t in m^3, or multiples of this unit.

In an attempt to formalize and rationalize the solution of the optimization problem in the case of multiple uses, two methods may be defined:

Weighting Method. This method includes several direct or indirect decision variables in the objective function, where examples include power generation flows, irrigation flows, flows for other consumptive uses, minimum and maximum levels for navigation, recreation, conservation, etc. In this case, the restriction equations considered are the physical characteristics and water balance. In this method, the objective function is a weighted type, where the weights of each objective are defined by the stakeholder.

Restriction Method. This method includes a single use in the objective function, for example, power generation, and considers the other uses in the restriction equations. One may then determine the exchange relations between uses (Pareto curves), varying the limits of fulfilling each objective as related to the other.

Considering both irrigation and power generation, the solution of the two methods proposed is presented below. In this case, an attempt is made to maximize the mean energy and mean discharge for irrigation throughout the period of analysis.

Objective function weighting method:

$$Max\left(\alpha.\sum_{i=1}^{m}\frac{\sum_{t=1}^{n}E_{i,t}}{n}+\beta.\sum_{i=1}^{m}\frac{\sum_{t=1}^{n}QI_{i,t}}{n}\right) \tag{7}$$

Objective function restriction method:

$$Max\left(\sum_{i=1}^{m}\frac{\sum_{t=1}^{n}E_{i,t}}{n}\right) \text{ with } QI_{i,t} = DI_i \tag{8}$$

where $QI_{i,t}$ = irrigation flow supplied by reservoir i during interval t; $DI_{i,t}$ = irrigation demand to be fulfilled by reservoir i during interval t; $E_{i,t}$ = energy generated by the plant of reservoir i during time interval t; α and β are the weighting parameters for energy and irrigation, respectively.

The parameters, α and β, are values that express the relative importance of each use. In a way, they indicate an order of preference of one use over another, i.e., a hierarchical order. These parameters are assigned subjectively by the stakeholders and/or managers. In this chapter, the parameters were subjected to a sensitivity.

When one of the uses of a given system is to generate electric energy, the problem of reservoir operation must be complemented by functions that rule energy production, as follows:

$$E_{i,t} = 9{,}81.10^{-3}.ng.nt.nh.HB_{i,t..}QT_{i,t} \tag{9}$$

where $E_{i,t}$ = the mean generation of reservoir i in interval t, in average MW, which is the energy corresponding to the mean power generated over a month or a certain number of months; ng, nt and nh = the efficiencies of the generator, turbine and hydraulic circuit (penstock and restitution), respectively; $HB_{i,t}$ = mean monthly gross head in reservoir i during interval t in meters (difference between water levels in the reservoir and plant tailrace); $QT_{i,t}$ = turbine discharge corresponding to reservoir i during interval t in m^3/s.

Next, the solution for the operation of a reservoir system to generate electric energy based on the initially proposed optimization problem is presented. In this case, an objective function is used, which seeks to maximize the mean energy of the system over the period of analysis.

$$Max\left(\sum_{i=1}^{m}\frac{\sum_{t=1}^{n}PRT_{i,t}.QT_{i,t}}{n}\right) \tag{10}$$

where

$$PRT_{i,t} = 9{,}81.10^{-3}.ng.nt.nh.HB_{i,t} \tag{11}$$

where $PRT_{i,t}$ = power production factor of plant i during month t in $MW/m^3.s^{-1}$.
The equation is subject to

$$VF_{i,t} = VF_{i,t-1} + \left[QA_{i,t} - QT_{i,t} - QV_{i,t} - QC_{i,t} \right].K - EV_{i,t} \tag{12}$$

$$Vminimum_i < VF_{i,t} < Vmaximum_i \tag{13}$$

$$PRT_{i,t}.QT_{i,t} < PI_i.ID_i \tag{14}$$

$$QTminimum_i < QT_{i,t} < QTmaximum_i \tag{15}$$

$$QCminimum_i < QC_{i,t} < QCmaximum_i \tag{16}$$

$$QT_{i,t} \text{and } QV_{i,t} > 0 \tag{17}$$

where $QV_{i,t}$ = spilled discharge from the plant corresponding to reservoir i during interval t in m^3/s; PI_i = installed capacity of plant i in MW; ID_i = availability index of generators in plant i, which defines the mean power capacity available over time, where the scheduled or forced downtime for maintenance and other reserves have been discounted.

The objective function represented in eq. (10) is non-linear because $PRT_{i,}$ is a non-linear function of reservoir volume and of the released flow (sum of turbine and spilled discharges). To obtain the value of $PRT_{i,t}$, the mean gross head over interval t should be calculated. This head is obtained from the difference between the water level in the reservoir (upstream level) and the downstream water level. The water level in the reservoir is calculated based on the elevation volume curve. The downstream water level is obtained from the rating curve (stage-discharge relation) of the plant tailrace. Both relations are represented by non-linear relations.

The optimization model formulated is stochastic in nature because the natural flows into the reservoir are random variables associated with time, whose future performance is unknown.

One way of treating the problem indirectly is through the implicit method. Based on a time series generation model, several sequences of synthetic natural inflows are generated, which are then used as input data to solve the optimization problem. The results obtained are statistically analyzed, and then the system operation rules, the levels to ensure fulfilling demands, and others are defined.

The generation of synthetic inflows was conducted using AR (autoregressive) and ARMA (autoregressive and moving average) models, which are the most commonly used models in hydrology and many other areas, with some adjustments depending on the type of time series modeling and application. There are also MA (moving average) and ARIMA (autoregressive integrated moving average) models.

The AR model of order p, which is usually referred to as AR (p), is presented by Salas (1993) as follows:

$$y_t = \mu + \sum_{j=1}^{p} \phi_j (y_{t-j} - \mu) + \varepsilon_t \tag{18}$$

where y_t = the random variable modeled, i.e., the time series under study; p = "lag" or order of the model, indicates the degree of temporal autocorrelation; ε_t = uncorrelated noise - the random variable normally distributed with mean zero and standard deviation σ_ε.

Because ε_t is normally distributed, y_t is normally distributed as well. The model parameters are μ, ϕ_1, $\phi_2,\ldots,\phi_p$ and σ_ε. The parameter, μ, can be estimated by the mean of y_t, and the other model parameters are estimated by the Yule-Walker equations. All these equations are presented in detail in the reference.

The ARMA models with p autoregressive and q moving average parameters, known as ARMA (p, q), are presented by the referred author, as the following equation:

$$y_t = \mu + \sum_{j=1}^{p} \phi(y_{t-1} - \mu) + \varepsilon_t - \sum_{j=1}^{q} \theta_j \varepsilon_{t-j} \tag{19}$$

The last term in eq. (19) is the moving average term. The other terms correspond to the portion of the autoregressive model, as shown in eq. (18). The moving averages parameters are θ_1, $\theta_2,\ldots,\theta_q$. A moving average model, MA (q), displays the corresponding terms, i.e., the last term in eq. (19). The equations for estimating the parameters of ARMA models are also presented in Salas (1993).

4. Case study

The São Francisco River basin, which is shown in Figure 1, has seven hydropower plants on its main course, two of which, Três Marias and Sobradinho, have a large accumulation capacity. These two reservoirs allow plurianual discharge regulation, i.e., they have filling and emptying cycles of over one year. The Itaparica reservoir located downstream from Sobradinho presents an annual regulation capacity, and the other reservoirs have a small water accumulation capacity. Downstream from Itaparica are located the power plants of the Moxotó-Paulo Afonso complex and the Xingó hydroelectric plant.

The model was prepared based on the structured language, GAMS, to solve mathematical programming problems. The solver used was MINOS, which can solve linear and non-linear programming problems.

The water uses in the case study were electric energy generation and irrigation, which are the main conflicting uses in the São Francisco basin.

The optimization problem also took into account other operational restrictions for each reservoir/plant.

The time interval adopted for calculation was on a monthly basis, where the horizon for the analysis period is 6 years (72 months), which was chosen according to the duration of the recorded critical period observed in most river basins in Brazil.

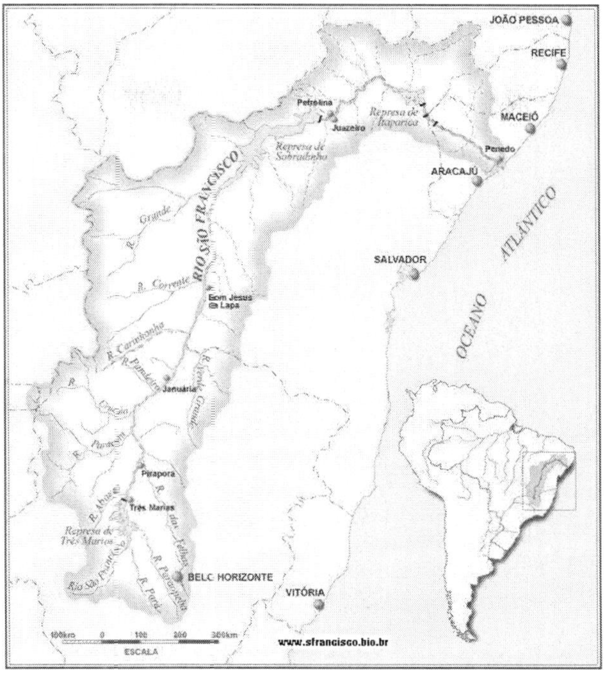

Fig. 1. Location of the São Francisco River basin in Brazil.

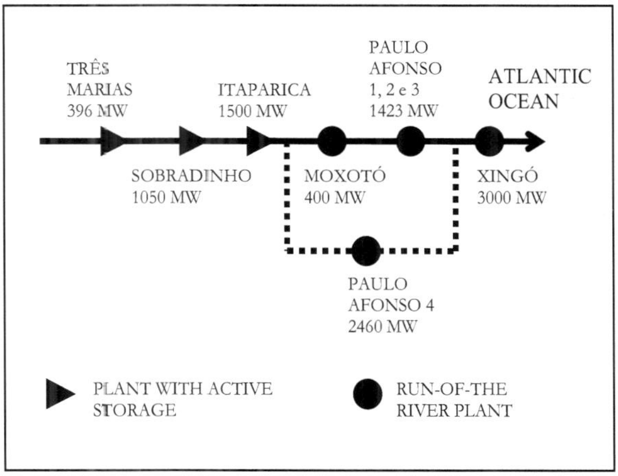

Fig. 2. Topological scheme of the São Francisco system.

Table 1 shows key information related to the hydropower plants and reservoirs of the São Francisco system.

Power plant	Distance to the mouth (Km)	Drainage area (Km^2)	Active storage (Hm^3)	Maximum water level (m)	Minimum water level (m)	Installed capacity (MW)
Três Marias	2220	50560	15278	572.50	549.20	396
Sobradinho	800	498425	28669	392.45	380.50	1050
Itaparica	310	587000	3548	304.00	299.00	1500
Moxotó	270	599200	-	251.50	251.50	400
Paulo Afonso 1,2 E 3	270	599200	-	251.50	251.50	1423
P.Afonso 4	270	599200	-	251.50	251.50	2460
Xingó	210	608700	-	138.00	138.00	3000

Table 1. Hydropower plants and reservoirs of the São Francisco system.

The model was developed in two modules. The first module treated the problem according to the restriction method. In this case, the objective function sought to maximize the energy generated by the system over the period of analysis. The second module focused on the problem according to the weighting method, where the solution included an objective function, which sought to maximize a weighting between the mean energy generated by the system and the total mean flow extracted from the system for irrigation.

5. Analysis for multiple uses

During this phase, a time series of mean monthly flows lasting six years was used, which consisted of pre-defined hydrological scenarios.

5.1 Selection of hydrological scenarios

The studies for the selection of hydrological scenarios were developed based on the Três Marias and Sobradinho flow series that were considered "key" series in this study.

The values of the correlation coefficients of the Três Marias series and the incremental Sobradinho series were on the order of 0.65 for both the annual series and the monthly series. This value indicates that there is a considerable spatial correlation between these two flow series that must be taken into account when formulating hydrological scenarios.

To select the scenarios, the moving averages of six consecutive years were calculated for the two flow series. Based on the joint distribution of these two variables, the cases that correspond approximately to the 1st, 2nd and 3rd quartile were selected, i.e., periods of six consecutive years whose moving averages were associated with joint cumulative probabilities, equal to 25%, 50% and 75%. These scenarios were called "dry", "median" and "wet", respectively.

5.2 Multiple use analysis: restriction method

The processing used to analyze the restriction method using the model was done based on the maximization of the mean energy generated in each time sequence considered. The flows diverted for irrigation were treated as constraints. The three hydrological scenarios selected plus the scenario corresponding to the critical period of the interlinked systems, which occurred during the first half of the 1950s, were considered.

The results obtained are reproduced in Figure 3, where the exchange relationship can be seen, i.e., the trade-off between irrigation and hydroelectric generation in the São Francisco system. The curves presented in this figure are the Pareto curves. The gradients of these curves indicate the trade-off value between these two uses. In the case of the dry scenario and the critical period, this gradient was on the order of -2 average MW of generation per m^3/s used in irrigation. For each m^3/s diverted for irrigation, approximately 2 average MW are lost, which over a year totals 17.5 GWh of power generation.

In the cases of the median and dry scenarios, these gradients are on the order of -1.3 and -0.8, respectively, as a consequence of more water availability contained in those scenarios. Based on these results, one can note the influence of the hydrological scenarios on the trade-offs between these two uses. In this case, because the irrigation flow is a restriction that must always be met, water availability regulates how much may be generated by the hydroelectric plants.

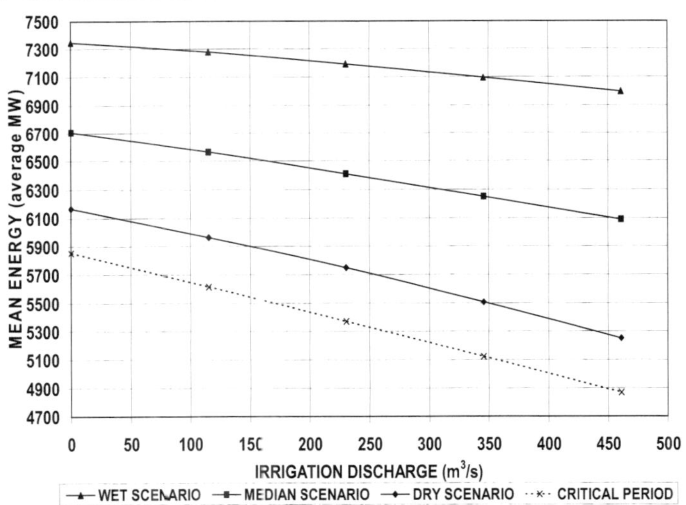

Fig. 3. Pareto curves with historical series - São Francisco River basin.

5.3 Multiple use analysis: weighting method

To evaluate the weighting method, the module for this type of formulation was used. To simplify the analysis, complementary weighting parameters, α and β, were selected, i.e., values whose sum is equal to the unit. However, there was no need for these parameters to be complementary; they only needed to express an order of preference of one use as compared with the others. In Figure 4, the results obtained for the dry scenario are shown.

According to these results, there was a region where the trade-off between irrigation and energy occurred more intensely. This region corresponded approximately to values of α between 0.25 (β = 0.75) and 0.60 (β = 0.40). The maximum and minimum limits were outside this region for the two uses. In other words, for values of α below 0.25 (β above 0.75), the upper limit of irrigation flow was found. In this case, the energy values were minimal, corresponding to approximately 5750 average MW. For α above 0.60 (β below 0.40), the minimum values for irrigation were found. However, these cases resulted in the maximum energy values, i.e., 6170 average MW. For the other hydrological scenarios, the results were similar.

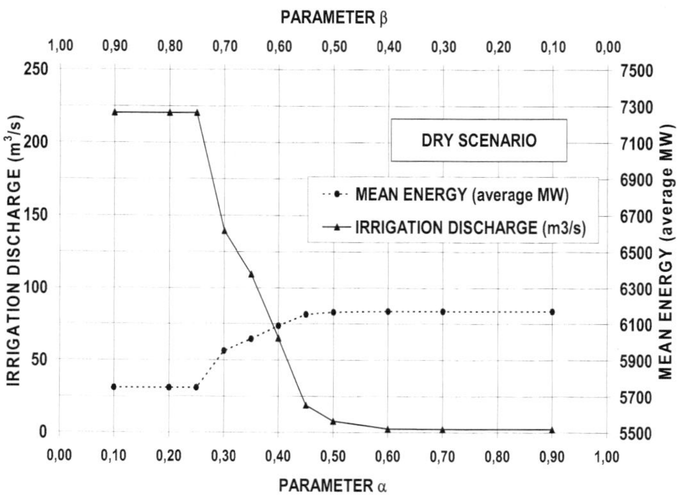

Fig. 4. Weighting method - dry scenario - São Francisco River basin.

6. Stochastic analysis of reservoir operation

Next, the treatment of the stochastic issue related to the random nature of natural flows into the reservoirs was evaluated. These evaluations were developed beginning with the formulation of the optimization problem, according to the restriction method, imposing the irrigation demand as a constraint to be met by the system. The objective function adopted was the maximization of the mean energy over the period of analysis.

Stretches of six-year synthetic series were generated. To evaluate the influence of the number of series in the results, sets of 200, 500 and 1000 series were considered.

Based on the processing of the optimization model with the three sets of synthetic series (200, 500 and 1000 series), the corresponding cumulative probabilities distributions of the mean energy, as shown in Figure 5, were calculated.

Additionally, 65 six-year series were generated based on the data of the historical series of mean monthly flows, available between 1931 and 2001. In this way, each 6-year series began each year of the historical trace up to the year 1995.

The probabilities distribution for the energy generated obtained from the history of flows is also presented in Figure 5, together with the distributions already defined based on the synthetic series.

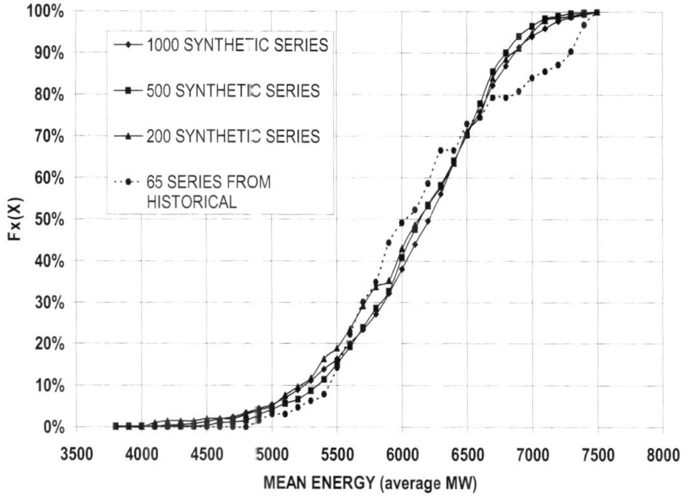

Fig. 5. Cumulative probabilities curves for energy obtained using the synthetic series and historical series - São Francisco River basin.

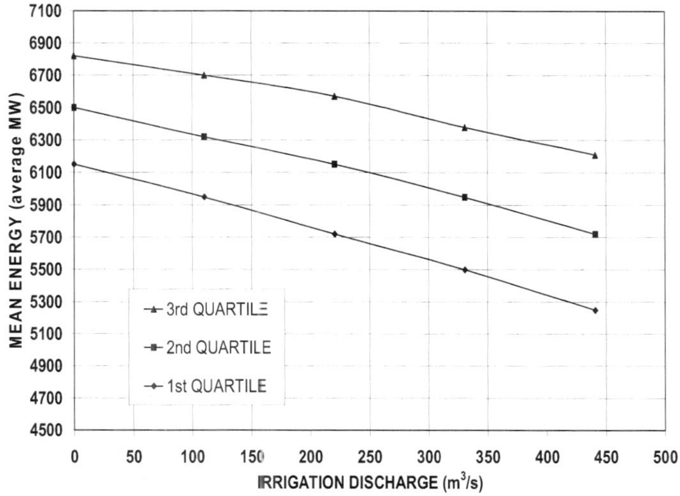

Fig. 6. Pareto curves with 500 synthetic series - São Francisco River basin.

The Kolmogorov-Smirnov test was applied to perform a statistical evaluation of the hypotheses of equality on the distributions. The results of the test showed that for 90% and 95% levels of significance, the null hypothesis, i.e., equality among the four distributions, cannot be ruled out. Thus, based on these results, it was accepted that any of the distributions obtained produced statistically similar results. However, based on the visual analysis of the curves in the previous figure, a greater discrepancy was observed between the distributions due to historical traces compared with those resulting from the application of the synthetic series, particularly at the extremes.

Pareto curves were also developed for the 1st, 2nd and 3rd quartiles for the case with 500 synthetic series. These curves are shown in Figure 6.

The gradients of these curves were -2.0, -1.7 and -1.3 average $MW/m^3.s^{-1}$ for the 1st, 2nd and 3rd quartile, respectively. The values of the gradients determined for the scenarios extracted from the historical series, as stated before, were -2.0, -1.3 and -0.8 average $MW/m^3.s^{-1}$ for the dry, median and wet scenarios, respectively. Comparing these numbers, for the dry scenario, the same trade-off values were observed. In the cases of the median and wet scenarios, the trade-offs obtained using the synthetic series were more marked than those corresponding to the historical series.

Based on the comparison of the Pareto curves obtained by the synthetic series to those of the historical series, it was found that the curves essentially coincided with those of the dry scenario. For the median and wet scenarios, the curves obtained using the historical series were greater than those obtained using the synthetic series. For the median scenario, it was found that the increment in energy generated was between 3% and 6%, according to the flow used for irrigation. For the wet scenario, the increment varied between 8% and 13%.

These results indicate that, for the dry scenario, the results are the same, regardless of whether the hydrological scenario is extracted from a historical series or a synthetic series. For the median and wet scenarios, the use of synthetic series has repercussions with more severe evaluations for both the trade-offs (steeper gradients) and the values of the generated energy.

7. Conclusions

Because most reservoir systems exhibit a competition among water uses, it is necessary to define several specific rules to properly distribute the available water resources. To meet the commitments, water planners and managers must consider a number of allocation alternatives using system modeling tools. Therefore, it is possible to define the proper criteria and procedures to balance the allocation rules. In this paper, the model proposed was based on non-linear programming optimization, which was developed using the mathematical programming language, GAMS. The optimization problems formulated were solved by the linear/non-linear programming solver, MINOS. These tools were used to develop a model to optimize multi-purpose reservoir systems operations.

The case study was based on the São Francisco River basin reservoir system located in northeast Brazil. The system consists of six reservoirs, from which three are operated with active-storage. The reservoir's main purpose is power generation, but irrigation, flood control, water quality control, recreation and environmental preservation are also important issues. The major conflicts arise between power generation and irrigation demands. Therefore, the trade-offs between these two uses were analyzed in more detail. The total energy produced by the system is approximately 53,000 GW-hour per year, and the average estimated irrigation demand in the coming years is approximately 230 m^3/s.

The trade-off analysis on the power generation and irrigation demands was formulated using the restriction and the weighting methods. In the first method, the objective function included only one water use, where the others uses were part of the restriction equations as constant or seasonal demands or constraints. In the second method, the competing water uses were included in the objective function and weighted by coefficients previously chosen by the water planners or managers.

The restriction method was simpler and more directly applicable. This method clearly showed the trade-offs between the competing uses. In other words, it was possible to evaluate the options and find a compromise between the water uses or users. However, it is a method that should only be applied when there are few uses (3 at most) due to the physical limitation in visualizing multidimensional graphic representations.

The weighting method enabled the analysis of a greater number of water uses. However, the weighting coefficients should be established between the competing uses, which introduce a certain degree of subjectivity into the analysis.

In the stochastic analysis, it was concluded that for the case studied, the number of synthetic series does not significantly influence the form of the probability distributions of the energy generated by the optimization model. However, comparing the results of the model obtained using the synthetic series and the historical series, it was observed that the model was sensitive to the synthetic series, particularly when the extremes of the probability distributions of energy were analyzed.

Comparing the Pareto curves defined based on the hydrological scenarios derived from the historical series and those defined based on the generation of synthetic series, the curves essentially coincided in the dry scenario. For the median and wet scenarios, there was a tendency to underestimate the curves obtained using the synthetic series compared with those based on the historical series.

8. References

Barros, M.T.L. (1989). *Otimização estocástica implícita de um sistema de reservatórios considerando múltiplos objetivos.* Doctoral thesis - University of São Paulo, São Paulo, Brazil (in Portuguese)

Barros, M.T.L. et al. (2003). Optimization of large-lcale hydropower systems operations. *Journal of Water Resources Planning and Management,* Vol.129, No.3, (May 2003) pp. (178-188)

Braga, B.P.F. et al. (1991) Stochastic optimization of multiple-reservoir-system operation. *Journal of Water Resources Planning and Management,* Vol.117, No.4, (Jul. 1991), pp. (471-481)

Braga, B.P.F. et al. (1998) Siscom: sistema computadorizado de apoio ao planejamento e operação de sistemas hidroelétricos. *Revista Brasileira de Recursos Hídricos,* Vol.3, No.4, pp. (89-101), (in Portuguese).

CEPEL - Centro de Pesquisa de Energia Elétrica. (2011). Manual de referência: modelo NEWAVE. Feb. 11th 2011. Available from http://www.cose.fee.unicamp.br (in Portuguese)

Francato, A.L. & Barbosa, P.S.F. (1997). Fatores determinantes das propriedades operativas ótimas de um sistema hidroelétrico. *Proceedings of 12th Brazilian Symposium on Water Resources*, Vitória, Espírito Santo, Brazil, November, 1997 (in Portuguese)

Georgakakos, A.P. & Yao, H.; Yu, Y. (1997). A control model for dependable hydropower capacity optimization. *Water Resources Research*, Vol.33, No.10, (Oct. 1997), pp. (2349-2365)

Kelman,J. et al. (1990). Sampling stochastic dynamic programming applied to reservoir operation. *Water Resources Research*, Vol.26, No.3, (Mar. 1990), pp. (447-454)

Labadie, J. (1998). Reservoir system optimization models. *Water Resources Update Journal*, Issue107, pp. (83-110)

Labadie, J. (2004). Optimal Operation of Multireservoir Systems: State-of-the-ArtReview. *Journal of Water Resources Planning and Management*, Vol.130, No.2, pp. (93-111)

Lopes, J.E.G. (2001). *Otimização de sistemas hidroenergéticos*. Master thesis, University of São Paulo, São Paulo, Brazil (in Portuguese)

Lopes, J.E.G. (2007). Modelo de planejamento da operação de sistemas hidrotérmicos de produção de energia elétrica. Doctoral thesis, University of São Paulo, São Paulo, Brazil (in Portuguese)

Lund, J.R. & Ferreira, I. (1996). Operating rule optimization for Missouri river reservoir system. *Journal of Water Resources Planning and Management*, Vol.122, No.4, (Jul./Aug. 1996), pp. (287-295)

Murtagh, B.A. & Saunders, M.A. (1995). *Minos 5.4 user's guide*. Systems Optimization Laboratory, Stanford University, Palo Alto, California, USA

Oliveira, R. & Loucks, D.P. (1997). Operating rules for multireservoir systems. *Water Resources Research*, Vol.33, No.4, (Apr. 1997), pp. (839-852)

Peng, C-S. & Buras, N. (2000). Dynamic operation of a surface water resources system. *Water Resources Research*, Vol. 36, No. 9, (Sep. 2000), pp. (2701-2709)

Pereira, M.; Campodónico, N. & Kelman, R. (1998). Long term hydro scheduling based on stochastic models, *Proceedings of International conference on electrical power systems operation and management - EPSOM'98*. Zurich, Switzerland, 1998

Philbrick Jr., C.R. & Kitanidis, P.K. (1999). Limitations of deterministic optimization applied to reservoir operations. *Journal of Water Resources Planning and Management*, Vol.125, No.3, (May/Jun. 1999), pp. (1 35-142)

Ponnambalam, K. & Adams, B.J. (1996). Stochastic optimization of multireservoir systems using a heuristic algorithm: case study from India. *Water Resources Research*, Vol.32, No.3, (Mar. 1996), pp. (733-741)

Salas, J.D. (1993). Analysis and modeling of hydrologic time series, In: *Handbook of hydrology*, D.R. Maidment, pp. (19.1-19.72), McGraw-Hill Inc., ISBN 0070397325, New York, USA

Tejada-Guilbert, J.A.; Stedinger, J.R. & Staschus, K. (1990). Optimization of value of cvp's hydropower production. *Journal of Water Resources Planning and Management*, Vol.116, No.1, (Jan. 1990), pp. (52-70)

Yeh, W.W-G. (1985). Reservoir management and operations models: a state-of-the-art review. *Water Resources Research*, Vol.21, No.12, (Dec. 1985), pp.(1797-1818)

Zambon, R.C. (2008). Planejamento da operação de sistemas hidrotérmicos de grande porte. Doctoral thesis, University of São Paulo, São Paulo, Brazil (in Portuguese)

10

Hydropower Scheduling in Large Scale Power Systems

Monica Zambelli, Ivette Luna Huamani, Secundino Soares, Makoto Kadowaki and Takaaki Ohishi
University of Campinas, Brazil

1. Introduction

Hydropower is the most important and widely-used renewable source of energy. Nearly one-fifth of the world's energy each year is supplied by hydroelectric power generation, which is more than solar, wind, biomass and all other renewable sources combined.

Brazil is the third largest producer of hydroelectricity in the world, preceded only by China and Canada. (Source: www.eia.gov). In 2009, hydropower accounted for 87 percent of Brazilian electric power generation, with smaller amounts coming from conventional thermal, nuclear, and other renewable sources. But managing a power system with over 110 GW of installed capacity, most of it coming from around 150 hydro plants, is a daunting task.

Hydro plants are located in 8 River Basins with specific hydrologic characteristics. Many of them are capable of storing water on reservoirs that can be used to regulate the river stream flow throughout the year, others are run-of-river plants and are subjected to seasonal river flows. Reservoir operations at a hydro plant affect the whole cascade downstream and the benefits of holding water for future use are not easy to estimate.

To illustrate these characteristics let us observe Itaipu Hydro Plant, with 14 GW of installed capacity, the hydro plant with the greatest generation in the world, located downstream from the Paraná River Basin on the frontier with Paraguay. Fig. 1 shows the average and standard deviation of its monthly inflows.

The seasonal behavior of the stream flow in the Parana River Basin is easily observed in Fig.1. The dry season goes from May to October, with average inflows around 6,000 m^3/s in August. The wet season goes from November to April, with average inflows around 16,000 m^3/s in February. Inflow variability is much higher during the wet season, as indicate the standard deviation values which are near 5,000 m^3/s in February and around 2,000 m^3/s in August.

Seasonal fluctuations are common in stream flow data and being able to smooth it and hedge unexpected events such as droughts and floods is a major concern in reservoir operation problems. Consumption of potable water, water usage for industrial and irrigation purpose, and also river flow control for navigation are some of the issues that constraint reservoir operation for hydropower generation.

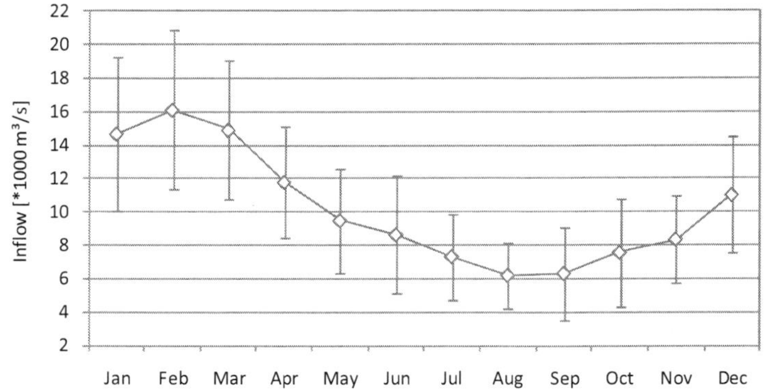

Fig. 1. Average and standard deviation of monthly inflows for the Itaipú hydro plant.

Upstream Itaipú, there are 57 hydro plants located on the several tributaries of the Parana River. The coordination of reservoir operation is extremely important since water is a limited resource and global hydropower generation can be increased if plants are dispatched concerning the whole river basin.

Moreover, stream flow profiles generally differ among River Basins according to the geographic region where they are located. For example, Fig.2 presents the average and standard deviation of monthly inflows for the hydro plants of Santo Antônio, in the Madeira River and Salto Caxias, in the Iguaçu River. The former is located in the northern region (N) of Brazil, flowing across the Amazon rain forest whereas the latter is located in the southern region (S).

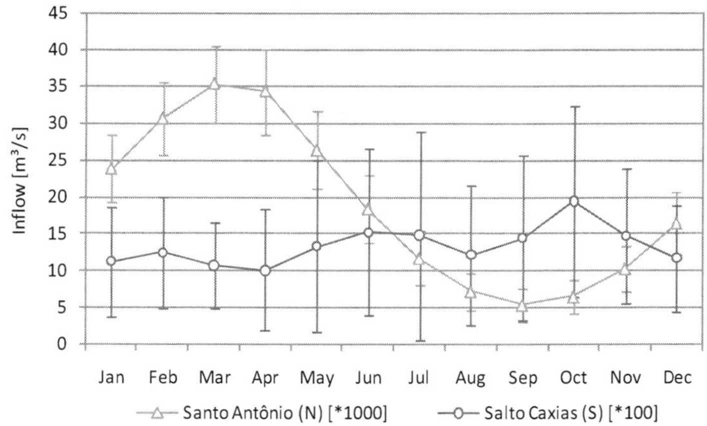

Fig. 2. Monthly inflow profile for two hydro plants located in the Northern (N) and Southern (S) regions of Brazil.

It is possible to observe that the seasonal profile is clearly defined for the northern river with very high inflows (10x greater than that of the southern river) and distinct wet and dry seasons. The minimum average value, in September, is 6.5 times greater than the maximum average inflow, in March. However for Salto Caxias, in the southern region, the inflows do not follow the same pattern. Average values range from 1,010 m³/s to 1,946 m³/s and standard deviations are high (over 55% of average) throughout the whole year, reaching 95% of average in the month of July.

Coordinated hydropower scheduling can take advantage from different seasoning so that power generation among regions can be complementary, if the transmission network supports the resulting power flow.

At this level, the basic question is how much water should be used to generate hydropower at the present and how much should be stored for future use. This decision is difficult due to the uncertainty of coming inflows. If the decision is to use increase use at present and the coming inflows are lower than expected, the future costs from thermal generation will be high. Otherwise, if the decision is to store more to the future and the coming inflows are high, there is risk of spillage, which means a waste of power resources. Stream flow forecasting is thus a fundamental issue to support system's operation, increasing benefits and reducing overall costs.

Above all this, one characteristic of hydropower systems deserves special attention: the conversion of water potential energy into electricity is a nonlinear phenomenon, typically expressed as a function of the hydraulic head and the rate of water flow. Nonlinear models should be used to estimate power plants' operation more precisely, especially when the planning horizon is straightened, and real time operation approaches.

Concerning the daily operation, the inflow variation is not an issue whereas meeting the load demand and attending the large set of power systems operating constraints turn it into a very complex scheduling problem. Consumers requirements must be met instant-to-instant because, at this level, the electric system is basically a huge electric circuit with alternating current. Besides the electric circuit laws, the system operation must consider many other aspects, such as forced outage (contingency), stability, voltage collapse, security constraints, multiuse of water and environmental requirements.

The total consumption varies instant-to-instant, but it presents a cyclic pattern as can be seen at Fig.3, which shows a week load, with one hour discretization. Every day there is a maximum load around 19:00h and a minimum load at day-break that reaches about 60% of the peak load.

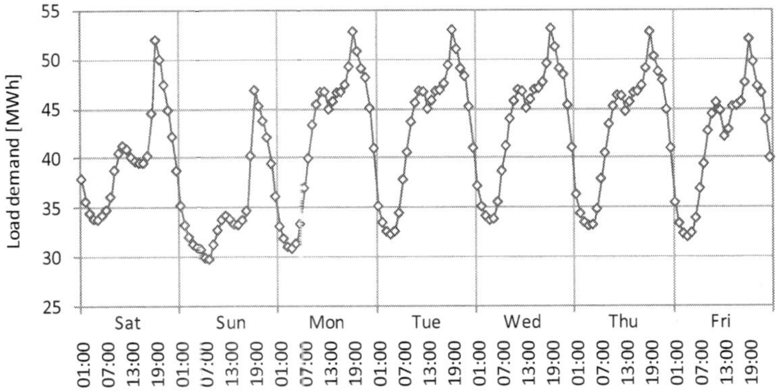

Fig. 3. Load demand curve for the Brazilian power system in a typical week.

In hydro dominant power systems the load tracking should be accomplished by hydro plants, which requires their generating units to be powered on and off along the day. The generating units in operation have great influence on turbine-generator efficiency so that deciding the commitment of hydro generating units constitutes a complex combinatorial optimization problem.

The government plays a substantial role in the management of the Brazilian electricity sector (BES). The operation planning of the Brazilian Interconnected System (BIS) is coordinated by an independent system operator (ISO) in accordance to operating rules agreed by the electric sector agents and supervised by a regulatory agency, within a tight pool market arrangement.

These rules aim to establish a coordinated hydrothermal scheduling (HS) for the whole system. The HS problem consists on the determination of a reliable and cost effective power plant dispatching that meets the load demands of the system at every time stage. This involves using the available generation resources optimally to reduce operating costs without compromising the security of the system operation in the future.

In practice, this task is accomplished by a set of optimization and/or simulation models, used to guide the HS of the BIS (Maceira, 2002). The models are connected in a chain that provides operating strategies for each instance of the HS problem, according to the time horizon it covers. These are:

a. Long-Term Hydrothermal Scheduling (LTHS)
b. Short-Term Hydrothermal Scheduling (STHS)

The LTHS problem consists on defining the generation scheduling decisions on a monthly basis for a planning period of several years (five years in the Brazilian case) that supply the load demands at minimum operating costs. The primary concern of this problem relies on uncertainties associated with the long planning horizon, especially regarding inflows. Some simplification in hydropower plants modeling can be acceptable and average values for turbine-generator efficiency can be taken for hydropower generation calculation instead of its nonlinear hill curve function, as it is considered in the Brazilian case.

The LTHS results are used to guide the operation on a short term basis. The monthly power generation decisions are disaggregated into weekly generation and the first week generation is established as a target to the STHS model which will provide a more detailed unit dispatch.

STHS aims to determine the thermal and hydro power outputs and the number of generating units dispatched at each hydro plant to meet the system load demand for each hour of the week ahead. The STHS is designed to optimize some performance criterion but also accomplish the average generation targets established by the LTHS, and respect the operational constraints of the power plants. Hydrologic aspects can be estimated more precisely so that the challenge at this step lies on the representation of systems nonlinearities and the attendance of the operating bounds, such as generation limits, ramp and spinning reserve limits, hydraulic system operation requirements, and transmission system capacities.

This chapter is organized in six sections. Sections 2 and 3 describe the LTHS and the STHS problems respectively, with some discussion on modeling and solution techniques. Section 4 presents the inflow forecasting models applied to the HS. Section 5 illustrates the power systems scheduling concepts in a case study comprising the Brazilian large scale power system. Finally, section 6 states the main conclusions of this chapter.

2. Long term hydrothermal scheduling

Long term hydrothermal scheduling (LTHS) for a multireservoir system consists on a quite complex optimization problem due to issues such as the long planning horizon to be analyzed (several years), the time dependence of decisions, the coupling of hydro plants in the same river basin, and the nonlinear relations involved in the hydro power generation functions and thermal costs. Above all this, the major concern in hydrothermal scheduling is the stochastic nature of inflows. Various approaches have been proposed to solve the LTHS problem and they can all be classified as either stochastic or deterministic according to the modelling of inflows (Labadie, 2004).

Stochastic approaches usually consider the uncertainty of water inflows on the basis of probability distribution functions and most of the applications use stochastic dynamic programming (SDP) as optimization tool (Stedinger et.al., 1984). SDP has been the most commonly used technique for the solution of the LTSH. Among its advantages there is the possibility to explicitly model the uncertainty of inflows and to represent important nonlinear relations inherent to the problem. However, for multireservoir systems, it requires some kind of simplification due to the intense computational requirements (Bellman, 1957).

One way of overcoming this problem is by aggregating multiple reservoirs to form a composite reservoir of energy (Arvanitidis and Rosing, 1970; Turgeon, 1980; Valdes et.al., 1995), and/or by piecewise linear approximation of nonlinear functions (Diniz et.al., 2008) (Dias et.al., 2010). This is the case of the methodology currently used in Brazil which is based on stochastic dual dynamic programming (SDDP) (Pereira and Pinto, 1991), using Benders decomposition (Pereira and Pinto, 1985) with nonlinearities in the power generation and future cost functions modelled as piecewise linear. This approach should be used carefully in large scale systems since it assumes a high correlation between inflows of the hydro plants within the composite reservoir which is not easy to observe in practice.

Deterministic approaches, on the other hand, take into consideration specific hydrological scenarios and provide solutions for individual plants. The stochastic aspects of the problem can thus be implicitly handled by the selection of such inflow scenarios and by the analysis of the optimal deterministic solutions associated with each one of them (Dembo, 1991) (Escudero et.al., 1996). The advantage of this approach, also known as implicit stochastic optimization or Monte Carlo optimization, is that nonlinear models can be directly applied even to large scale hydropower systems. The primary disadvantage is that operational policies are only optimal for the assumed hydrologic time series and deriving a general optimal operation rule may not be straightforward, but a more detailed model describing the operation for individual hydro plants is possible even for very large scale power systems.

Therefore, considering that this chapter is focused on large scale hydropower systems, the LTHS will be defined based on deterministic optimization in a framework of scenarios. The model named Optimal Dispatch for the Interconnected Brazilian National system (ODIN) is based on adaptive model predictive control (MPC) (Camacho and Bordons, 2004), an optimization framework widely applied in real-time control and industrial processes which can provide high quality suboptimal solutions for the LTHS problem with acceptable computational effort.

2.1 Problem formulation

In the LTHS optimization problem the primary goal is to supply the total load demand at minimum expected operating costs for a planning period of several years.

In a hydro dominant power system this source is used primarily for the load attendance but costs associated with hydroelectric power generation are considered to be negligible in relation to those of thermal generation. Therefore the operation can be evaluated by the thermal generation z using the cost function ψ_j which represents generation fuel costs associated to non-hydraulic sources j dispatched additionally to attend the load demand. Costs related to importing electricity from neighboring markets and energy deficits can be modeled in a similar way.

The objective function can thus be written as in Eq.1, and aims to define the hydro plants' releases q that minimize the expected costs of operation with respect to the inflows y, for a planning period with T stages. An interest rate λ_t is applied to calculate the present value of the monthly operating costs. J is the number of thermal plants in the system.

$$\min_{q} \underset{y}{\mathcal{E}} \sum_{t=1}^{T} \left\{ \lambda_t . \sum_{j=1}^{J} \psi_j(z_{j,t}) \right\} \tag{1}$$

The load demand should be attained by the power sources available in the system as stated in Eq.2, where z and p are total thermal and hydro power generation, respectively. G is the power generation of small generation companies not explicitly controlled by the ISO, including hydro plants with less than 30MW of installed capacity and alternative energy sources, such as wind, solar and biomass.

$$z_t + p_t + G_t = D_t \qquad \forall t \quad (2)$$

Total power generation provided by thermal sources in a stage t is given by the sum of individual plants j constrained by their limits for operation, as stated in Eq. 3 and 4.

$$z_t = \sum_{j=1}^{J} z_{j,t} \qquad \forall t \quad (3)$$

$$Z_{j,t}^{\min} \leq z_{j,t} \leq Z_{j,t}^{\max} \qquad \forall j,t \quad (4)$$

In Eq.4 the minimum thermal generation at a stage t is defined by operational limits or imposed by contracts with fuel suppliers. The upper limit is determined by the generation capacity of the plant, which is the installed capacity discounting maintenance and unexpected outage factors.

Total hydro power generation is calculated in Eq.5 as the sum of the energy provided by each individual plant i in a set of I hydro plants during a stage t.

$$p_t = \sum_{i=1}^{I} p_{i,t} \qquad \forall t \ (5)$$

Hydro power generation of a single plant i is a nonlinear function of the water head $h_{i,t}$ and the discharge $q_{i,t}$, as expressed in Eq.5, where k is a constant factor representing the product of water density, gravity acceleration and a conversion factor that gives the energy production in MW. η_i is the turbine/generator efficiency that depends on water head and discharge. Although very important in STHS modelling, for LTHS the average efficiency can be considered for the sake of simplicity.

$$p_{i,t} = k.\eta_i.h_{i,t}.q_{i,t} \qquad \forall i,t \ (6)$$

The water head, in turn, is a nonlinear function of average reservoir storage x^{avg}, water discharged through the turbines q, and total water released from the reservoir u, as expressed in Eq. 7.

$$h_{i,t} = h_{Fi}(x_{i,t}^{avg}) - h_{Ti}(u_{i,t}) - h_{Li}(q_{i,t}) \qquad \forall i,t \ (7)$$

The forebay $h_F(x^{anv})$ and tailrace $h_T(u)$ elevations are calculated by 4th degree polynomial functions and the penstock head loss $h_L(q)$ is determined by a quadratic function of the discharge, but a percentage of the nominal water head or a constant can be used alternatively.

The system dynamics is stated in Eq.8 which describes the water balance in the hydro plants' reservoir.

$$x_{i,t} = x_{i,t-1} + \left(y_{i,t} + \sum_{k \in \Omega_i} u_{k,t} - u_{i,t} - ev_{i,t} - U_C \right) .\gamma_t \qquad \forall i,t \ (8)$$

The reservoir storage $x_{i,t}$, of hydro plant i at the end of stage t, is thus given by the sum of the storage at the previous stage plus the total water flow received by the plant during that stage, converted to storage unit by a factor γ_t. The total inflow is determined by the sum of incremental average inflow $y_{i,t}$ plus the total water released from the set of plants Ω_i located immediately above hydro plant i in the same river basin, minus the water released from plant i. Evaporation $ev_{i,t}$ is represented as a nonlinear function of the reservoir storage, and amounts of water taken from the reservoir for alternative purposes U_C are also considered.

The average reservoir storage used in the forebay elevation function (Eq. 7) is thus stated in Eq.9.

$$x_{i,t}^{avg} = \frac{x_{i,t-1} + x_{i,t}}{2} \qquad \forall i,t \ (9)$$

Total water release u is composed by the sum of water discharge q through the turbines plus the water spilled v over the spillways.

$$u_{i,t} = q_{i,t} + v_{i,t} \qquad \forall i,t \ (10)$$

Operating constraints are expressed by Eq. 11 to 14

$$X_{i,t}^{\min} \leq x_{i,t} \leq X_{i,t}^{\max} \qquad \forall i,t \quad (11)$$

$$u_{i,t} \geq U_{i,t}^{\min} \qquad \forall i,t \quad (12)$$

$$q_{i,t} \leq q_{i,t}^{\max}(h_{i,t}) \qquad \forall i,t \quad (13)$$

$$v_{i,t} \geq 0 \qquad \forall i,t \quad (14)$$

Lower and upper bounds on variables are imposed by the physical operational limits of the hydro plants, as well as the limitations associated with multiple uses of water. For example, the lower bound for reservoir and release can vary over time to allow navigation, water supply, irrigation and recreation. The upper bound for reservoir can be imposed for purpose of dam safety and flood control. The upper limit for the discharge in Eq. 13 is also a nonlinear function of water head.

2.2 Solution technique: model predictive control

The MPC approach corresponds to an operational policy for LTHS problems based on an open loop feedback control framework. The stochastic aspects of the problem are implicitly handled by the use of expected inflow values and an accurate representation of the generating plants' operational characteristics is possible since optimal dispatch for individualized plants at each stage is obtained from a deterministic nonlinear optimization model.

The decision-making process runs under a simulation model in which the discharge decisions are implemented. This means that for each stage of the simulation procedure, the forecasting and optimization models should be executed over an optimization horizon in order to obtain the discharge decisions to be implemented for the first stage of each optimization.

The feedback control scheme is assured since for each stage the optimization model updates the discharge decisions as a consequence of the new inflow forecasting sequence and of the new initial reservoir storage resulting from the previously simulated water balance.

Previous tests with this approach, focusing specifically on the uncertainty of inflows, evaluated the results for single reservoir systems where dimension is not an issue (Martinez and Soares, 2002; Zambelli et.al. 2009). The approach has shown results equivalent to those of standard methods based on stochastic dynamic programming.

An outline of the MPC operational policy for the LTHS problem is shown in Fig. 4 where, for a given stage t of the simulation horizon, the hydro system is observed and the reservoir storage levels x_{t-1} are taken as the initial condition for the deterministic optimization model that must solve the LTHS problem for a given optimization horizon T^*.

The optimization regards a series of predicted values for the unknown parameters to be considered, in this case, water inflows, determined by the Predictor module, based on past observed values.

The Optimizer module then provides optimal releases for each hydro plant for the optimization horizon but only the discharge decision for the first stage q^*_t is selected and submitted to a simulation model. The latter calculates the consequences of such decision in terms of storage and generation considering the inflow being simulated and makes the necessary corrections, if needed, according to formulation (1)-(14). Corrections are frequently needed due to differences between the predicted inflow series $\bar{y}$ and the simulated inflow series y.

In the following stage $t+1$, the storage level of the reservoirs resulting from simulation is observed, and a new forecasting of inflows is provided based on the latest available information. This procedure of "forecast-optimize-update" is repeated until the end of the planning horizon T.

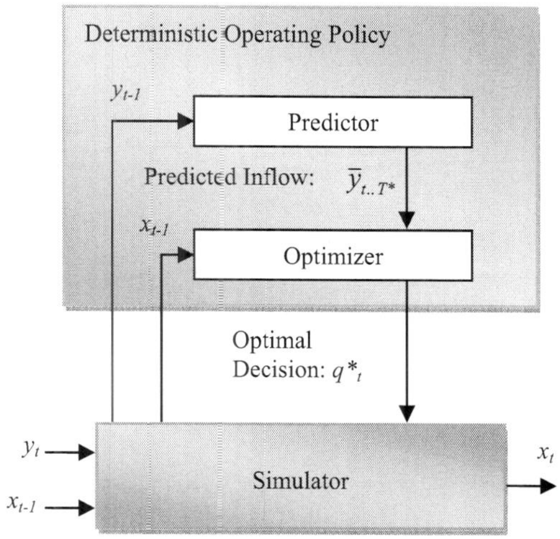

Fig. 4. Model predictive control scheme.

One important aspect affecting the performance of the MPC approach is the boundary conditions of the optimization model in terms of the final storage of the reservoirs and the optimization horizon. Results from the optimization model over the whole period of historical inflow records assuming perfect foresight of inflows indicate that the reservoir storages at the beginning of each dry season are almost always full. With this in mind, the optimization horizon implemented by the MPC approach reported in this paper adopts a rolling horizon of at least 13 months and at most 24 months ahead, depending on the current monthly stage. Full reservoir storage is imposed as a boundary condition at the end of this horizon, adjusted to match the month of April as it is the end of the wet season in almost all the Brazilian river basins. These parameters were estimated based on successive simulation tests with various considerations and have shown to maximize the approach performance.

In the MPC approach the deterministic nonlinear optimization model (1)-(14) can be solved using specialized optimization techniques such as network flow algorithms with capacitated arcs (Oliveira and Soares, 1995) or interior point methods (Azevedo, et. al. 2009). The former was used in the case study presented in section 5. The thermal part of the problem (3)-(4) is determined afterwards by an economic dispatch algorithm (El-Hawary and Christensen, 1979).

3. Short term hydrothermal scheduling

Short-term hydrothermal scheduling (STHS) considers a planning period of one week on an hourly basis. At this step, the two major decisions are the start-up and shutdown of generation units and the generation levels of online units at each time interval. The first decision is called Unit Commitment (UC) and the second is called Generation Schedule (GS). In a hydro-dominant power system, the start-up and shutdown of generating units are concentrated on hydroelectric plants since they should provide the load tracking in order to keep the thermoelectric units operating on a flat dispatch.

The approach presented in this chapter for the STHS is based on optimization models and heuristic procedures and considers some specific characteristics of hydro-dominant systems that make it different from the approaches designed for thermal-dominant power systems, either in terms of performance criterion and problem constraints. In this section the performance criterion adopted is presented as a composition of hydro efficiency loss functions and start-up/shutdown costs of the generating units.

3.1 Performance criterion

3.1.1 Efficiency loss functions for hydroelectric plants

One important characteristic of the operation planning chain for hydro-dominant power systems is that the LTHS establishes the generation target for each hydro plant during the next week. Therefore, an adequate objective function for the STHS should be to generate this target using the lowest possible amount of water. Thus the optimization criterion adopted in this chapter considers the hydro generation efficiency through loss functions expressed in MW (Soares and Salmazo, 1977).

The goal is to represent the generation loss at each hydro plant as a function of its power output. For a given forebay elevation, the increase in generation is accomplished by increasing discharge at each generating unit in operation. This implies on variations of tailrace elevation, penstock head loss and turbine-generator efficiency. The following analysis presents in details the aspects that influence the generation efficiency of a hydro generating unit.

By efficiency of a hydro generating unit it is meant the ratio between power output and discharge input. In mathematical terms, from Eq. 6 and 7, the production function of a hydro plant is given by Eq. 15.

$$p = k.\eta.\{h_F(x) - h_T(u) - h_L(q)\}.q \tag{15}$$

where the indexes were dropped out for the sake of simplicity. The generation efficiency of a hydro plant, given by the ratio between power output and discharge input, also called plant productivity, depends on the turbine-generator efficiency η, the tailrace elevation h_T, and the penstock head loss h_L, and consequently depends on storage x, discharge q and release u.

Fig. 5 shows a typical turbine-generator efficiency curve, also called *hill curve,* of a generating unit. As can be seen, the turbine-generator efficiency depends on net water head and discharge.

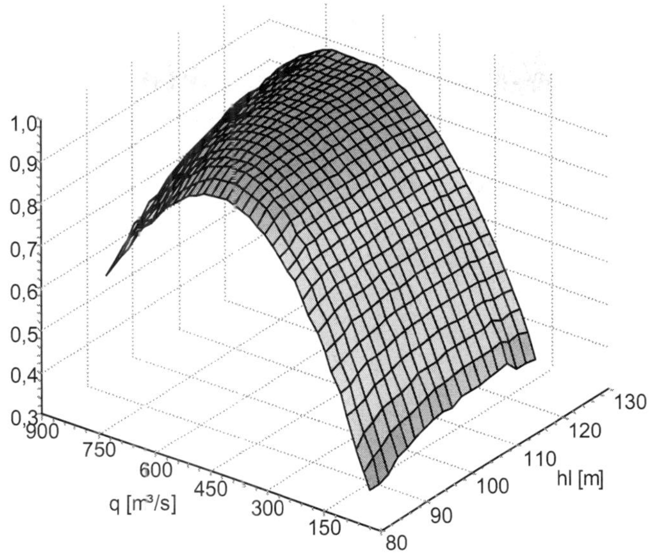

Fig. 5. Turbine efficiency curve (hill curve) of a hydro generating unit.

The hydro efficiency will be measured by the power loss in generation due to variations on turbine-generator efficiency and net water head. These variations will be calculated in MW in order to be compared on the same basis (Soares and Salmazo, 1997). The hypothesis of identical generating units is assumed, which is true for almost all hydro plants. Thus, for a given number of generating units in operation, the power loss in generation due to variations on tailrace elevation can be calculated as Eq. 16.

$$p_T = k.\eta.\left\{ h_T(u) - h_T^{ref} \right\}.q \tag{16}$$

where h_T^{ref} is the tailrace elevation assumed as a reference. In a similar way, the power loss in generation due to variations on penstock head loss can be computed as Eq. 17.

$$p_L = k.\eta.\left\{ h_L(q) - h_L^{ref} \right\}.q \tag{17}$$

where h_L^{ref} is the penstock head loss assumed as a reference. Finally, the power loss in generation due to variations on turbine-generator efficiency can be calculated as Eq. 18, where η^{ref} is the turbine-generator efficiency assumed as a reference.

$$p_\eta = k.\left\{ \eta(q) - \eta^{ref} \right\}\left\{ h_F(x) - h_T(u) - h_L(q) \right\}.q \tag{18}$$

Fig. 6 shows the loss curves for a given hydro plant with a total of 4 generating units and 1192 MW of installed capacity. Part A details each one of the losses in generation and the total loss with 1 unit on. Part B presents the total loss curves for one up to four generating units.

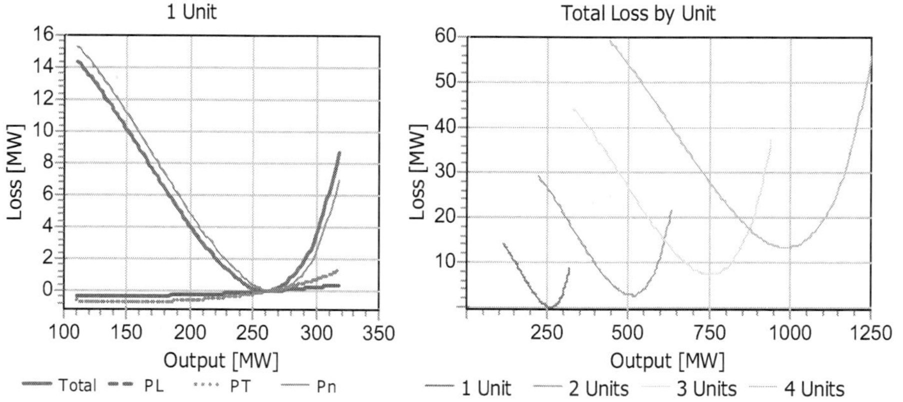

(A) (B)

Fig. 6. Loss curves for a hydro plant with A) one and B) several generating units.

In Fig. 6 (A) the three types of generation loss are depicted as well as the total one. It can be noted that the minimum total loss occurs at around 250 MW of output, which is the point adopted as a reference for Eq. (16)-(18). Therefore, at this point the total loss is minimal by construction, and corresponds to the maximal productivity of the hydro plant. It is interesting to observe that the best point in terms of turbine-generator efficiency is around 260 MW whereas the best point in terms of productivity is around 250 MW. This difference is due to the impact of tailrace elevation and penstock head loss over plant's productivity.

In Fig. 6 (B) it is possible to notice that a certain desired output can be provided by multiple unit dispatches, thus the minimum loss criterion helps to define the optimal solution.

3.1.2 Start-up and shutdown costs in hydroelectric plants

Load demand presents significant variations during each day and week. As load increases a larger number of generating units should be committed to power generation and the opposite occurs when the load decreases, once generation should track the load. These frequent start-up and shutdown of generating units should be avoided as much as possible since they represent an increase on tear and wear of the units, reducing their operational life.

For dominant thermal systems the load tracking is most performed by thermal plants and the corresponding start-up and shutdown costs of such units as well as operational constraints such as minimum down and up times of units, rump up and down limits must be considered. For hydro dominant systems, on the other hand, the load tracking can be exclusively performed by hydro plants whose operational constraints and start-up and shutdown costs are less restrictive than for thermal ones. Nilsson e Sjelvgren (1997) estimate that in Sweden the start-up and shutdown costs of hydro generating units are around 3 US$ per MW. This value was adopted in the case study presented in this chapter.

3.2 Problem formulation

The STHS is formulated as a mixed integer nonlinear programming problem whose main goal is to minimize operating costs along a certain planning horizon T on an hourly basis.

The objective function includes the start-up and shutdown costs c_i^s of hydro generation units, the generation loss of hydro plants f_i and the generation costs of thermal plants f_j, as expressed in Eq. 19.

$$Min \sum_{t=1}^{T} \left\{ \sum_{i \in I} \left[c_i^s \left| n_{i,t} - n_{i,t-1} \right| + c^{MW} f_i(n_{i,t}, p_{i,t}) \right] + \sum_{j \in J} f_j(z_{j,t}) \right\} \quad (19)$$

where $n_{i,t}$ is the number of generating units on operation at hydro plant i and stage t; z and p are thermal and hydro power generation, respectively; and c^{MW} is the energy price.

The load demand D should be attained at each stage by the summation of hydro and thermal generation, as in Eq. 20.

$$\sum_{i \in I} p_{i,t} + \sum_{j \in J} z_{j,t} = D_t \qquad \forall t \quad (20)$$

Eq. 21 imposes that each hydro plant should meet the generation targets m_i, provided by the LTHS.

$$\sum_{t=1}^{T} p_{i,t} = m_i \qquad \forall i \quad (21)$$

Spinning reserve requirements are established by eq. 22 where $r_{k,t}$ is the spinning reserve constraint k at stage t for the set of plants $R_{k,t}$, where this constraint applies, and the set of stages T_r, when it holds.

$$\sum_{i \in R_{k,t}} (p_i^{\max} - p_{i,t}) \geq r_{k,t} \qquad t \in T_r; k = 1..n_{r,t} \quad (22)$$

Eq. 23 defines the ramp limits where $s_{k,t}$ is the maximal ramp at constraint k and stage t for the set of plants $S_{k,t}$, where this constraint applies, and the set of stages T_s, when it holds.

$$\left| \sum_{i \in S_{k,t}} (p_{i,t} - p_{i,t-1}) \right| \leq s_{k,t} \qquad t \in T_s; k = 1..n_{s,t} \quad (23)$$

Eq. 24 and 25 represent the limits on hydro and thermal generation, respectively. The hydro generation limits vary according to the number of committed generating units.

$$p_i^{\min}(n_{i,t}) \leq p_{i,t} \leq p_i^{\max}(n_{i,t}) \qquad \forall i,t \quad (24)$$

$$Z_j^{\min} \leq z_{j,t} \leq Z_j^{\max} \qquad \forall j,t \quad (25)$$

The limits on the number of generating units associated with a given hydro generation can be defined as in Eq. 26.

$$n_i^{\min}(p_{i,t}) \le n_{i,t} \le n_i^{\max}(p_{i,t}) \qquad \forall i,t \quad (26)$$

The water balance at each reservoir is expressed by Eq.27 where the reservoir storage $x_{i,t}$, of hydro plant i at the end of stage t, is given by the sum of the storage at the previous stage plus the total water flow received by the plant during that stage, converted to storage unit by a factor γ_t. The total inflow is determined by the sum of incremental average inflow $y_{i,t}$ minus the water released from plant i ($u_{i,t}$), plus the total water released from the set of plants Ω_i located immediately above hydro plant i in the same river basin, considering the number of stages $\theta_{m,i}$ for water displacement between plants m and i.

$$x_{i,t} = x_{i,t-1} + \left(y_{i,t} + \sum_{m \in \Omega_i} \left(u_{m,t-\theta_{mi}} \right) - u_{i,t} \right) \cdot \gamma_t \qquad \forall i,t \quad (27)$$

Constraints in Eq. 28 and 29 establish limits for storage and discharge, respectively.

$$X_{i,t}^{\min} \le x_{i,t} \le X_{i,t}^{\max} \qquad \forall i,t \quad (28)$$

$$q_i^{\min} \le q_{i,t} \le q_i^{\max} \qquad \forall i,t \quad (29)$$

Finally, Eq. 30 defines the initial reservoir storages and number of generating units and Eq. 31 imposes that the number of generating units available at the hydro plants is an integer.

$$n_{i,0}; x_{i,0} \quad given \qquad \forall i \quad (30)$$

$$n_{i,t} \in N \qquad \forall i,t \quad (31)$$

3.3 Solution technique

The detail representation of the generating units in the operation of the hydro plants, which requires the consideration of integer and continuous variables, turns the STHS into a mixed integer nonlinear optimization problem whose solution is quite difficult for large scale systems such as the Brazilian one. In order to overcome this difficulty, a solution technique based on optimization-simulation decomposition is proposed (Kadowaki et. al., 2009). This decomposition is motivated by the fact that most of the hydraulic constraints (28) and (29) are not active at the optimal solution. This suggests a relaxation approach by which the hydraulic constraints (27), (28) and (29) are relaxed during the optimization phase, performed by the optimization model, but are considered during the simulation phase, performed by the simulation model.

The optimization model determines the number of generating units in operation at each hydro plant and stage and their respective generation schedule, as well as the generation at each thermal plant and stage that attains the load demand, the generation targets of the power plants and the limits on ramp, spinning reserve, generation and number of available generation units. After solving the optimization model, the solution obtained ($p_{i,t}$*, $n_{i,t}$*) is simulated at the simulation model in order to identify violations on the relaxed constraints. If violations are identified new constraints are included in the optimization model optimization in order to eliminate them, and this procedure is repeated until all the hydraulic constraints are satisfied.

Figure 7 shows the iterative procedure that implements the optimization-simulation decomposition approach.

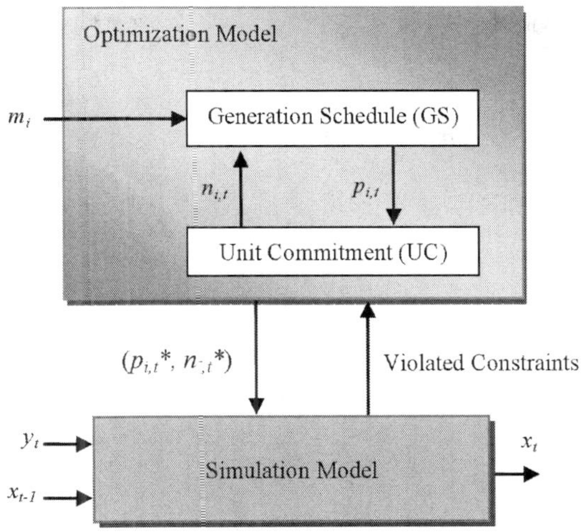

Fig. 7. Relaxation approach for solving the STHS problem.

The optimization model, by its turn, is also decomposed into two sub-problems: the UC sub-problem, that determines the number of generating units in operation at each hydro plant and stage for a given power generation, and the GS sub-problem, that determines the generation at each hydro plant and stage for a given number of generating units in operation. These sub-problems are solved iteratively until convergence is achieved when no more changes occur in the solution. The UC sub-problem is an integer nonlinear optimization problem for each hydro plant and can be efficiently solved by dynamic programming. The GS sub-problem, by its turn, is a continuous nonlinear programming problem that can be efficiently solved by Newton method, on a relaxation framework with respect to the inequality constraints.

4. Inflow forecasting

Forecasting of river inflows is an important input for planning and scheduling of hydroelectric power system, converting the cleaner energy stored in water reservations into electric energy and planning and managing water resources as effectively as possible (Al-Zùbi, et. al., 2010).

In spite of the economic and environmental aspects involved the fact that most of the electricity in Brazil comes from hydroelectric power plants makes the development of accurate inflow forecasting models essential.

These models need to be able of dealing with the dynamics, uncertainty and nonlinearities inherent to natural phenomena. The necessity of models with the capability of processing uncertainties is reinforced by the presence of missing data and the production of wrong values due to technical or human error.

In that sense, several researches have been devoted to the formulation and development of accurate inflow forecasting models. In general, the river flow models are assigned to one out of three broad categories: deterministic, conceptual or parametric.

Deterministic models describe the rainfall-runoff process using physical laws of mass and energy transfer. Conceptual models provide simplified representations of key hydrological process using a perceived system. Parametric models use mathematical transfer functions to relate meteorological variables to runoff (Dawson & Wilby, 2001).

Deterministic and conceptual models require a large amount of high-quality data associated with hydrological, meteorological and natural geographical characteristics as well as human activities. In general, all the information required by deterministic and conceptual models - such as soil moisture and evapotranspiration- are not available for all the basins under consideration, especially when we are dealing with a large scale system such as the Brazilian one.

On the other hand, parametric models do not foreshadow a detailed understanding of the basin's physical characteristics, nor does it require extensive data pre-processing (Zhang, et. al., 2009).

Following this line of modelling, different approaches for inflow forecasting based on computational intelligence have emerged as an alternative for building inflow time series forecasting models. They are particularly powerful in situations where it is difficult to determine the physical process or when it is not possible to obtain a physical interpretation of the mathematical model representation (Price, 2008).

The main attribute associated to neural networks are the ability of modelling non-linear mapping between variables involved in several areas, including hydrology, where the most common structures are neural networks (Maier & Dandy, 2000; Bowden, et. al., 2005; Othman & Naseri, 2011) and radial basis functions (Jayawardena, et. al., 2006; Lin & Wu, 2011).

During the last decades, several proposals based on fuzzy systems and hybrid models have also found increasing applications in hydrology (Nayak, et. al., 2004; Luna, et. al., 2009), possible (Al-Zùbi, et.al., 2010). Fuzzy systems are useful to model uncertainties presented in hydrological variables, increasing flexibility for modelling the nonlinear relationships; and when combined with nonlinear optimization techniques, they appear as a very promising approach, obtaining structures that can be interpreted on the basis of IF-THEN rules.

Zambelli et. al. (2009) used an offline Fuzzy Inference System (FIS) for predicting annual inflows that are disaggregated into monthly samples used for long-term hydropower scheduling. Luna et. al. (2009) used an adaptive version of the FIS for the daily inflow forecasting of several basins and hydroelectric plants, considering precipitation information and the last inflows registered as input variables. The results of the adaptive FIS outperformed the ones achieved by a conceptual model for a short term forecast horizon, whereas the combination of both results outperformed the independent ones for longer lead times.

Therefore, this work makes use of Takagi-Sugeno FIS (Takagi & Sugeno, 1985), for monthly, weekly and daily inflow forecasting. The learning algorithm is based on the unsupervised clustering algorithm Subtractive Clustering (SC) (Chiu, 1994) and the offline version of the Expectation Maximization (EM) algorithm (Jacobs, et. al., 1991). The model structure and optimization algorithm is fully detailed in (Luna, et. al., 2010).

4.1 Monthly inflow forecasting

Two approaches were considered for obtaining the monthly inflow forecast sequences necessary for the PC approach.

4.1.1 Monthly models (FIS-M)

The seasonality of the monthly flows under study suggests the use of twelve independent models, as the technique traditionally used for the BES based on periodic autoregressive models (Maceira and Bezerra, 2007).

Therefore, the first approach adopted in this work consists of adjusting twelve different FIS models, one for each month of the year. Generally, these models are optimized by considering a one-step-ahead forecast error, which results in the degradation of performance when applied to a long-term forecasting task, although their performance is relatively good for one-step-ahead inflow forecasting.

Besides, the model complexity is in part limited by the size of the historical records available, due to the large amount of data necessary for the adjustment of all the model parameters.

In order to determine adequate models considering both accuracy and complexity, models were selected by the evaluation of the Bayes Information Criterion (BIC), which considers not only the reduction of the Root Mean Square Error (RMSE) but also the model complexity represented by the number of parameters to adjust.

Inflow data is normalized to the unit interval. Input variables were selected from a set of possible inputs composed by the last six lags of the time series. Input-output data set was split up into two subsets, the training set used for model optimization and the testing set used for validation and testing. Validation and testing sets were kept the same because of the limited duration of the historic inflow time series available.

Hence, monthly models were obtained through the following procedure

1. Build the input-output patterns considering a subset of possible inputs;
2. Define the initial set of fuzzy rules via the SC algorithm and the input-output patterns built previously;
3. Optimize the model parameters by the EM algorithm and
4. Evaluate the BIC penalization function.

The model with the lowest BIC was the chosen one. For a multi-step ahead forecasting task, these models were fed back with previous forecast results.

4.1.2 Annual models (FIS-A)

The second approach is based on the reduction of the long-term forecasting error by using a top-down forecast strategy (FIS-A). Top-down forecasting (TD) is extremely useful for improving the accuracy of detailed forecasts, since errors are compensated and variations can cancel each other out (Lapide, 2006).

The FIS-A approach predicts the aggregation of the twelve future monthly inflow samples (the aggregate inflow for the next year from the current month) by adjusting a unique model on an annual basis.

Input and model selection was also performed considering the BIC penalization function, following the procedure adopted by the FIS-M approach.

Consequently, the forecast results were disaggregated into the respective monthly estimates. This disaggregation was performed using the historical contribution factors of each month of the year, based on long-term average values.

4.2 Weekly inflow forecasting

Weekly inflow forecasting is essential for an adequate dispatch of generation of hydropower plants in the SIN. Weekly inflow time series for every hydropower plant in the SIN is composed by fifty two weeks a year. Figure 8 shows historical mean and standard deviation of weekly inflow time series for Furnas UHE, where we can observe a high variability during humid periods (at the beginning and at the end of the year).

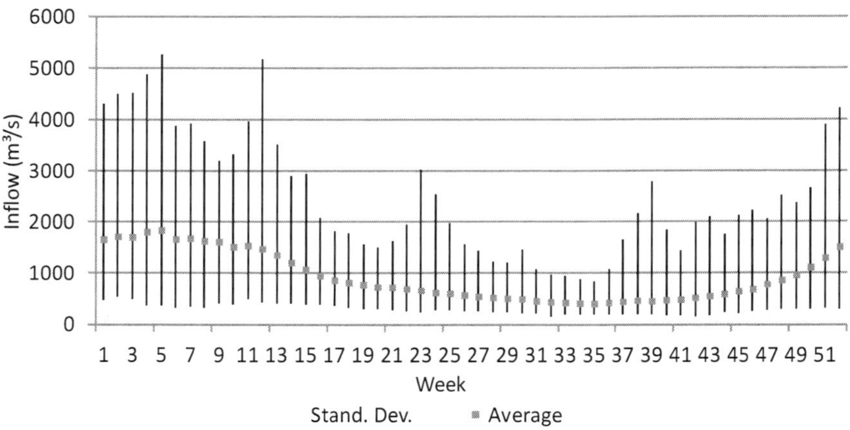

Fig. 8. Weekly historical average and standard deviation for Furnas UHE.

Even though each week has its own characteristics, it is not possible to adjust a model for each week, mainly because of the short historical series and the large scale of the problem addressed in this work. Therefore, we chose the setting of a single model for each UHE. The model adjusted following the procedure adopted by the FIS-M approach was used for a multi-step ahead forecasting task, considering a lead-time varying from one up to five weeks ahead.

5. Case study

In order to illustrate the application of the concepts described in the previous sections a case study simulating the monthly operation schedule from June, 2011 to December, 2015 for the whole Brazilian power system was considered. It comprises 147 hydro plants (95.6 GW of power capacity) and 144 thermal plants (32.4 GW of power capacity) with real operative constraints and the expansion plans for a planning horizon of 55 months ahead.

The MPC approach is usually implemented and simulated on a monthly basis in a framework of scenarios. Inflow series are taken from historical database, extending from

1936 to 2010, comprising 75 scenarios of 55 months each. In this paper, however, due to limitations in space, only two scenarios were considered: a favourable one (A) corresponding to the scenario beginning on June, 2005 and an unfavourable one (B) corresponding to a scenario beginning on June, 1952.

All constraints presented in the formulation of the optimization problem (1)-(14) have been considered in the simulation. The forebay h_F and tailrace h_T elevations were represented by 4th degree polynomial functions and the penstock head loss h_L by a linear function.

System data used for the case study were taken from official data source, accessible online (www.ons.org.br), and reflect the system configuration by June/2011.

The expected values of inflows used by the optimization model within the MPC framework were given by the FIS forecasting model described on section 4.1 and considered two strategies: monthly forecast (FIS-M) and annual forecast (FIS-A). In the latter the monthly values are derived from the relative contribution of long-term average values of historical inflows for each month to its annual value.

5.1 LTHS results

The simulation results for both inflow scenarios and considering the two inflow forecasting approaches presented on section 4.1 are summarized in table 1.

Favorable scenario (A)

	Cost [Million US$]	Hydro Gen. [MW]	Termal Gen. [MW]	Final Storage [MW-month]
FIS-M	7.307,22	53894,30	647,70	217347,00
FIS-A	7.526,52	53778,10	764,00	228997,00

Unfavorable scenario (B)

	Cost [Million US$]	Hydro Gen. [MW]	Termal Gen. [MW]	Final Storage [MW-month]
FIS-M	65.610,75	43837,5	10704,5	50478,8
FIS-A	48.459,70	44022,0	10519,8	45783,3

Table 1. Simulation results from MPC approach.

As can be seen, there are no significant differences on the favourable scenario with respect to the forecasting inflow approach considered. While the cost is slightly lower (3%) for the FIS-M approach, the final storage is also slightly lower (5%), resulting on quite similar performances. For the unfavourable scenario, however, the expressive higher performance of the FIS-A approach in terms of cost (26%) is not compensated by its slightly lower final storage (9%). This is an interesting feature of the proposed approach since it is on the most critical inflow situations, where the operational costs are higher, that the annual inflow approach provides better performances.

The evolution of the system's stored energy with both approaches and inflow scenarios is presented in Fig. 9. The stored energy is calculated by the sum of the useable water on the reservoir of each hydro plant pounded by the average cumulative productivity of this plant, which in turn is the summation of the efficiencies of all hydro plants located downstream.

It is possible to notice that the FIS-A approach yields to higher stored energy in both inflow scenarios. In general, the system's reservoirs go down in the dry season as a consequence of using the water to regulate river's flow. During the wet season the reservoirs recover to near full levels. This behavior is observed for the favorable scenario (A) as the stored energy presents peaks and valleys around the months of April and November, respectively, which constitutes the beginning and ending of the wet season in the majority of Brazilian's River Basins.

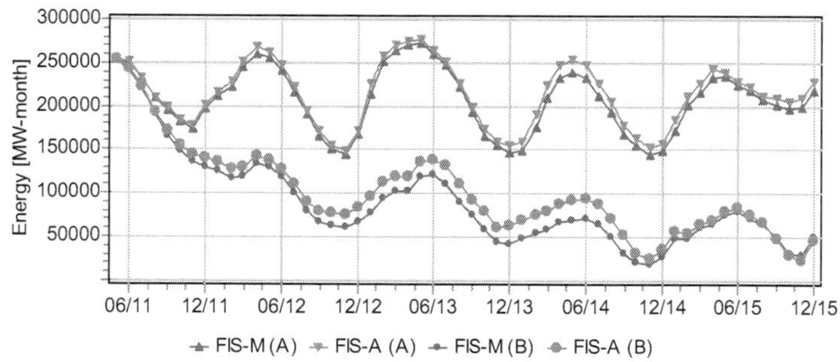

Fig. 9. Stored Energy evolution.

However, in the unfavorable scenario (B), the reservoirs are unable to recover and the stored energy is reduced year by year. Differences between the monthly and annual approaches are more expressive and reach 24.2 GW-month of stored energy in May, 2014. This is a consequence of the better inflow forecast with FIS-A since optimal decision intend to preserve the water reserves but inflow errors are compensated in simulation by increasing the plant's releases.

In Fig. 10 the operating costs are presented with both approaches and inflow scenarios. A cut was done in December, 2013 to allow rescaling in the remaining period.

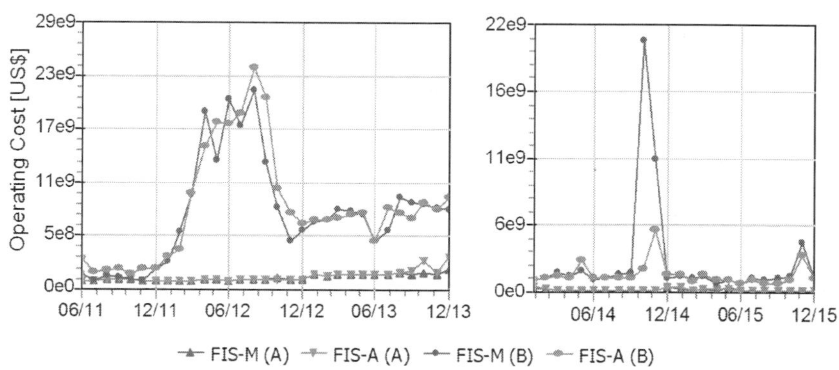

Fig. 10. Operating costs evolution.

In the favorable scenario (A) the operating costs are very close and in the first two years they suggest that thermal dispatch is near minimal. Nevertheless in the unfavorable scenario (B) the costs are higher and the FIS-M approach presented a maximum of 22 billion dollars in October, 2014 related to the lack of power supply to meet the demand. FIS-A approach not only avoids this deep energy shortage but also presents smoother cost variations which indicates less volatile energy prices.

The higher costs incurred by this approach in the first year are a consequence of the anticipation of thermal dispatch to save water and gain efficiency in the hydro plants. This phenomena known as *head effect* is related to the fact that operating with higher water heads the plants can deliver more power with the same (limited) water supply, a feature which is only possible to reproduce with nonlinear optimization models.

Three cascaded hydro plants were selected to present further detail of the individual operation with FIS-A for the favorable inflow scenario. Plants 1 and 2 have accumulation reservoirs whereas Plant 3 is a run-off-river plant. Plant 1 is the most upstream and plant 3 the most downstream in this river. Fig. 11 shows the reservoir storage trajectory for the three cascaded hydro plants.

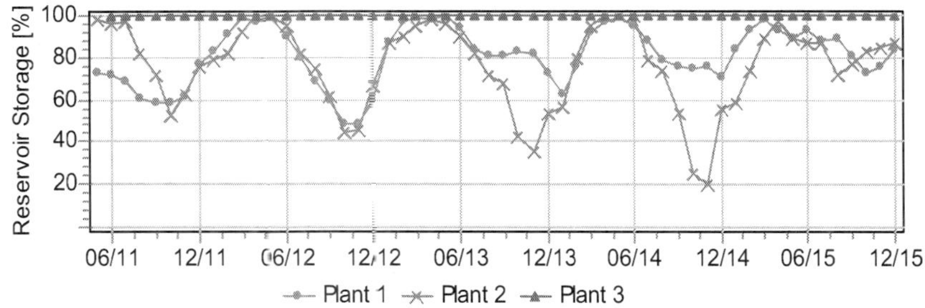

Fig. 11. Reservoir storage trajectory for three cascaded hydro plants.

The forecasted inflow series for Plant 1 is presented in Fig. 12 along with the simulated one. It can be seen that the forecasted series is quite acceptable, although there are peaks at the beginning of years 2012 and 2013 that were not identified by the forecasting model.

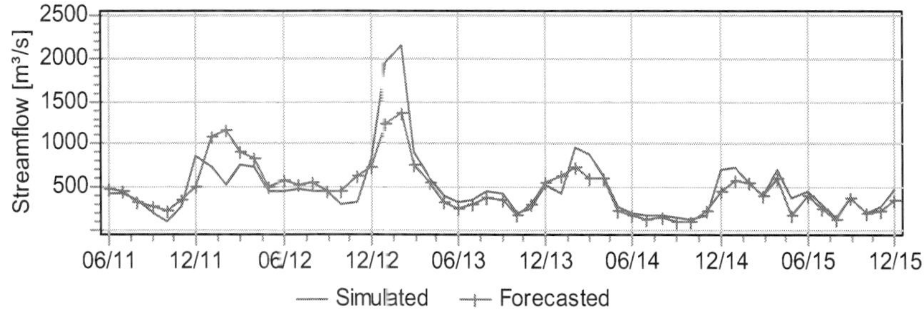

Fig. 12. Simulated and Forecasted stream flows for Plant 1 with FIS-A in favorable scenario.

Still it is important to notice that in the LTHS with MPC approach the forecasting model tries to hit the annual trend, which is more relevant than the specific monthly inflows. This is due to the fact that the discharge decisions of the optimization model are much more sensitive to the total annual inflow than to the specific values of each month (Zambelli et al. 2009).

5.2 STHS results

Next step on the operation planning chain, STHS is fed with the power generation of the first week ahead, disaggregated from the monthly generation determined by the LTHS model for each individual plant. The results of the STHS model will be presented for the same three cascaded hydro plants. Their generation targets in this case study were 493.60; 930.10; and 373.00 MW, respectively.

As shown in the LTHS problem, Plants 1 and 2 have accumulation reservoirs for annual flow regulation and therefore, in a week time period do not vary significantly. Plant 3, in turn, is a run-of-river plant in a long term planning horizon but presents reservoir head variations in short term planning.

In the first iteration of the relaxation procedure, the simulation model identifies violations of reservoir's maximum level in the optimal solution for Plant 3 from the third day until the end of the week. A new constraint was then added and the optimization model solution was updated. Reservoir storage of Plant 3 before and after the additional constraint is presented in Fig. 13 where the violation is eliminated. It is important to note that the generation targets remained satisfied.

STHS inputs and outputs for the selected plants are presented in Table 2.

	Generation Target [MW]	Initial Forebay Elevation [m]	Final Forebay Elevation [m]
Plant 1	493.6	652.62	652.45
Plant 2	930.1	517.71	519.67
Plant 3	373.0	430.67	431.30

Table 2. STHS simulation results.

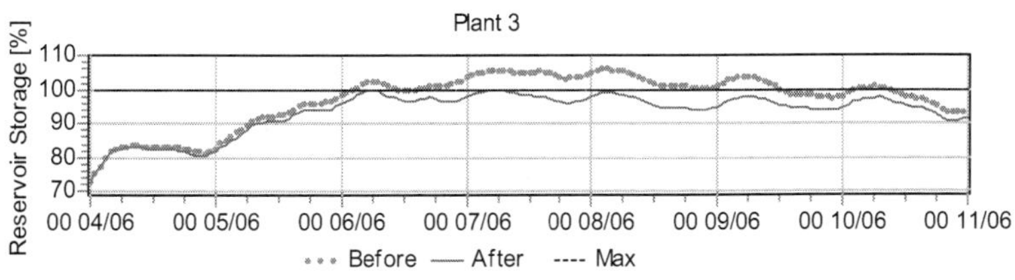

Fig. 13. Reservoir storage for Plant 3 with constraint violation.

The UC and GS solutions solutions change to fulfill the operating constraints, not only for the referred plant, but all cascaded plants can be affected. UC and GS solutions before and after the additional constraint are presented in Figs. 14 and 15, respectively for the three selected hydro plants.

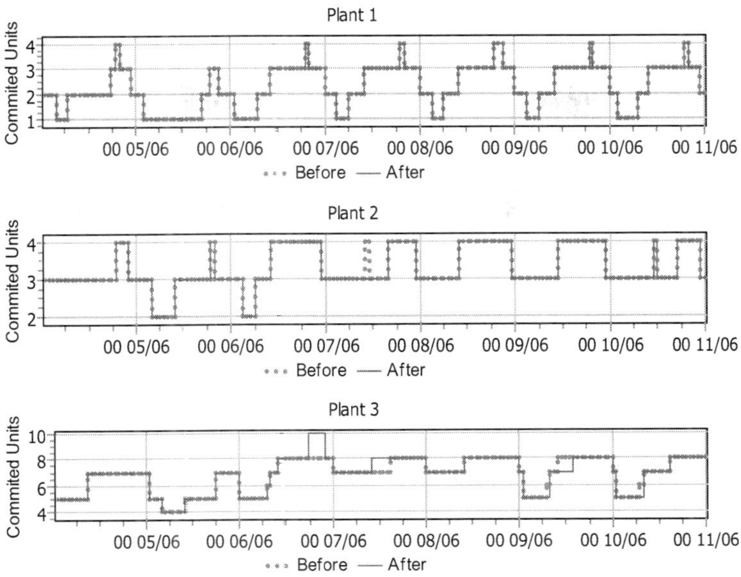

Fig. 14. Unit Commitment solution before and after the new constraint.

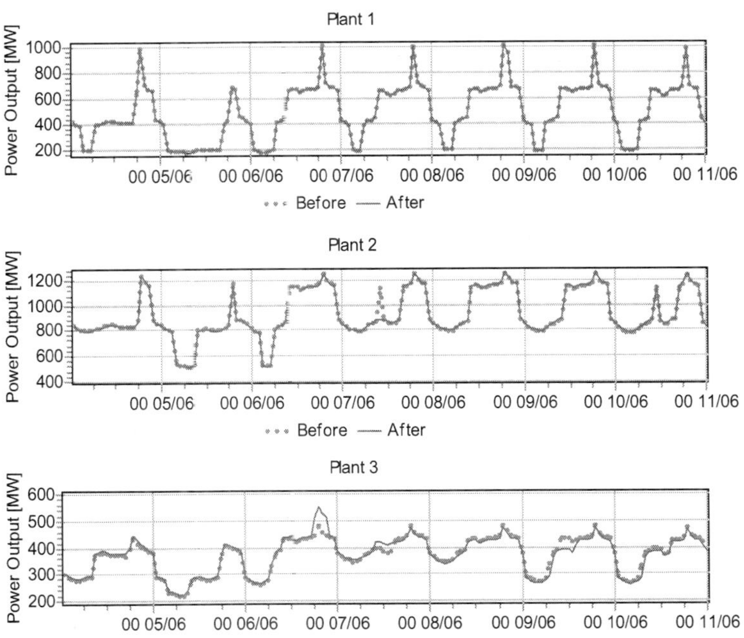

Fig. 15. Generation Schedule before and after the new constraint.

As can be seen, the differences in the scheduling of the selected plants are quite small. Plant 1 was not affected by the new constraint whereas in Plant 2 the fourth unit was not committed at 10h, day 07/06 which consequently reduced the generation at this time. On the other hand, in Plant 3 where the violation occurred the scheduling has slightly changed by increasing the generation on the first three days and reducing it on the last four days so that its generation target remains satisfied.

The results for the whole Brazilian system show that most of the hydraulic constraints relaxed are not active in the optimal solution, justifying the adoption of the relaxation procedure proposed for the STHS. A case study without the start-up and shutdown costs has also been performed. The results show that in this case the number of start-up and shutdown increase from 1,610 to 2,346 whereas the generation loss reduces from 311,667.7 to 309,178.8 MWh. Thus, the proposed approach allows the decision maker to choose the desirable trade-off between these two objectives by fixing the appropriate start-up and shutdown cost.

6. Conclusion

This chapter has presented an approach for hydropower scheduling in large-scale hydro-dominant power systems. An operation planning chain composed of two steps have been suggested. The first step corresponds to the long term hydrothermal scheduling (LTHS) that determines the generation at each plant and month over a planning horizon of up to 5 years ahead. The first month decision is then disaggregated into weekly intervals, and constitutes the generation targets for the next planning step. The second step corresponds to the short term hydrothermal scheduling (STHS) that determines the generation at each plant and hour over a planning horizon of 168 hours ahead.

To solve the LTHS problem a Model Predictive Control approach has been proposed by which a deterministic nonlinear optimization model is executed at each month to provide the generation decisions and then disaggregated into weekly intervals. To feed the optimization model with forecasted inflows, a Fuzzy Inference System (FIS) approach has been implemented in two different ways: one on a monthly basis (FIS-M) as it is usually adopted in the literature, and one on an annual basis (FIS-A), further disaggregated on a monthly basis proportional to the historical average values.

To solve the STHS problem an optimization- simulation decomposition approach based on relaxation has been suggested. According to this approach, the hydraulic balance constraints and limits are relaxed resulting on a mixed integer nonlinear optimization problem that minimizes the generation loss and start-up/shutdown costs at hydro plants while attaining the generation targets established by the LTHS. Then, the generation scheduling and the number of hydro generating units dispatched are provided to a simulator model that calculates the corresponding hydraulic variables and identifies possible violations. This violations yield new constraints to be added to the optimization model and the procedure is repeated until all hydraulic violations are eliminated.

A case study with the Brazilian power system composed of 147 hydro plants comprising 95.6 GW of power capacity and 144 thermal plants comprising 32.4 GW of power capacity illustrates the proposed approach. Two inflow scenarios were considered in the LTHS, one favorable and one unfavorable. In both scenarios the FIS-A approach has presented better results, especially in the unfavorable case when 26% of cost savings have been obtained.

This result reflects two facts: that inflow forecasting errors are lower on an annual basis than on a monthly basis, and that the generation decisions are more dependent on the total future inflows than any particular month value.

With respect to the STHS problem, the case study reported shows that the relaxation procedure concerning the hydraulic constraints is quite efficient since most of these constraints are not active in the optimal solution. Furthermore, the few violated constraints can be easily eliminated by the addition of appropriate linear constraints in the optimization model.

Further research is going on to improve the proposed methodologies. In particular other optimization techniques are being developed to substitute or to validate the heuristic approach used to solve the mixed integer nonlinear programming problem in the STHS.

7. Acknowledgment

The authors cordially acknowledge the whole team working in the support decision system at the laboratory for coordination of power systems operation. Special credit should be given to M.Sc. student M. Lopes for providing the inflow forecasts for the case study and Ph.D. student J. Borsoi for making the official data available for the case study.

8. References

Al-Zùbi, Y., Sheta, A. and Al-Zùbi, J. (2010). Nile River Flow Forecasting Based Takagi-Sugeno Fuzzy Model, Journal of Applied Sciences, 10: 284-290.

Bowden, G. J., Maier, H. R. andDandy, G. C. (2005). Input determination for neural network models in water resources applications. Part 2. Case Study: forecasting salinity in a river," Journal of Hydrology, no. 301, pp. 93–107.

Camacho, E.F. and Bordons, C. (2004). Model Predictive Control. Springer, Berlin.

Chiu, S. (1994). A cluster estimation method with extension to fuzzy model identification, Procs of The IEEE International Conference on Fuzzy Systems, vol. 2, pp. 1240–1245.

Dawson, C. and Wilby, R. (2001). Hydrological modelling using artificial neural networks, Progress in Physical Geography, vol. 25, no. 1, pp. 80–108.

El-Hawary, M. E. and Christensen, G. S. (1979). Optimal Economic Operation of Electric Power System. Academic Press.

Gwo-Fong Lin & Ming-Chang Wu (2011). An RBF network with a two-step learning algorithm for developing a reservoir inflow forecasting model, Journal of Hydrology, In Press, Corrected Proof, Available online.

Jacobs, R. Jordan, M., Nowlan, S. and Hinton, G. (1991). Adaptive mixture of local experts. Neural Computation, vol. 3, no. 1, pp. 79–87.

Jayawardena, A.W., Xu, P.C., Tsang, F.L. and Li, W.K. (2006). Determining the structure of a radial basis function network for prediction of nonlinear hydrological time series, Hydrological Sciences Journal 51 (1), pp. 21–44.

Jun Zhang; Chun-tian Cheng, Sheng-li Liao, Xin-yu Wu & Jian-jian Shen (2009). Daily reservoir inflow forecasting combining QPF into ANNs model. Hydrol. Earth Syst. Sci. Discuss., 6, 121-150.

Kadowaki, M. Ohishi, T., Martins, L.S.A. and Soares, S. (2009) Short-term hydropower scheduling via an optimization-simulation decomposition approach. 2009 IEEE Power Tech Conference.

Lapide, L. (2006). Top-down & bottom-up forecasting in S&OP. Journal of Business Forecasting 25(2). pp. 14-16.

Luna I., Maciel, L., Lanna, R. da Silveira, F. and Ballini, R. (2010). Estimating the Brazilian central bank's reaction function by fuzzy inference system. Communications in Computer and Information Science, vol. 81, pages 324--333. Springer.

Luna, I.; Soares, S., Lopes, J.E.G. and Ballini, R. (2009). Verifying the use of evolving fuzzy systems for multi-step ahead daily inflow forecasting. Procs.of the 15th International Conference on Intelligent System Applications to Power Systems, pp. 1–6.

Maceira, M.E.P and Bezerra, C.V. (1997) Stochastic Streamflow Model for Hydroelectric Systems, 5th International Conference on Probabilistic Methods Applied to Power Systems - PMAPS, Vancouer, Canada.

Maceira, M.E.P., Terry, L.A., Costa, F.S., Damázio, J.M. and Melo, A.C.G. (2002). Chain of optimization models for setting the energy dispatch and spot price in the Brazilian system. Proceedings of the Power System Computation Conference - PSCC'02, Sevilla, Spain.

Maier, H. R. & Dandy, G. C. (2000). Neural networks for prediction and forecasting of water resources variables: Review of modelling issues and applications. Envir. Modelling and Software, 15, 101–124.

Nayak, P., Sudheer, K. Rangan, D. and Ramasastri, K. (2004). A neuro-fuzzy computing technique for modeling hydrological time series, Journal of Hydrology, vol. 291, pp. 52–66.

Nilsson, O. and Sjelvgren, D. (1997) Hydro unit start-up costs and their impact on the short term scheduling strategies of Swedish power producers. IEEE Transactions on Power Systems, vol. 12, no. 1, pp. 38–44.

Othman, F. and Naseri, M. (2011). Reservoir inflow forecasting using artificial neural Network, International Journal of the Physical Sciences Vol. 6(3), pp. 434-440.

Price, R. (2008). Lecture on knowing the context between hydroinformatics and flood modelling, UNESCO-IHE Institute for Water Edutation, Tech. Rep.

Soares, S. and Salmazo, C. T. (1997) Minimum loss predispatch model for hydroelectric power systems," IEEE Transactions on Power Systems, vol. 12, no. 3, pp. 1220–1228.

Takagi, T. and Sugeno, M. (1985). Fuzzy identification of systems and its applications to modeling and control, IEEE Transactions on Systems, Man and Cybernetics, no. 1, pp. 116–132.

Zambelli, M.; Luna, I. and Soares, S. (2009). Long-Term hydropower scheduling based on deterministic nonlinear optimization and annual inflow forecasting models. Procs. of the PowerTech Conference, pp. 1–8.

11

Fuzzy Scheduling Applied on Hydroelectric Power Generation

Carlos Gracios-Marin et al.[1*]
[1]Benemérita Universidad Autónoma de Puebla,
Facultad de Ciencias de la Electrónica, Puebla,
México

1. Introduction

New industrial processes are evolution in terms of novel requirements to be accomplished. Interoperability, Open and Dynamic structures and Fault tolerance characteristics are some of them described by Shen and Norrie (1999).

In the environment of fault tolerance concept, it is desirable to reduce the effects of bad decision in the scheduling of activities/resources and to make good decision (adaptability) when one or more resources in the process fail.

In many application areas in which a malfunction of the system can cause significant losses or even endanger the environment or human life, a fault analysis model is required to evaluate the performance and can anticipate possible faults in the process defined by Zhuo and Venkatesh (1999).

Examples of such areas are in transport, process control and instrumentation with devices. The systems which are used in such or similar application areas are expected to exhibit always an acceptable behaviour. This property of a system is often referred to as **dependability**. Any departure from the acceptable behaviour is considered a system **failure**. Failures are caused by **faults**, which can arise in different phases of the manufacturing system lifecycle.

Most of the techniques which have been devised for fault analysis are targeted towards hard-wired systems and do not match the characteristics of software. A crucial difference between hardware and software system is that a program can neither break nor wear-out. Software faults can always be traced back to mistakes, which have been made during software specification, design or implementation.

To detect and remove faults, the software can be verified and validated against the requirements specification (model). The weakest point of this procedure is the requirements

* Gerardo Mino-Aguilar[1], German A. Munoz-Hernandez[1], José Fermi Guerrero-Castellanos[1], Alejandro Diaz-Sanchez[1,2], Esteban Molina Flores and Eduardo Lebano-Perez[1,3]
[1] *Benemérita Universidad Autónoma de Puebla, Facultad de Ciencias de la Electrónica Puebla, México*
[2] *INAOE.- Tonantzintla, Puebla, México*
[3] *UPAEP.- Puebla, México*

specification. Any fault or ambiguity in the specification can result in a fault in the process implementation. Fault avoidance is another approach to increase process dependability.

Most of the measures applied throughout the development process attempt to make the development more strict and formal. It is important that the process of adding rigor and formality could start from the very beginning, i.e. from developing a formal requirements specification which defines the space of all behaviours, which can be exhibited by the software. In the next step the unacceptable, e.g. dangerous, behaviours can be identified and defined in terms of the same formalism. Finally, one can check whether an unacceptable behaviour can be deduced from the specification. If this is the case, the specification can be modified and the analysis repeated.

A special case arises, when the formal model which underlies a software specification is discrete. In such a case the space of all behaviours is discrete, and a definition of an unacceptable behaviour can be reduced to a definition of unacceptable states. The evaluation of the software behaviour can be conducted as verification whether or not such states belong to the state space of the specification.

2. Background of concepts

2.1 Intelligent fault diagnosis

Actual industrial processes must accomplish the new requirement by high performance characteristics. Interoperability, Open and Dynamic structures and Fault tolerance characteristics are examples of these requirements, described by Shen and Norrie (1999).

The fault tolerance concept, consider as desirable to reduce the effects of bad decision in the scheduling of activities/resources and to make good decision (adaptability) when one or more resources in the process fails.

In many applications in which a malfunction of the system can cause significant losses or even cause danger to the environment or human life, a fault analysis model is required to evaluate the performance and can anticipate possible faults in the process. Examples of such areas are in transport of material, process control and instrumentation with devices.

The systems which are used in such or similar application areas are expected to exhibit always an acceptable behaviour. This property of a system is often referred to as dependability. Any departure from the acceptable behaviour is considered a system failure. Failures are caused by faults, which can arise in different phases of the process system lifecycle.

By definition, a fault represents an unexpected change of system function, although it may not represent a physical failure. The term failure indicates a serious breakdown of a system component or function that leads to a significantly deviated behaviour of the whole system. The term fault rather indicates a malfunction that does not affect significantly the normal behaviour of the system considered in first time by Rzevski (1989) and Nasr (2007) actually.

An incipient (soft) fault represents a small and often slowly developing continuous fault. Its effects on the system are in the beginning almost unnoticeable. A fault is called hard or abrupt if its effects on the system are larger and bring the system very close to the limit of acceptable behaviour. A fault is called intermittent if its effects on the system are hidden for discontinuous periods of time. Although a fault is tolerable at the moment it occurs, it must be diagnosed as early as possible as it may lead to serious consequences in time.

A fault diagnosis system is a monitoring system that is used to detect faults and diagnose their location and significance in a system. The system performs the following tasks:

1. Fault detection - to indicate if a fault occurred or not in the system
2. Fault isolation - to determine the location of the fault
3. Fault identification - to estimate the size and nature of the fault

The first two tasks of the system - fault detection and isolation - are considered the most important. Fault diagnosis is then very often considered as Fault Detection and Isolation (FDI). A fault-tolerant control system is a controlled system that continues to operate acceptably following faults in the system or in the controller. An important feature of such a system is automatic reconfiguration, once a malfunction is detected and isolated. Fault diagnosis contribution to such a fault-tolerant control system is detection and isolation of faults in order to decide how to perform reconfiguration.

2.2 Diagnosis based on analytical models

The model based fault diagnosis can be defined as the determination of the faults in a system by comparing available system measurements with a priori information represented by the system's analytical/mathematical model, through generation of residuals quantities and their analyses.

When an analytical model is used to represent any system under diagnosis is that it cannot perfectly model uncertainties due to disturbances and noise. The differences provoked by the non-complete description of the model, cause the residual values, which are instruments to indicate faults. By Palade and Jain, a robust FDI scheme represents a FDI scheme that provides satisfactory sensitivity to faults, while being robust (insensitive or even invariant) to modelling uncertainties.

The principal challenge in designing a robust FDI scheme is to make it able to diagnose incipient faults. The effects of an incipient fault on a system are almost unnoticeable in the beginning, thus effects of uncertainties on the system could hide these small effects.

A fault diagnosis task consists of two main stages: residual generation and decision-making. Residual generation is a procedure for extracting fault symptoms from the system, using available input and output information. A residual generator represents an algorithm used to generate residuals. Decision-making represents examining the residual signals in order to establish if a fault occurred and isolate the fault.

Residual evaluation techniques are divided into threshold decisions, statistical methods, and classification approaches. In accord of Jain, residual evaluation techniques, offers the adequate conditions to classify the faults in a machine problem. Considering the work of German Munoz, the fusion of Artificial Neural Networks and Fuzzy Logic, can be applied to model, simulate and control, industrial processes like Power Generation, to predict fault in the scheduling and maintenance of turbo generators.

Fuzzy logic tools can also be applied for residual evaluation in the form of a classifier as to reduce the variability in the decision-making to prevent faults in machines as well. One possibility is the combination of this qualitative approach with a quantitative residual generating algorithm. This idea is an improvement at the Recursive Decision Feedback Extension (RDFE) method presented by Graciós et al (2005).

3. An alternative fault tolerance model

Considering the last concepts explained, an alternative method is proposed a Fuzzy Filter to improve the decision-making of a Recursive Decision Feedback Extension, presented by Graciós, Munoz, Diaz, Nuno-de-la-Parra & Vega-Lebrúm (2009), which were applied to evaluate the possibility of fault and control of a real power generation plant. The fuzzy filter design is develop using the method described by Jain as follows.

3.1 Method proposed

The method proposed is based on the Recursive Decision Feedback Extension presented by Graciós (2009) and the Fuzzy Residual Evaluation embedded in a filter to diagnostic the soft fault generated in the behaviour of continuous function machines. In this paper, the principal solution is evoked at turbo generators in a hydro-plant.

To develop the improvement over the Recursive Decision Feedback Extension, we will invoke the concepts presented in early articles. Graciós, Vargas and Díaz (2005), have demonstrated that Fuzzy Logic (FL) and Artificial Neural Networks (ANN), can be used to transform *quantitative* knowledge (now residuals) into *qualitative* knowledge. This knowledge can be considered further information to evaluate the correct performance of the process for decision-making. Muñoz-Hernández et al (2009), presented a novel application of Neural PID to establish an adequate solution to control the behaviour in power and frequency for a hydro-plant. Perhaps the model is almost complete, the performance with the neural part, improved the result in simulation.

Considering the results obtained by the simulator, it is necessary to obtain a robust FDI because the decision-making in the output of the RDFE scheme is hard. The FDI can resolve the problem to "soft" the decision in terms of the type of faults that present in the generators.

```
PORT>>Get [P1(t), P2(t), P3(t), P4(t)] for each machine (i)
Z [P1(t), P2(t) P3(t), P4(t)]= [P1(k), P2(k), P3(k), P4(k)]
Fuzzyfing [P1(k), P2(k), P3(k), P4(k)]

For i,j=1 to k (data number)

        A(u_ij)=B(1) (f_1j(u), f_2j(u), f_3j(u), f_4j(u));
        Intell_Sch=Max [Min [B(f_1j)(f_1j(u_1)) ,B(f_2j)(f_2j(u_1)), B(f_3j)(f_3j(u_1)), B(f_4j)(f_4j(u_i))]];
        PORT <<FF (Mesch);
End
```

Fig. 1. Improved RDFE algorithm.

Describing the method, the inputs of the Fuzzy Filter requires of the three basic components:

- Signals Fuzzified.
- Inference algorithm.

- Classification of the fault indication.

The process of fuzzification have realised by RDFE. The inference algorithm is described by Munoz-Hernandez et al (2009), using the results obtained by neural PDF strategy. The weights of the ANN applied in the scheme, are used to refine the decision-making in the RDFE scheme.

Several improvements have been developed in the application technique. First the decision algorithm is soft by the use of the Filter, which is designed with the classification generated in the last paper. The fuzzification is well-defined in the concept outlined of the DFE implementation procedure obtained by Jain & Martin (1998).

Jones and Mansoor (See Muñoz-Hernández:2009), have contributed in the state of the art of Instrumentation and Control of hydro-plant subsystem considering the possibilities in presence of faults. These concepts were applied in the design of the filter to present the classification of faults as a output of the method.

The FF function is a pseudo process to evaluate the hard decision obtained by the RDFE algorithm and the Filter "softens" the decision-making in the control strategy. PORT is an added function to MAP the digital values that arrives from and to the process using the adequate and confidence sensors a actuators.

3.2 The algorithm

The structural and functional description of the algorithm is based on the contribution presented by B. Köppen-Seliger and P. M. Frank. Their work is included in the book of Fusion of intelligent schemes (see Jain & Martin: 1998), where the question: *is it possible to distinguish between all defined faults using the given rules?* is answered.

To answer this question in the case of turbo generators in a hydro-plant it is necessary to prove whether or not a distinction between the faults can be made.

The assumptions of the premises presented by Köppen-Seliger (in the same book of Jain & Martin: 1998), are used to discriminate the possible faults in terms of 4 parametric behaviours for each turbo generator. The original algorithm was modified to represent the more possible faults for lead to the following description:

f_k occurs, but none of the other faults

To handle this with the algorithm, each rule has to be transformed into a fuzzy switching function defined by Jain. If the result of the fuzzy switching function is equal to 0.5 then the rule is fuzzy-consistent. If the result is < 0.5 the rule is fuzzy-inconsistent. That means that for fuzzy-inconsistent rules the compatibility degree is < 0.5 for all x.

The complete design algorithm has been summarized a recent work presented by Graciós (2009) in local conferences with good results and each step was developed as follows:

Step 1. Define the number of faults which are of interest.

In this step, the number of faults was obtained using the simulation platform described by German and collaborators in another In-tech Chapter of book where several simulations were realised to create an adequate Database for the fault definition.

Step 2. For each residual component, two fuzzy sets have to be defined as an initial definition. These two fuzzy sets are *normal* and *not normal*. However, there is no definite rule about the suitable number of order partition, the higher the precision will be. But it will take more time in computing processing and more complicated form in modelling. Considering the results by the simulator, for the hydro plant is necessary 3 partitions only to improve and soft the decision making scheme.

The fuzzy sets obtained from the step 1 were defined using Fuzzy Time Series (FTS) as follows: let

$$X_t^{u(i)} = \{x[t,u(i)] \text{ for } t = 1, 2, \ldots, n\} \in R, \quad (1)$$

Be a time series, Ω be the range of X_t^i for each *i* parameter under measure of the *u* machien and P_j, for j=1, 2, 3, ... n,

$$\bigcup_{j=1}^{r} P_j = \Omega \quad (2)$$

be an ordered partition Ω. Let L_j denote linguistic variables with respect to the ordered partition set. For t= 1, 2, ..., n, if $\mu_{i,j}$, is the grade of membership (possibility of fault) of X belongs to L_j , satisfies u. Then $FX_t^{u(i)}$ is said to be the Fuzzy Time Series for each u(i) machine parameter and written as:

$$FX_t^{u(i)} = \mu(1)X_t^{u(2)}/L_1 + \mu(2)X_t^{u(2)}/L_2 + \mu(3)X_t^{u(3)}/L_3 + \mu(4)X_t^{u(4)}/L_4 \quad (3)$$

where / is used to link the linguistic variables with their membership in FX, and the + operator indicates, rather than any sort of algebraic addition, that the listed pairs of linguistic variables and membership collectively.

In this approach, triangular membership function was selected for easy transformation process when the cacalculating corresponding membership functions of linguistic variables in FTS.

Step 3. The rules are transformed into fuzzy switching functions. As a simple example for this transformation consider the following rule:

If *Res*1 is normal and *Res*2 is negative or If *Res*1 is positive and *Res*2 is negative then *f*1 the corresponding fuzzy switching function is given as the definition is based on the assumption that the first index indicates the residual and the second index indicates the fuzzy set of this residual.

Step 4. Based on the fuzzy sets defined in Step 3 and the faults defined in Step 1, the resulting number of rules has to be generated.

Step 5. Prove whether or not it is possible to distinguish between *all* faults. That means that the fuzzy switching functions have to be checked for phrases described by Equation (5). This procedure must be performed for all faults. This leads to the following scheme:

In this formula to check for *faultk*, just the terms for *k* to *p* are considered, because the previous steps checked that fault *k* is fuzzy consistent with respect to *fault*1, . . . , *faultk*-1.

Step 6. If the distinction is possible, all faults can be detected *and* isolated and the procedure is terminated.

Step 7. If a perfect distinction with this choice of the distribution of the fuzzy sets is not possible, one or more fuzzy sets have to be modified. This means, for example, that instead of the fuzzy set *"not normal"* two fuzzy sets *"slightly deviating"* and *"strongly deviating"* may be discriminated. The fuzzy set that has to be changed is a result of the reduced switching function. It is the fuzzy sets that lead to a fuzzy-inconsistency.

Step 8. Now carry out the algorithm again and repeat the procedure until a unique distinction is possible.

This algorithm ensures that all faults are detectable *and* distinguishable. If a perfect distinction of the residuals is not possible, the algorithm indicates these inconsistencies. This helps the operating personnel to evaluate signals giving consideration to the inconsistency. To prove the algorithm it was applied to a part of a wastewater plant.

There are two possible uses of the supporting algorithm for the design of the Fuzzy Filter in the residual evaluation process. The first possibility starts with an *empty* rule base. That means that the designer has to generate a *complete* rule base using this algorithm. This ensures that the generated rule base is consistent and complete with respect to the fault detection scheme. Therefore, all rules have to be consistent and unique in order to represent each fault under consideration. The suggested algorithm automatically checks whether or not these conditions are fulfilled.

The second possibility starts with a given, possibly inconsistent, rule base. The task is now to check which part of the given rule base is inconsistent and/or incomplete. This part of the rule base should be modified as described in the next section. It should be mentioned that these two possibilities use the same algorithm; just the initial conditions of these two possibilities are different. To use this algorithm, the so-called *Fuzzy Switching Functions* have to be defined in order to simplify the procedure.

4. Case of application

The hydropower generation process is a well-known industrial process that converts mechanic force into electric power. The conversion pathway of the process is shown in Figure 2. The output of the process is electrical energy where power and phase parameters are evaluated by performance index.

To run such an energy conversion process under efficient conditions, which means to reduce the probability of fault in the influence of electrical net, some requirements have to be met. To include all effects, the model of the process has to be highly nonlinear and of high order. Because of these facts some researchers have tried to simplify the model by reducing the number of subsystem parts in the process. The complete model contains three subsystems for the whole plant:

- Guided Vanes
- Hydraulic
- Electric

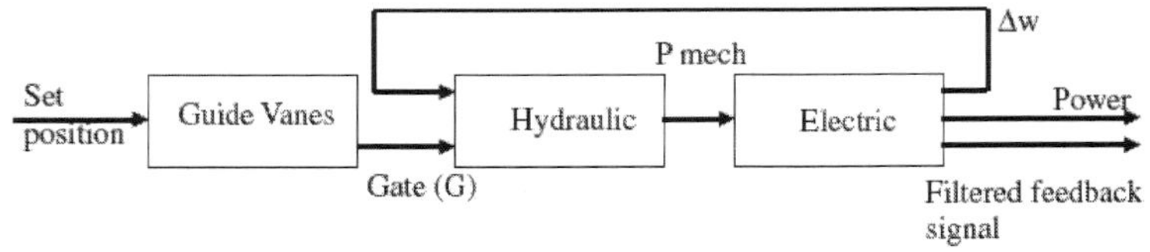

Fig. 2. The three subsystems of the hydroelectric plant.

Because electric subsystem is the most significant part some models contain just similar form of model. The complete model of the electrical subsystem (Fig. 3) is based on the 'swing' equations [20], and includes the effect of synchronizing torque. The first-order filter is included in the feedback loop for noise reduction. The models are expressed in the per-unit system, normalized to 300MW and 50 Hz.

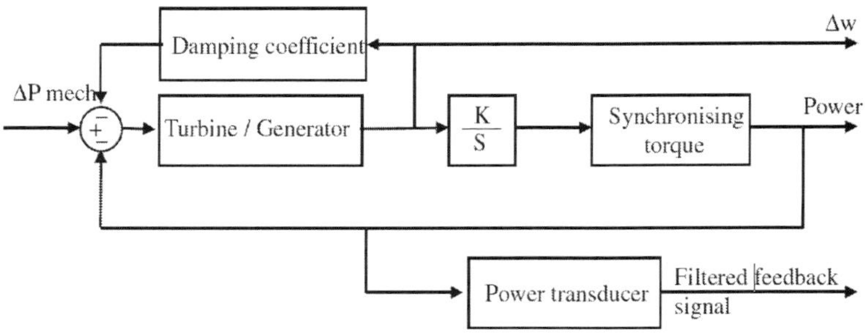

Fig. 3. Electric subsystem of the hydroelectric plant.

This model was implemented in Simulink©. It was designed to be scalable, allowing different behaviours to be selected according to the objective of the study. For the purposes of this work, a multivariable linear model was chosen.

4.1 Design of the fuzzy filter for residual calculation

The design of the fuzzy filter for the qualitative residual evaluation is based on the following assumptions:

- The structure of the fault diagnosis scheme is based on the topology described in Figure 4. This includes a nonlinear model of the process as well as a linear model for the observer-based residual generator.
- For the design of the fuzzy filter, only a qualitative description of the faults is needed.
- Both the quantitative residual vector and the qualitative description of the fault behaviour are used as inputs to the fuzzy filter in order to detect and isolate the faults (Fig. 4.).

As is described in the design method, the fuzzy filters are tuned using an array of Fuzzy Time Series to define the membership grades to identify and isolate the faults.

Steps 1 to 8 are developed as the original RDFE scheme is presented in [9], however the refine of the decision for each decision-making parameters is made coupling the Fuzzy Filters (FF) at the out of the scheme with the FEED evaluation from the B(u_i) adjusts.

Step 9. Calculate the FF {$B\ (u_{i)})$ } for i =1, 2, 3, 4.

The FF was implemented using MATLAB and tested using SIMULINK and the Simulator used in [10] to refine the decision in the environment.

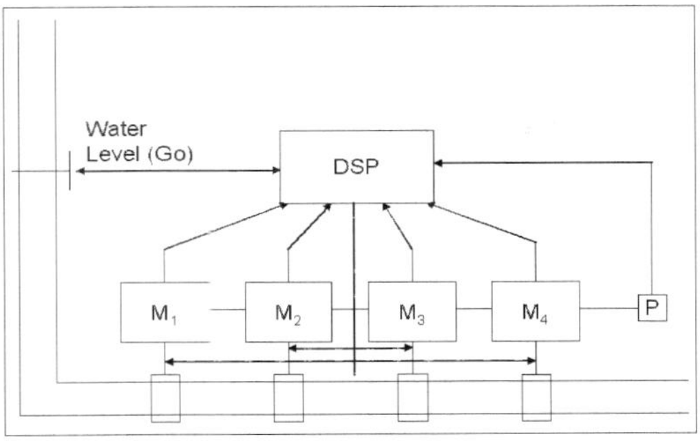

Fig. 4. Schematic for hydro-plant.

The simulated process is presented schematically in figure 4. The initial values of each performance parameter (f_j) by each machine (u_i) and their corresponding membership function values $B\ (f_i)\ (f(u_i))$ is given in the Table 1.

u/f	P1	P2	P3	P4
M1	0.86	0.91	0.95	0.93
M2	0.98	0.89	0.93	0.90
M3	0.90	0.92	0.85	0.96
M4	0.91	0.83	0.91	0.87
Fuzzy Parametrization				
M1	0.6	1	1	1
M2	1	0.9	1	1
M3	1	1	0.5	1
M4	1	0.3	1	0.7

Table 1. Initial Average Performance and Fuzzy Membership.

When the procedure of fuzzification is developed, we get $A\ (u_1) \approx 0.6$, $A\ (u_2) \approx 0.9$, $A\ (u_3) \approx 0.5$ and $A\ (u_2) \approx 0.3$. So, for this particular case M_2 has the best performance of the four machines, where in this actual condition, can be considered has good functioning condition.

Now consider we propose a performance curve index to define the best schedule for the 4 machines in terms of the 4 parameters defined and then a novel algorithm can be developed for a recursive version for the DFE strategy.

Figure 5 is the result of the scheduling strategy without the Fuzzy Filter, reported by Graciós et al(2009), however, figure 6 shows the "soften" result in the scheduling of the performance for the 4 turbo generator considering the Fuzzy Filter designed.

Consider that several time intervals are found scheduling instability due the time of calculation for the tune filter scheme. This is a part to be improved in a future work.

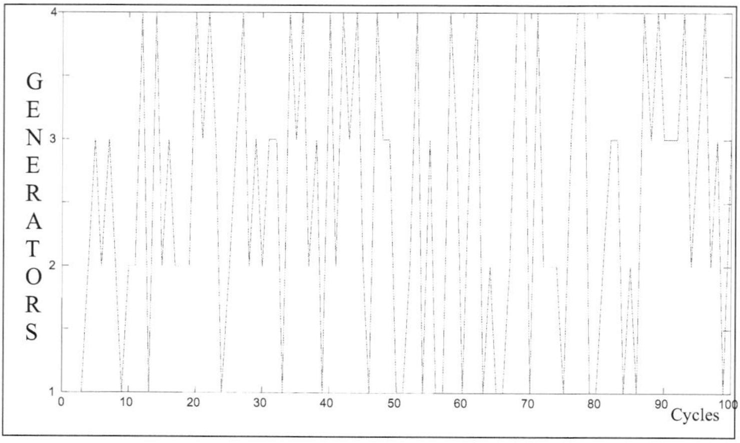

Fig. 5. Simulation of the scheduling for the plant without Fuzzy Filter.

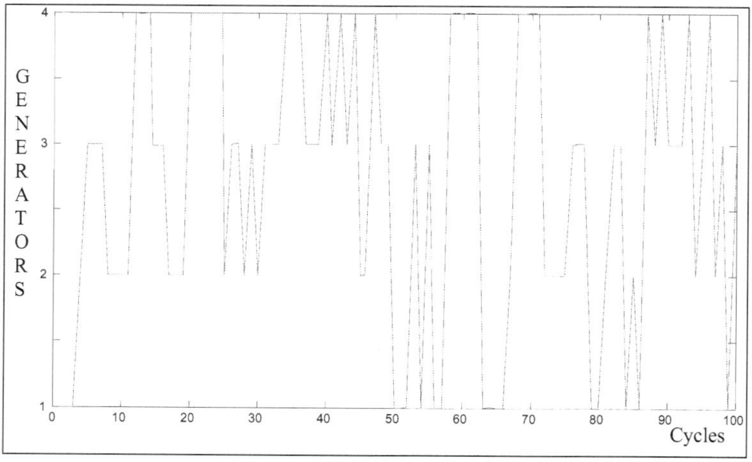

Fig. 6. Simulation of the scheduling for the plant with Fuzzy Filter.

This model was implemented in Simulink©. It was designed to be scalable, allowing different behaviours to be selected according to the objective of the study. For the purposes of this work, a multivariable linear model was chosen.

5. Conclusions

The theory of Fuzzy decision using Feedback extensions was introduced to determine a novel fuzzy scheduling scheme to be applied in the meet of production goal by a programmed performance. This type of schedule represents a novel alternative to transform typical industrial process on "intelligent process" inserting AI agents in the regulation/control of activities for each resource.

It lets to improve the requirements of agility and fault tolerance for actual processes, where the agent can be developed using rapid prototyping architectures like FPGA as well.

With the inclusion of soften filters in the decision of performance evaluation, the graphs can shown, the improvisation of the stability decision in the performance of each turbo generator in a typical plan of 4/6 in Dinorwig hydroplant.

The use of the actual simulator presented by Muñoz-Hernandez et al (2009) in the validation of the performance represents an adequate form off line to protect the real system.

The soft parts developed by the fuzzy filters, reduce in a 40% the computational time for the scheduler considering a good value of 33.3 ms for each time of scheduling cycle.

The equations to calculate the membership for each fuzzy filter were described in a programmable device obtaining a total delay time of 1 ms, which is affordable with the implementation of this type of application.

The practical results compared using Matlab© showed best performance in the function of the machines processes validating the possibility to recursive a basic feedback extension definition.

6. Acknowledgment

Carlos Arturo Gracios Marin wants to give thanks to Profs. Dewi I. Jones, Sa'ad Mansoor and German Muñoz by their advised, in the redaction of this work. My particular and great grateful, at the colleagues of the Academic Group of Power traction, quality and Generation of the Puebla Autonomous University (BUAP) and in special for Prof. Fernando Porras, Francisco Portillo and my special consideration to the support of Education and Postgraduate Offices of this Institution because without them, this publication has not been possible. Finally, I want to express my personal acknowledgment to Dr. Agüera (Rector of the University) and De la Peña Mena (Scientific Director of CONACYT) for your financial support in this project.

7. References

Cassandras, C. & Lafortune, S. (2008). Introduction to Discrete Event Systems, Second Edition, Springer Science +Business Media LLC, ISBN 978-0-387-33332-8.

Dagli, C. H. (1994). Artificial Neural Networks for Intelligent Manufacturing, First Edition, Chapman & Hall, U. K.

Graciós-Marín, C. A. ; Munoz-Hernandez, G. A. ; Diaz-Sanchez, A. ; Nuno-de-la-Parra, P. ; Estevez-Carreon, J. & Vega-Lebrúm, C. A. (2009). Recursive decision-making feedback extension (RDFE) for fuzzy scheduling scheme applied on electrical power control generation, *International Journal of Electrical Power & Energy Systems*, Vol. 31, Issue 6, pp. 237-242, Elsevier, U. K.

Graciós-Marín, C. A. ; Vargas-Soto, E. & Díaz-Sánchez, A. (2005). Describing an IMS by a FNRTPN definition: a VHDL approach, *International Journal on Robotics and CIM*, Vol. 21, Issue 3, Elsevier.

Jain, L. C. & Martin, N. M. (1998). Fusion of Neural Networks, Fuzzy Systems and Genetic Algorithms: Industrial Applications, ISBN 0849398045, CRC Press, CRC Press LLC.

Munoz-Hernandez G. A.; Gracios-Marin C. A.; Diaz-Sanchez A.; Mansoor S. P. & Jones D. I. (2009). Neural PDF Control Strategy for a Hydroelectric Station Simulator, In: *Automation Control - Theory and Practice,* A D Rodić (Ed.), ISBN: 978-953-307-039-1 Intech, Available from http://sciyo.com/articles/show/title/neural-pdf-control-strategy-for-a-hydroelectric-station-simulator.

Nasr, E. A. & Kamrani A. L. (2007). Computer-Based Design and Manufacturing: An Information-Based Approach, ISBN 0-387-23323-7, Springer Science+Business Media, LLC.

Palade, V. ; Bocaniala, C. D. & Jain, L. (2006). Computational Intelligence in Fault Diagnosis Advanced Information and Knowledge Processing, ISSN 1610-3947.

Rzevski, G. (1989). Artificial Intelligence in Manufacturing, Computacional Mechanics Publications. *Springer-Verlag, Proceedings of the fourth International Conference on the Applications of Artificial Intelligence on Engineering,* Cambridge, U. K. July 1989.

Shen, W. & Norrie D. H. (1999). *Agent-Based Systems for Intelligent Manufacturing: A State-of-the-Art Survey,* Knowledge and Information Systems, an International Journal, 1(2), pp. 129-156.

Zhou, M. & Venkatesh, K. (1999). Modelling, Simulation and Control of Flexible Manufacturing Systems.- A Petri Net Approach, *Series in Intelligent Control and Intelligent Automation Vol. 6,* World Scientific.

12

Integration of Small Hydro Turbines into Existing Water Infrastructures

Aline Choulot[1], Vincent Denis[1], and Petras Punys[2]
[1]*Mini-Hydraulics Laboratory (Mhylab),*
[2]*Water & Land Management Faculty,*
Lithuanian University of Agriculture,
[1]*Switzerland,*
[2]*Lithuania*

1. Introduction

Climate change due to CO_2 emissions has been defined as the major environmental challenge to be faced nowadays by the International Community. The European Directive 2009/28/EC of 23 April 2009 on the Promotion of Renewable Energy aims at achieving by 2020 a 20% share of energy from renewable sources in the EU's final consumption of energy. Each EU Member State adopted a national renewable energy action plan (NREAP) setting out its national targets for the share of energy from renewable sources consumed in transport, electricity, heating and cooling in 2020.

The public awareness on environmental topics has improved significantly, leading to a European environmental awareness. One of the latest manifestations of this awareness is the European Water Framework Directive (2000/60/EC), aiming at an overall protection of water. But this Directive tends to be in contradiction with the above mentioned directive, slowing down the development of hydropower including small hydropower plants (SHP). However, there is no doubt about the benefits of converting energy by SHP plants that means climate change mitigation and security of energy supply. Then, it implies regional development and employment. On a local level, SHP integration into the local environment, optimal use of water resource and mitigation measures are now key words for SHP design and implementation, which can lead to creation of positive impacts on the local ecosystem (Chenal et al. 2009).

Multipurpose hydro schemes, which lead to energy recovery in existing infrastructures thanks to hydropower plants, are one of the rare issues that may perfectly respect both the "Renewable Energy Directive" and the "Water Framework Directive". In addition, it can offer a solution to many potential issues discussed on water policy when it comes to sustainable management of the resource in sectors like agriculture, inland navigation, wastewater treatment or drinking water supply. In other words there is a significant market niche of this "sleeping" hydro potential. Even extremely small water infrastructures can generate hydropower - including the systems that deliver water to homes or subsequently scrub it of pollutants. Anywhere there is excess head pressure in a infrastructure dealing with water; there can be a good opportunity to generate electricity.

This chapter is meant to answer two main questions:

- Where are the potentials stemming from a water infrastructure?
- How (technically) can energy be recovered by a small turbine or unconventional small hydropower plant?

To answer these questions the overall objectives were to:

- Identify potentials for non traditional hydropower installations,
- Review main steps for development of a multipurpose project,
- Provide typical recommendations for installing SHP plants into existing infrastructures,
- Summarise good practices of these technologies based on cases studies,

Main findings of this study are based on a specific Swiss experience and the expertise of Mhylab (Mini-Hydraulics Laboratory). The SHAPES project outcome - Energy recovery in existing infrastructures with small hydropower plants (ESHA et al., 2010) is here used extensively, with some to the most relevant cases studies, collected all over the European Union and Switzerland. Table 1 presents these 16 case studies, with their main characteristics (nominal discharge, gross head, electrical output and electrical production), while their description and main peculiarities will be developed through different sections as referred in the last column of this Table 1.

Moreover, a variety of information resulting from a range of publications in open sources, conference proceedings, internet resources and case studies on the application of energy recovery were collected and analysed.

2. Overview of small hydropower

Hydropower plants are divided into two main areas: the "large" and the "small" ones. At present time there is no satisfying definition to determine if a hydropower plant is small or large. This differentiation depends on a multitude of criteria, such as the output of the scheme and its size or technical or economic characteristics.

The criterion currently used for defining small hydropower plants is that of output, but many variants are in use. Eurelectric, the European Commission, ESHA (European Small Hydropower Association) as well as several other countries have defined a scheme of less than 10 MW as being small (Chenal et al., 2009).

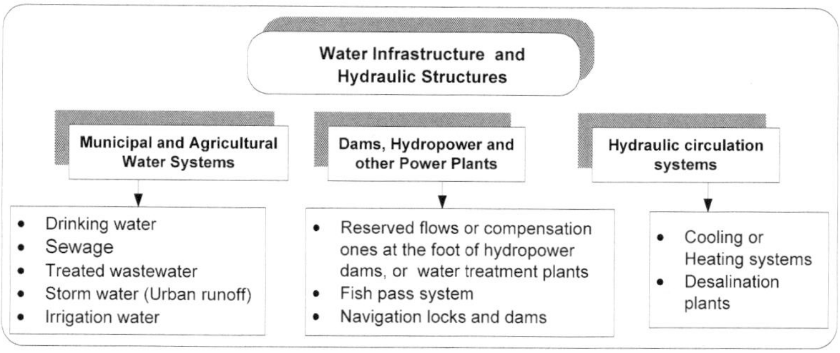

Fig. 1. Components of the water industry covered by this analysis.

Existing infrastructures	Power plant name and Country	Nominal discharge (m^3/s)	Gross head (m)	Electrical output (kW)	Electrical production (GWh/year)	Section
Drinking water network	La Zour, CH	0.30	217	465	1.8	3.2.1
	Mühlau, AT	1.60	445	5750	34.0	3.2.1
	Poggio Cuculo, IT	0.38	28	44	0.36	3.2.1, 5.3.2, 5.3.5
Irrigation network	Armary, CH	0.09	105	68	0.45	3.2.2
	Marchfeldkanal, AT	6.00	2	70	0.50	3.2.2
	Rino, IT	0.78	446	2800	14.00	3.2.2
Raw wastewater network	Le Châble, Profray,CH	0.10	449	380	0.85	3.2.3, 5.3.2, 5.5
Treated wastewater network	Seefeld, AT	0.25	625	1192	5.50	3.2.3
	Nyon, CH	0.29	94	220	0.70	3.2.3, 5.3.2
Hydropower dam and reserved flow	Llys y Fran, UK	0.16	25	29	0.22	3.2.5, 5.3.2
	Le Day, CH	0.60	27	126	0.58	3.2.5
Hydropower dam and fish pass	Aire-la-Ville, CH	2.00	21	348	2.72	3.2.6, 5.3.2
Navigation lock	L'Ame, FR	10.80	2	145	0.65	3.2.7
Desalination plant	Tordera, ES	0.11	685	720		3.2.8
Cooling system	Sangüesa, ES	1.16	11	75	0.50	3.2.9
	Skawina, PL	23.30	8	1560	6.39	3.2.9

Table 1. Selected European case studies of multipurpose schemes (ESHA et al., 2010).

Here the chapter deals with small hydropower plants that can operate as auxiliary installations into municipal and agricultural water systems, hydraulic structures, power plants, desalination plants, heating or cooling systems, while guarantying their primary functions (Fig. 1).

3. Where are the potentials?

3.1 Potential estimation

Hydropower depends on two main parameters: the head (or the pressure), and the discharge. Therefore any process implying a water discharge, steady or not, and an unused pressure, is a potential energy source.

Nowadays and worldwide, the multipurpose schemes operating in the water industry equipped with small hydropower plants are limited. For example, no one has been identified in the Baltic countries. Moreover there is a lack of data in Europe concerning the operating and remaining potential, apart from Switzerland, as shown in Table 2. Can it be then implied that the remaining potentials can be interesting?

Water network type	Potential type	Number of sites	Output (MW)	Production (GWh/year)	Electricity consumption equivalent households
Drinking water	Operating	90	17.8	80	17780
	Remaining	380	38.9	175	38890
Untreated wastewater	Operating	3	0.4	1,4	310
	Remaining	86	7.1	32	7110
Treated wastewater	Operating	6	0.7	2.9	640
	Remaining	44	4.2	19	4220

Table 2. Hydropower schemes in the water industry in Switzerland: operating and remaining potential (Chenal et al., 1994; SFOE, 1995).

To better promote energy recovery within water networks, the Swiss Federal Office of Energy has produced guidelines for installing SHP plants (SFOE, 1996). It can be noted that the Swiss software tool TURBEAU can help in estimating the cost efficiency of the identified potentials (Boillat et al., 2010).

3.2 Typical potential sites

These potentials, for which electricity generation is not their primary priority, but the second, are so called **multipurpose schemes**. This implies the integration of the power plant in the existing infrastructure while guaranteeing its primary function. For example, for a drinking water network, the primary priority is to supply in quantity and quality the needed water; whilst for a desalination plant, it is to generate drinking water from sea water. Most of the time the respect of the primary function will imply the setting of a by-pass of the turbine as mentioned in chapter § 5.7.

As multipurpose schemes are characterized by a wide range of water quality, from drinking water to wastewater, there is a need for an overview of different techniques.

3.2.1 Drinking water network

A simple drinking-water network can be described as follows (Fig. 2):

- a spring at altitude,
- a forebay,
- a penstock,
- a reservoir,
- a water supply network.

From the elevation of the sources, and as the pressure at the consumers cannot generally exceed 4 bars, there can be an excess of pressure in the networks to recover. The main idea here is to replace the pressure breakers, used traditionally to waste the excess pressure, by turbines so as to generate electricity.

Different energy recovery possibilities can be identified and defined by the turbine positions:

- on a reservoir:

Water passes through the turbine before being accumulated in a reservoir. This method is the most flexible, as it permits disconnection of the turbine operation from the water supply network to guarantee at any time the primary function of the existing infrastructure.

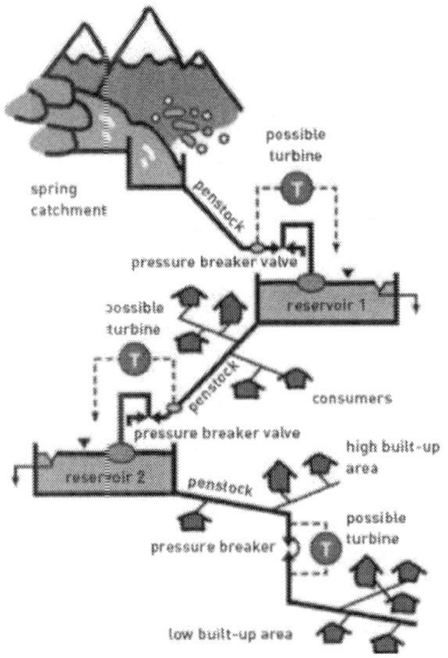

Fig. 2. Layout of a drinking-water network and possible positions of the turbines.

- within the supply network:

Water passes through the turbine and carries on its way through the pipe. This setting means that a pressure defined by the network requirements has to be maintained at the turbine outlet, which reaction turbines and counter pressure Pelton can achieve (cf. section § 5.3).

- before restitution to the environment:

Excess water that is not supplied to the consumers passes through the turbine before restitution to the environment.

When the drinking water source is underground and has to be pumped to the reservoir, no turbine integration will be possible.

- Case study, La Zour, Switzerland[1]: The drinking water system of Savièse commune had to be upgraded in anticipation of population growth, increases in per-capita water

[1] All the characteristics (nominal discharge, gross head, electrical output, electrical production) of the case studies are presented in Table 1.

consumption, and glacier retreat. In the scope of this project, two small hydro schemes (250 kW and 330 kW) were commissioned in 2001, La Zour scheme in 2004 (cf. Photo 1 and Photo 29) and a fourth one in 2009. The performances of the three first hydro plants are to the expected level. The fact that the commune has recently ordered a fourth turbine demonstrates the technical and economic attractiveness of these kinds of SHP developments.

- Case study, Mülhau, Austria: The plant that collects water in a tunnel more than 1.6 km long (the average time the water takes to pass through the rock mass and into the tunnel is estimated at 10 years), supplies drinking water for the major part of Innsbruck. With a generating capacity of 6 MW (cf. Photo 2), it is one of the biggest drinking water power plants in Austria.
- Case study, Poggio Cuculo, Italy: The Poggio Cuculo water treatment plant, which supplies drinking water to Arezzo main reservoir, operates with three different raw water discharges supplied by a large upstream reservoir, depending on the electricity price: 280 l/s during the day, 360 l/s during the winter night and 380 l/s during the summer night.
  As the difference of levels between an intermediate reservoir and the water treatment plant is 28 meters, a turbine has been set as a by-pass of the former regulation valve (cf. Fig. 7). This means that the raw water discharges through the hydro turbine before entering the water treatment works for processing.

Although the pipeline related head loss is considerable for the 3 operational discharges (the efficiency of the penstock is 45% only for 380 l/s), the existing pipe work could not be changed for administrative and cost reasons. However, thanks to a runner with 8 adjustable blades (cf. Photo 28) and a variable turbine rotation speed, the turbine can be operated with good hydraulic efficiency under any of the three operating discharges. Moreover the turbine has become the discharge regulation device for the reatment plant inlet, thanks to automation of the runner blade adjustment.

The water treatment plant consumes more than 2 GWh/year of electrical energy, to be compared to the 0.36 GWh/year generated by the small hydropower plant.

Photo 1. La Zour: the setting of the runner and the generator.

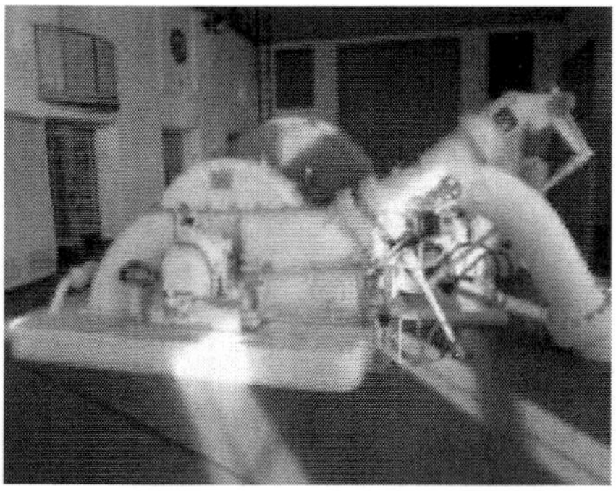

Photo 2. Mülhau: Drinking water turbine (2-nozzle Pelton turbine).

One of the first hydropower plants on the drinking-water network in Europe was erected on the drinking water pipe running down to Lausanne, Switzerland, in 1901. The power plant, Sonzier, still operates nowadays, with an output of 1.6 MW and a yearly production of 6.6 GWh, or the electricity consumption of 1470 European households[2].

On the agenda of a recent SHP conference organised in Lausanne (Switzerland, 2010) , one of many items discussed was the multipurpose hydro schemes under which electricity generation in drinking water supply networks were deeply analysed. A number of case studies were presented (Krasteva, 2010; Toader et al., 2010; Bischoff, V. & Salamin, 2010). Conception and design of a micro-hydro in a water supply system are discussed in Ramos et al. (2010). A US based company proposed a turbine that can be used instead of the pressure-reducing valves found throughout municipal water systems (Bodin, 2008). Rather than overcoming the resistance of a valve's spring-loaded diaphragm, the energy of the water drives the turbine. A similar project was realised at another water supply system with installation of a hydropower system by replacing a pressure reducing valve (White, 2011).

3.2.2 Within an irrigation network

The potentials available within an irrigation network are similar to the ones on a drinking water network. The SHP project has to be flexible enough to maximise the electricity production the whole year and not only during the irrigation period (Giacopelli & Mazzoleni, 2009).

- Case study Armary, Switzerland: Historically, the Armary, a small water stream, was used to irrigate the lands of Allaman castle. Before the hydro scheme implementation, the farmers used diesel driven pumps to irrigate their fields during the summer season. In 2006, a penstock was installed as a by-pass to the stream, still fed with a reserved flow, connected to a turbine and to spraying devices in the fields (145 hectares) for irrigation.
  The turbine discharge regulation is the water level of the forebay. Using this parameter allows the turbine to operate automatically even during the irrigation season. When the farmers are irrigating their fields, the forebay level drops causing the turbine discharge to be reduced or even stopped. As the turbine is equipped with two jets, it operates with good efficiency even on low part-flow discharges.

In this way, water is available for the farmers at the pressure directly suitable for their spraying equipment (10 bars). Therefore, pumping is no longer necessary, which has reduced CO_2 related emissions. Water is also available all year round for the hydro plant (cf. Photo 3).

- Case study Marchfeldkanal, Austria: The existing irrigation channel system is about 20 km long and comprises 8 weirs equipped with flap gates to regulate the water level. The highest weir was selected to implement a small hydropower plant upon (cf. Photo 4). All the irrigation operational requirements have been safeguarded. The system is an unusual one in that it uses a so-called "hydraulic coupling". Both turbines

---

[2] The average electrical consumption of a European household is estimated here at 4,500 kWh/year.

are connected indirectly to a unique generator via oil hydraulic pumps. The hydraulic pumps drive a hydraulic motor, which then drives the electrical generator.
The purpose of the hydraulic coupling is to replace the two-speed increasers and two generators by two pumps, one motor/generator and an oil pressure unit. The hydraulic circuit gives freedom to locate the motor/generator at a distance of 10 meters from the turbines, on the bank of the water course. The first advantage of this arrangement is that the size of the complete installation is substantially reduced. The second advantage is that the location of all the electrical equipment is on the bank well clear of flooding and easily accessible. Due to the additional stages in the energy conversion process, losses are increased, something that was underestimated at the start of the project. The overall efficiency may be between 60 - 70%. The annual output (0.5 GWh/year) is due to the considerable discharges available in the channel, which is itself fed by the Danube River.

Photo 3. Armary: the power house.

Photo 4. Marchfeldkanal: the turbines.

Photo 5. Rino: recreation area around the basin.

Photo 6. Rino: a view of the power station.

- Case study Rino, Italy: The multipurpose use of water in an Alpine Park (hydroelectric production + tourist attraction + irrigation) makes the Rino hydroelectric plant (cf. Photo 6) an interesting example of how to balance the temporary use of natural resources with considerable environmental constraints. The small basin permits transfer part of the daily production from the off-peak hours to the peak ones. This has been designed to be an attractive place for the tourist activities (angling, picnic, recreation).

The plant was designed to exploit the variation of water levels in the basin, which is kept between precise limits in July and August so that it can be utilised for angling. The tourist use of the basin has been improved by the construction of a recreation area nearby (wood, picnic sites, fountains, toilets block) (cf. Photo 5). The tail race of the hydroelectric plant supplies screened de-silted and regulated water to a sprinkler irrigation plant.

The success of this project, being in a park environment, shows that carefully designed small hydro development is compatible with sensitive management of the environment and with other enterprises (such as agriculture and tourism). The aim of the project was not only to respect these activities but, when possible, to enhance them.

3.2.3 Wastewater treatment plant

There are two possibilities to generate electricity from wastewaters (Fig. 3). The first one is before the wastewater treatment plant (WWTP). In such case, the wastewater network of a built-up area will lead to a forebay equipped with a thin trash rack and a rack cleaner. The wastewater is then led through a penstock to the WWTP, situated at a lower elevation, where it passes through the turbine before being treated through the usual process.

The turbine has to be set as close as possible to the elevation of the treatment basin to maximise the head.

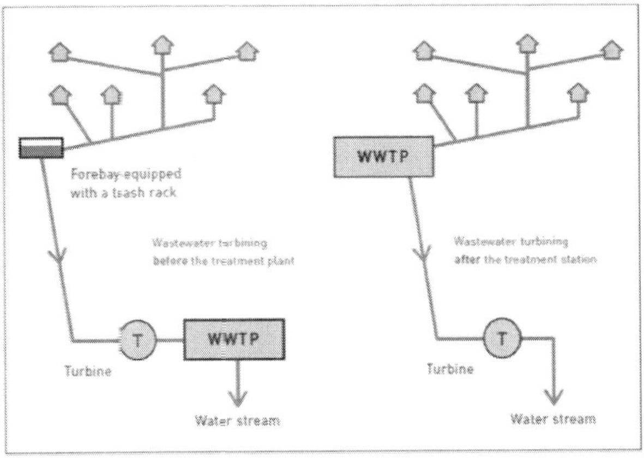

Fig. 3. Turbine setting before and after the wastewater treatment plant (WWTP).

Case study Le Châble Profray, Switzerland: The wastewater from the outlets of the Verbier ski resort is collected in a storage basin of 400 m^3, equipped with a 6 mm trash rack to remove floating material. This basin is now also used as a forebay for a hydro scheme where the power house is located at a distance of 2.3 km below within the treatment plant (cf. Photo 7). After passing through the hydro turbine (cf. Photo 8,Photo 37 andPhoto 38), the wastewater discharges into the treatment plant inlet before finally being re-introduced to a nearby water stream. A bypass is incorporated to guarantee the wastewater treatment operation, whether or not the hydro plant is operational, and for times when the plant operational discharges need to be greater than the turbine maximum discharge.

Photo 7. Le Châble Profray: The valley and the wastewater treatment plant where the turbine is set.

Photo 8. Le Châble Profray: The turbine and its runner and the alternator, during the erection.

The second possibility is after the WWTP. In this case, the treated water that comes out of the WWTP is led down through a penstock to a turbine before being discharged to a lake or a water stream. To maximize the head, the turbine will then be close to this restitution.

For some sites, the hydropower project can lead to improving the cost efficiency of a longer penstock to reach a water stream where dilution can be more significant.

- Case study Seefeld, Austria: To reach the Inn River, the treated wastewater from Seefeld wastewater treatment plant needs to be pumped to pass over a hill and then discharges to the hydropower plant. After the turbine (cf. Photo 10), the water passes through a de-foaming plant and then is discharged into the Inn River, meeting the dilution criteria for treated wastewaters. To guarantee these discharges, a permanently available bypass with energy dissipation is installed. The turbine and its bypass are integrated in a central process control system for automatic operation.
  The project feasibility is justified by the site topology. The hill between the sewage plant and the Inn River is a relatively small percentage of the over gross head available (head for the pumps: 94 m / head for the turbine: 625 m). Note that the electricity generation from this scheme exceeds both the pump energy consumption (1.5 GWh/year) as well as the wastewater treatment plant consumption (0.5 GWh/year) so that excess local generation can be exported onto the grid network. Additionally, by discharging the treated wastewater into a larger receiving stream, the local ecology is improved. A creative approach has been to the architecture of the power house: a water droplet shape creates a thought provoking image for the general public (cf. Photo 9).
- Case study Nyon, Switzerland: In the 1990's, due to a lack of space near Geneva Lake, the new wastewater treatment plant (WWTP) of Nyon City was built 110 meters higher on the plateau. Since then wastewaters are collected in a basin close to the lake, pre-treated, and then pumped to the WWTP where they are treated. Then they pass through a turbine before their discharge to the lake.

The electricity production (0.7 GWh/year) represents half of the pumps consumption, and the third of the water treatment one.

Photo 9. Seefeld: the power house.

Photo 10. The turbine using treated wastewaters.

It can be noted that both possibilities can be technically implemented. As Samra project in Jordan is an example of electricity production from wastewaters before and after the water treatment plant (cf. Photo 11-14). This project at the time was one of the largest of its kind in the world considering the output (2 x 830 kW and 2 x 807 kW). The electrical energy balance can also be pointed out: 90 % of the electrical consumption of the wastewater treatment plant is covered by these hydropower plants and an anaerobic digestion process (Denis, 2007; 2008).

Using wastewater flows to make power is a relatively new idea, but not unprecedented. As for drinking water networks, hydropower production from wastewater flows is also popular in Switzerland (Chenal et al., 1994; SFOE, 1995).

Vienna's main wastewater treatment plant is one of the biggest and most technically advanced sewage treatment facilities in Europe and this requires an enormous input of energy. The concept made use of the existing gradient between the plant outlet and the receiving water - the Danube Canal, along which some 6.5 m^3/s of purified effluents are discharged from the treatment plant per day. Based on the current amount of effluents and a level difference of 5 m between headwater and tail water, the use of a turbine typically designed for small hydropower plants presented itself as a viable option (some 400 kW capacity) (Hahn, 2009).

The US based the Low Impact Hydropower Institute's (LIHI) highly certified the Massachusetts Water Resource Authority's Deer Island hydroelectric project at its WWTP (LIHI, 2009). Once treated wastewater is disinfected, it is discharged into effluent channels and transmitted through to two corresponding hydro turbines (each 1 MW Kaplan).

Australia's North Head Sewage Treatment Plant started up a 4.5-MW small hydro unit that harvests power from treated wastewater falling down a 60-meter shaft. Along with a methane gas cogeneration unit that was also recently installed, this plant now generates nearly 40% of its own power (Patel, 2010).

Photo 11. As Samra hydropower plant and wastewater treatment plant inlet structure.

Photo 12. As Samra hydropower plant on the treated wastewater.

Photo 13. Two 5 nozzle Pelton turbines set on the raw wastewaters of Amman City, As Samra plants (Jordan) (H = 104 m, Q = 2 x 1.25 m^3/s, P= 2 x 830 kW, E=12.5 GWh/year, 2007).

Photo 14. Two Francis turbine set on the treated wastewaters of Amman City, As Samra plants (Jordan) (H = 42 m, Q = 2 x 2.3 m^3/s, P= 2 x 807 kW, E=8.6 GWh/year, 2007).

3.2.4 Within a urban runoff collection system

The type of potentials available within a runoff collection system is similar to the ones on a drinking water network. The main issues are the particles carried by the water through the turbine and irregularity of the discharges, which can be managed by accumulation.

3.2.5 On a reserved flow or compensation discharge

In most developed countries, water withdrawal from a river goes by the definition of an environmental body of a minimal flow to be maintained in the river, the amount and variability depending on national laws. This flow, called reserved, environmental or compensation discharge, is discharged to the rivers at the foot of weirs or dams built for hydropower schemes or water treatment works. Thus this implies a loss of electricity for the hydropower schemes (Pelikan, 2005). But an energy recovery is possible by setting a SHP plant at the foot of the weir or dam to use this reserved flow and the difference of levels between the upstream water level in the basin and the level of the water restitution to the river.

- Case study Llys y Fran, United Kingdom: In the United Kingdom, abstractors of water normally have an abstraction license from the Environment Agency, that defines a compensation flow to be maintained in the river at all times. Llys y Fran water treatment scheme, located near the Preseli Mountains in Pembrokeshire, is composed of a dam built on a river (cf. Photo 15) to accumulate water that will be then treated before consumption. As a compensation discharge of 160 l/s is required, and thanks to the difference of levels between the reservoir water levels and the foot of the dam, a turbine has been set that generates around 0.2 GWh/year.
  The existing hydro scheme commissioned in the early 1970s was underutilised, mainly because of a lack of automation The main issues dealt with working on an operational site where the priority lay with delivering raw water for treatment, whilst at the same time, making sure that the compensation discharge was not affected. In 2008, the hydro plant operation was refurbished and automated, whilst the compliant grid connection was facilitated.

Photo 15. Llys y Fran: the dam.

Photo 16. Le Day: The foot of the dam where the small power plant will be set.

- Case study Le Day, Switzerland: Le Day dam (cf. Photo 16) was built in the 1950s on the Orbe River to feed the underground power plant of Les Clées (27 MW) and Montcherand (14 MW). At the foot of the dam are located the valve chamber and the penstock that leads to Les Clées power plant. In Switzerland, from the federal law on water power use (from 1916 and revised in 2008), to let a reserved flow at the foot of dams becomes mandatory five years at the latest after the concession expiry. Although the concession is here valid until 2034, the operator applies already the recommendations from the cantonal water authority by letting a reserved flow of 400 l/s to the water stream. Recently the authority has defined again the reserved flow regarding the seasons. Finally, it will be 600 l/s from July to September and 300 l/s the rest of the year, which represents the same annual amount of water as the current situation. The project is then to use this reserved flow and the gross head between the back water level and the foot of the dam to produce electricity.

As the head varies between 17 and 27 meters, a Kaplan turbine (cf. § 5.3.1 and 5.3.2) with variable speed will be set. The hill chart of the turbine is here an essential tool as it permits to optimise the production by guaranteeing high performance and operation with cavitation erosion for the two discharges and head variations. This project has then two positive impacts: it permits to recover a part of the green electricity production lost by the large power plant while favouring the local ecosystem.

3.2.6 On a fish pass system

Fish passes and bypass systems at hydropower plants can cause losses in electricity generation from a few percent to more than 10%. Modern technologies as well as unusual design solutions allow to transform the water energy lost as reserved flow in a new resource available downstream of weirs and dams of existing hydro power plants (Papetti &Frosio, 2010; Rizzi et al, 2010).

To help fish to locate and navigate their way to the fish pass entrance, an additional discharge is necessary at its entrance downstream. The idea is to exploit this discharge and the head in the dam with a small hydro scheme, by arranging for an intake upstream of the dam with a penstock pipe routed parallel to the fish pass, and the turbine discharging near the entrance to the fish pass.

- Case study Aire-La-Ville, Switzerland: The Verbois large hydropower plant (100 MW, 466 GWh/year) is sited on a dam across the river Rhône near Geneva. The maximum head achievable in the dam is 21m. In 1999 a fish pass (cf. Photo 17) was installed (the longest of Switzerland with 350m), comprising 107 pools, supplied by a discharge of 710 l/s. To help fish to locate and navigate their way to the fish pass entrance, an additional discharge of 2 m^3/s was deemed to be necessary at its entrance downstream all year round. A proposal was made to exploit this discharge and the head in the dam with a small hydro scheme, by arranging for an intake upstream of the dam with a penstock pipe routed parallel to the fish pass, and the Francis turbine (cf. Photo 18) discharging near the entrance to the fish pass. Since 2003, the upstream fish migration has been guaranteed for 26 species, while the production of electricity has been facilitated.

Photo 17. Aire-La-Ville: Verbois fish pass and the SHP.

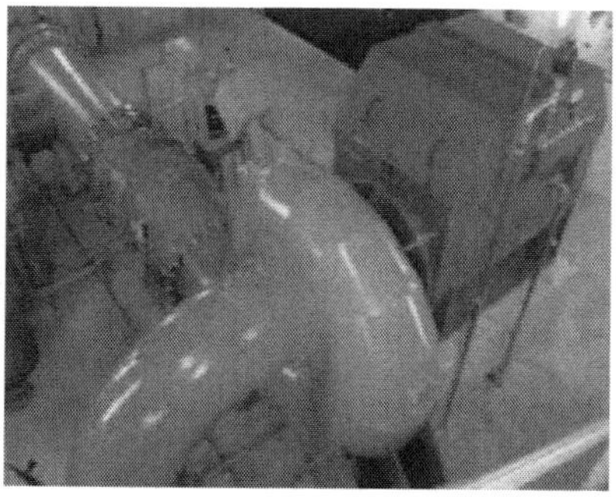

Photo 18. Aire-La-Ville: the Francis turbine set on the attraction discharge, close to the fish pass entrance.

3.2.7 In a navigation lock or dam

Navigation locks and dams cause water level fluctuations. Energy recovery consists then in using the difference of water levels by setting the turbine into the channel, even during the filling and emptying of the locks. As the flood passage capacity has to be maintained, the machine will have either to be set as a bypass of the channel, or to be lifted higher than the upstream flood level.

- Case study L'Ame, France: The Mayenne River is navigable and equipped with 16 locks & dams. The l'Ame project is the second fitted with a very-low-head turbine (Kaplan type) on this river (cf. Photo 19 and Photo 20). A program to equip the 14 remaining locations is being developed.

The main challenge in this case was to fit in 19th century infrastructures with a small visual impact and high fish friendliness due to the presence of silver eels.

Photo 19. L'Ame: downstream global view of the dam and the turbine.

Photo 20. L'Ame: upstream view of the turbine.

During ship locks operation depending on their construction and frequency of passage of the vessels 0.01 to 1% of annual flow volume must be available. This represents a loss in electricity generation if inland navigation is associated with a hydropower plant. To recover this type of energy, a pilot project was installed in a ship lock at Freudenau hydropower plant in Vienna, Austria (Wedam et al, 1999). The 5 MW capacity module is designed to generate power during both the filling and emptying of lock operation. It is composed of 25 small and identical units of 200 kW each, arranged within a frame in the shape of a matrix (Wedam et al., 2004; Schlemmer et al. 2007). An alternative technology to recover energy lost for ship locks operation has been developed in the US. There is an opportunity to install low head hydro for over 230 locks and dams with auxiliary locks in the U.S (Krouse, 2009).

3.2.8 In a desalination plant

Desalination plants use reverse osmosis to separate water from dissolved salts through semi-permeable membranes under high pressures (from 40 to 80 bars).

The residue of liquid water containing salt, still at high pressure can be passed through a turbine in order to recover part of the energy used for the initial compression.

- Case study Tordera, Spain: Tordera desalination plant generates drinking water for Maresme Nord and for La Selva, situated on the North coast near Barcelona. The plant takes sea water from wells, which implies that less water is taken from the aquifer and sea intrusion can be stopped. The reverse osmosis is the process used to separate water from dissolved salts through semi-permeable membranes under high pressures. Here four groups are set (cf. Photo 21 and Photo 22), each one composed of a pump, a motor and a 1-jet Pelton turbine on the same axis. The pumps are used to increase the water pressure (up to 70 bars) so that the water (without salt) can cross the membranes, while the turbines recover the energy from the concentrate outlet of the reverse osmosis, inferring smaller motors. Finally 10 to 20 hm^3 of drinking water are generated per year.

Photo 21. Tordera: the four groups.

Photo 22. Tordera: a dismantled Pelton turbine.

Potentials for development of hydro-powered Red Sea water desalination in Jordan are discussed in Akash &Mohsen (1998). A paper dealing with a global environmental analysis of the integration of renewable energy—wind energy, photovoltaic energy and hydro-power—with different desalination technologies is given by Raluy et al. (2005).

3.2.9 In a cooling or heating system

Cooling or heating systems can present a pressure difference that can be recovered by hydro turbines. A system designed by Frederiksen et al. (2008) recovers excess pressure from a district heating system to direct-drive the circulation pump within the building (typically rated around 1 kW) and a small generator. This not only maintains the hot water circulation, but also provides enough power to run the electrical control system so that the heating continues to operate even when there is a fault in the electricity network. Wollerstrand et al. (2009) gives a similar case of a small turbine set for energy recovery that can drive (directly or not) the circulation pump. Bansal & Marshalla (2010) investigated the feasibility of recovering lost energy from typical bio-gas upgrading facilities by means of a hydraulic turbine, and presented analysis of different types of hydraulic power recovery turbines.

- Case study Sangüesa, Spain: This hydropower project was part of a scheme to improve the cooling system at the Sangüesa Biomass plant. Condenser cooling needs a back pressure to operate, which necessitates a tower of balance of 10.5 meter high (cf. Photo 23). It can be noted that as the biomass plant and the turbine (cf. Photo 25) operate together, the turbine needs operate in continuous service for around 8'000 hours/year.

- Case study Skawina, Poland: The hydropower plant (HPP) was planned together with the thermal one (ThPP). The ThPP uses cooling water from Laczany Channel that bypasses a 20 km long segment of Vistula River and serves also for navigation purposes. After passing through the cooling system of the ThPP, water is led to the HPP (cf. Photo 24) by two concrete channels. The final portion of these channels is open with side walls used as spillways. The plant is equipped with a single hydraulic unit (Kaplan turbine and generator).After leaving the HPP, water is discharged through a 30 m long tailrace channel to Skawinka river.

Photo 23. Sangüesa: the biomass plant and its tower of 10.5 meters.

Photo 24. Skawina: the powerhouse during turbine overhaul.

4. How to start and develop a multipurpose scheme project

4.1 Main calculations

Here is a brief reminder on basic calculation. For more details, the reader is referred to the Guide on how to develop a small hydropower project (ESHA et al., 2005).

The electrical output power, P, of a hydropower plant is defined by:

$$P = \rho \cdot Q \cdot g \cdot H \cdot \eta_c \cdot \eta_t \cdot \eta_e \cdot \eta_{tr} \qquad [W]$$

With:	ρ	=	specific weight of water ≅ 1000 [kg/m³]
	Q	=	discharge [m³/s]
	g	=	acceleration due to gravity [m/s²]
	H	=	gross head [m]
	η_c	=	penstock efficiency ≥ 90% at nominal discharge [-]
	η_t	=	turbine efficiency 88% ≤ η_t ≤ 94 % at nominal discharge [-]
	η_e	=	generator efficiency ≥ 92 % at nominal discharge [-]
	η_{tr}	=	transformer ≥ 97 % [-]

The efficiencies mentioned above correspond to the present state of the art for a scheme that uses optimally the water resource.

Whereas for rivers, the yearly production (kWh/year) can usually be estimated by multiplying the maximal electrical output by 4500 hours/year, it is not possible to define this factor for multipurpose schemes. Regarding the collected case studies, the operation at full load varies between 2200 and 8700 hours/year.

4.2 Recommended steps for developing a SHP project

The following Table 3 lists the recommended steps of a SHP project from site identification to commissioning. Due to cost efficiency constraints, it may be reduced for sites which output is lower than 15 kW.

	Steps	Goal
1	Site identification	To define the main site characteristics and specificities and to involve the main entities concerned by the existing infrastructure (cf. § 4.3)
2	Preliminary analysis	To evaluate the technical, environmental and economic (with an accuracy of circa 30 %) feasibility of the project: is it worth going further?
3	Feasibility study	To evaluate the technical, environmental and economic (with an accuracy of circa 25 %) feasibility of the project and define the final solution
4	Implementation project	To achieve the specifications for the whole design of the SHP plant (equipments and civil works), and the final plans with a focus on the water quality and on the integration into the existing infrastructure (cf. § 4.4 and 5.1)
5	Public information	To reduce the risk of future public opposition
6	Public inquiry	To obtain the necessary authorisations peculiar to each country
7	Call for tenders and final design	To achieve a call for tenders to equipment suppliers and civil engineering firms, to propose the award, to achieve the final drawings of the schemes
8	Implementation and commissioning	Turbine manufacturing, civil works, erection on site

Table 3. Recommended steps of a SHP project in an existing infrastructure, for an output higher than 15 kW.

4.3 Site identification

As mentioned in the previous table, the first step to start a multipurpose project consists in creating collaboration between the infrastructure owner and SHP specialists and collecting information. Here is a first checklist:

- Definition of the primary function of the existing infrastructure,
- Maps and drawings,
- Head or pressure definition:
  - What is the upstream water level?
  - What is the downstream water level?
  - What are their yearly evolutions?
- Pipes characteristics: length, internal diameter, nominal pressure, roughness, age, state, head losses regarding discharges,
- Hydrology:
  - Are there any flow meters in the water network?
  - Definition of the flow duration curve with daily data, the compilation on 10 years being an optimum (cf. Fig. 4),

- Are there any seasonal variations?
- For water networks: evolution of the inhabitants
- For drinking water networks: sources discharges, number of consumers, consumption data and their evolution,

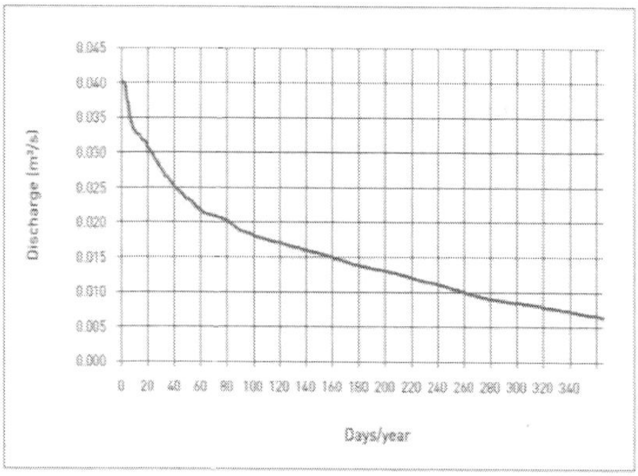

Fig. 4. Example of a flow duration curve.

- Water quality, as defined in section § 5.1
- Evolution of the existing infrastructure (projects? extension?)
- Where could the power house be set?
- Is there a grid close to the existing infrastructure?

Each SHP project is specific to the scheme where it has to be integrated. It is mainly defined by a nominal discharge, a gross head and head losses in the infrastructure as detailed in the Table 4. Then, the yearly average evolutions of the discharges and heads will lead to the production calculation.

Topic	Symbol	Units	Definition
Nominal discharge	Q	m^3/s	The nominal discharge depends on the flow duration of the site, so as to optimise the production all over the years (cf. Fig. 4).
Gross head	H	m	The gross head is defined by the difference in levels between the upstream water level at the collecting chamber or reservoir or penstock forebay and the downstream water level (at the reservoir, at the treatment plant).
Head losses	Hr	m	Head losses are a loss of energy within the infrastructure (penstock, channels) (cf. § 5.2).

Table 4. Main parameters to define a hydropower site.

4.4 Main requirement: integration to the existing infrastructure

Once the feasibility study has demonstrated the project viability, the implementation project will lead to define the whole design of the SHP plant, with a focus on the integration to the existing infrastructure. In other words, the SHP plant must not impact on the primary function of the site. Table 5 gives a list of basic recommendations.

Infrastructure requirements	Recommended technique
Water quality	The SHP plant must not impact on the water quality, unless it leads to its improvement, while optimising the equipment efficiencies and lifetime (cf. § 5.1).
Discharges at the turbine outlet	The turbine is designed from the flow duration curve of the scheme (cf. figure 3) so as to optimise the production. A bypass is set to reach the infrastructure discharge requirements at any times. Storage is avoided, apart when required for the existing infrastructures (cf. § 5.3.2 and 5.8).
Pressure at the turbine outlet	For heads > 60 meters, if the needed turbine outlet pressure has to be higher than the atmospheric one, the Pelton turbine is at a higher elevation, or a counter pressure Pelton turbine is set (cf. § 5.3.5).
Flexibility	The turbine has high efficiencies for the optimal range of pressure and discharges, defined by the existing scheme (cf. § 5.3.2).

Table 5. Technical recommendations for the integration of the SHP plant into the existing infrastructure.

4.5 Economic aspects specific to multipurpose schemes

The selected case studies show a wide range of investment: from €90,000 to 3,945,000, showing how each multipurpose project is specific. However, a few common principles can be mentioned.

First, the economic calculations distinguish the investments due only to the hydropower plant from the ones due to the primary function of the existing infrastructure. For example, a 100 mm diameter penstock can be sufficient for a water network, but as it may result in high head losses (cf. § 5.2), a 150 mm diameter pipe will be necessary for the hydropower project. Then only the cost difference between both penstocks (supply and setting) will be considered in the economic analysis of the SHP project.

Then, maintenance and operation costs will be reduced with sustainable equipment especially designed for the site. If the generator is connected to the national grid, the selling price will depend on the small-hydropower regulation proper for each country. Finally, by creating a source of income, a hydropower project can be a good opportunity to improve the existing scheme.

5. Technical recommendations for SHP plants set in existing infrastructures

The first recommendation, as for any projects, is the design as a whole at an early stage. In addition to this general principle, this section has the objective to list a selection of technical recommendations for multipurpose schemes, with a focus on integration to the existing infrastructures.

5.1 Water quality and its impacts on the scheme design

A SHP plant must not impact on the water quality, unless it leads to its improvement, while optimising the equipment efficiencies and lifetime. Especially while defining the penstock and turbine, attention will be paid on the mechanical resistance and manufacturing easiness of the selected materials but also on their corrosion and abrasive behaviour. Table 6 lists some technical consequences of the water characteristics on the SHP plant design.

Water quality	Recommended technique	Existing infrastructure									
		Drinking water network	Irrigation water network	Raw wastewater network	Treated wastewater network	Runoff collection system	Reserved flow	Fish pass	Navigation lock	Desalination plant	Cooling / heating system
Gravels and stones	Setting of a grid at the forebay	X	X	X		X	X	X	X		
Sand particles	Setting of a de-silted set before the forebay	X	X	X		X	X	X	X		
	Pelton runner built with mounted buckets to unset and replace the buckets										
Drinking water	All parts in contact with water in stainless steel	X									
	Electrical actuators to replace all oil ones	X									
Chlorinated water	Sacrificed anodes to prevent from erosion	(X)		X	X						
Salt	All parts in contact with water in a high quality stainless steel									X	
Organic wastes (bacteria)	Increase of the penstock internal diameter, to limit head losses due to the deposits on the wall created by bacteria			X	X						
Fat	Fat removing system at the forebay			X							
Fibrous and filamentous matter (plants, strings, ...)	Setting of a screening system equipped with a trash rack at the forebay to limit the wastes that enter the penstock and the turbine.		X	X		X			X		
	Suppression of all obstacles where the materials could accumulate. For Pelton turbines, it means no x-cross liner for the nozzles and no deflector.			X							
	Progressive flow speed increase within the turbine, to avoid trash accumulation			X							
	Integration of hand holes in the casing to clean the machine			X							
	For small Kaplan and diagonal turbines, special cleaning programme based on the closure of the downstream valve.		X	X			X		X		
Wastewater	All parts in contact with water in stainless steel			X							

Table 6. Technical recommendations due to water quality on SHP plant design.

It can be noted that the following infrastructures use water which quality is similar to rivers:

- irrigation water network
- reserved flows or compensation ones at the foot of hydropower dams, or of water treatment plants
- fish pass system
- navigation locks and dams

For cooling/heating systems, a priori the water quality does not imply a specific design for the turbine. Nevertheless, its temperature has to be considered.

5.2 Penstock and head losses

At the start of a SHP project in an existing infrastructure, a first issue is to define if the existing penstocks and channels are suitable to electricity production, which implies mainly to check their mechanical resistance (nominal pressure for a penstock) and head losses.

In general, head losses are acceptable if at nominal discharge they are lower than 10 % of the difference in levels, or in other words if the penstock efficiency is higher than 90 %. Indeed, this corresponds to the present state of the art for equipment that uses optimally the water resource.

To sum up, head losses in a penstock depend on:

- Its shape: singularities as elbows or forks tend to increase head losses
- Its internal diameter
- Its wall roughness and its evolution due to its degradation or/ and to wall deposits.

It may be recalled here that energy loss due to friction in a penstock can be estimated as being inversely proportional to its diameter to the power of five. For instance, a diameter increase of 20% leads to a head losses decrease of 60%.

When considering a wastewater network, the pressure due to the difference of levels between the forebay and the treatment plant (WWTP) has to be reduced, which tends to select a penstock with a small diameter. Thus this will transport wastewaters while wasting the pressure useless for the treatment process. On the contrary, if the objective is to produce electricity, the pressure has to be maximal where the turbine will be set. Therefore, a penstock with a larger diameter will be selected to minimise head losses.

When dealing with raw or treated wastewaters, a possible deposit of polluting loads on the penstock walls due to organic wastes has to be considered. Observations show that this deposit can easily exceed 1 to 2 mm.

Table 7 presents how important the choice of the penstock diameter is, and points out its clogging impact. Calculations have been achieved using Colebrook formula for an 860 m length penstock, a discharge of 280 l/s and a gross head of 115 m. The results are expressed as the penstock energy efficiency, ratio between the gross and net heads.

Penstock diameter (mm)	Polluting load scale thickness (mm)	Head losses (m)	Penstock energy efficiency (%)
312	0	22.7	80.3
312	2	44.2	61.6
380	0	8.5	92.6
380	2	15.5	86.5

Table 7. Head losses in a penstock regarding its diameter and clogging.

Turbine type	Operation range	Multipurpose schemes									
		Drinking water network	Irrigation network	Raw wastewater network	Treated wastewater network	Runoff collection system	Reserved flows and compensation discharges	Fish bypass system	Navigation locks and dams	Desalination plants	Cooling/heating systems
Pelton	60 -1000 m	X	X	X	X	X				X	
Francis	20 - 100 m	X	X	X	X	X	X	X		X	
Diagonal (Deriaz)	25-100 m	X	X	X	X	X	X	X		X	
Kaplan	1.5 - 30 m	X	X	X	X	X	X	X	X	X	X
Reverse pump	< 30 kW	X	X	X	X	X	X	X	X	X	X

Table 8. The five main types of turbine (see also Photos 25-29).

As shown by the above-mentioned values, a small diameter change (+ 21 %) does not only result in reducing head losses (and thus the production loss), but also in reducing the dependency from the clogging thickness. It can be noted that 312 mm and 380 mm are standard diameters, and that excavation and setting costs will be similar for both variants. Moreover, the energy efficiency of the 380 mm penstock without clogging fulfils the SHP performance requirements.

Finally, as for the whole hydropower area, head losses in pipes or channels have to be considered in the cost efficiency of a multipurpose project. Indeed, only a technical and economic calculation, based on the production gain and the cost difference between the variants, will permit to select the optimal equipment.

5.3 Turbines

5.3.1 Main types of turbines

The above table 8 presents the five main types of turbines. It shows that they are suitable to all multipurpose schemes (considering that dams and locks higher than 60 meters are rare for SHP to set a Pelton turbine). Reverse pumps (cf. Photo 26) are often found in drinking and other water networks, when the available output is lower than 30 kW, thanks to their affordable price (Williams, 2003; Williams, 2010, Budris, 2011; García et al., 2010; Steller et al, 2008; Sulzer Pumps, 2011). Another advantage of using pumps within the water industry is that a pump is a familiar piece of equipment, and maintenance requirements are well known (Orchard & Klos, 2009). However, as seen in section § 5.3.2, they are generally not suitable to multipurpose schemes.

Photo 25. Francis turbine with a spiral casing (case study Sangüesa).

Photo 26. Reverse pump set on the treated wastewater (case study Nyon).

Photo 27. The Diagonal turbine set in Mhylab's test bench.

Photo 28. Kaplan runner with 8 blades during manufacturing, to be set within a drinking-water network (case study Poggio Cuculo).

Photo 29. The Pelton runner and its 3 jets (case study La Zour).

5.3.2 Flexibility and performances

The SHP plant operation must not impact on the primary function of the existing infrastructure. Thus, the turbine has to be as flexible as possible regarding the available pressures and discharges, while guaranteeing high performances on the largest operation ranges (Table 9).

The turbine design is based on the site flow duration curve (cf. Fig. 4), a crucial tool to optimize the production and the viability of the project. Indeed, the discharges can evolve with the spring hydrology and/ or with human activities.

Turbine type	Discharge control device	Minimal discharge
Pelton	One to five adjustable nozzles	At least 15% of the nominal discharge of one nozzle
Francis	Adjustable guide vanes	Circa 50 % of the turbines' nominal discharge
Diagonal and Kaplan	Fixed or adjustable guide vanes, adjustable runner blades	At least 20 % of the turbines' nominal discharge
Reverse pump	No device	85 - 90 % of the machines' nominal discharge

Table 9. The five main turbines and their flexibility.

The case study le Châble-Profray set on raw wastewaters is an interesting example of over-dimensioned project. The first project, in 1993, was based on a nominal discharge of 240 l/s that considered the sudden discharge changes due to storms and snow melting, and also the important population increase due to the winter touristic activities. Therefore, the turbine was only working a few days per year at its nominal discharge. Moreover, during the dry season, the limited available discharge implied to be stored at the forebay to allow electricity production. This storage resulted in an important generation of decanted deposits. An accumulation of grease at the surface was also observed, leading to form a crust that had to be regularly removed. Furthermore, such wastewater storage makes the further treatment more difficult. Finally, the new turbine was designed for 100 l/s, leading to a production increase of 45% (0.85 GWh/year instead of 0.58 GWh/year), although the nominal discharge is 2.4 times lower.

Some multipurpose schemes deal with steady discharges, as for the following case studies:

- Aire-la-Ville, dealing with an attraction discharge for fish to find the entrance of the upstream migration system,
- Llys y Fran, dealing with a compensation discharge for water treatment schemes

Then, SHP plants at the foot of large hydropower dams generally work with a steady reserved flow. However, the case study Le Day deals with a reserved flow that doubles during the summer season.

For the case study Poggio Cuculo, the turbine works with three different drinking water discharges throughout the year depending on the season and if it is day or night. This variation is due to the price of the electricity consumed by the water treatment plant.

High performances depend on the site definition and on the whole design of the SHP plant. Therefore the project manager is recommended to go through all the analysis steps listed in Table 3 in collaboration with small hydropower specialists, and to ask the suppliers to justify the efficiencies of their equipment.

As shown on Table 9 and Fig. 5, Pelton, Diagonal and Kaplan turbines are especially recommended for their flexibility regarding discharges.

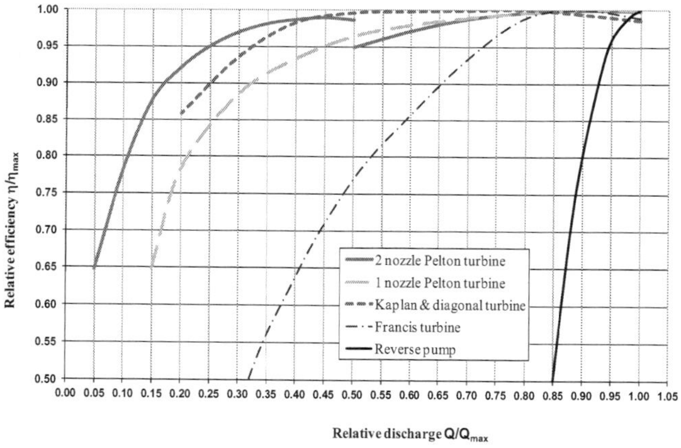

Fig. 5. Relative efficiencies regarding the discharges for Pelton, Diagonal and Kaplan, and Francis turbines, and reverse pump.

On the contrary, **a reverse pump** is not recommended regarding its lack of flexibility due to the absence of a regulation device, leading to:

- a cyclical operation:
  - it infers numerous starts and stops, leading to an untimely wear of the equipment,

- it requires a buffer reservoir designed for at least one operation hour,
- a problematical synchronisation,
- a specific design to operate with high performances as a turbine, which reduces its low investment advantage.

The case study Nyon commissioned in 1993 is composed of a reverse pump (cf. Photo 26) especially designed for the site. As it works with a fixed discharge, the frequent automatic operations to start up and shut down the reverse pump (circa 18 times per 24 hours) require especially sturdy drive systems that are relatively expensive. For example, the upstream butterfly valve has already been changed due to strong cavitation. Moreover, the neighbours complain about the noise and the vibrations due to these operations. Finally the operator has launched a study to replace the reverse pump with a Pelton turbine, with the objective to gain flexibility, reduce noise and vibrations and increase production.

5.3.3 Drinking water quality and turbines

To demonstrate that turbines can respect water quality, or in other words that drinking water can pass through the turbine before being consumed, a comparison with pumps can be achieved, as shown in Table 10.

	Pumping station	Turbine station
Inlet valve	yes	yes
Discharge regulation device	**no**	**yes**
Runner linked to a rotating shaft	yes	yes
Shaft gaskets	yes	yes
Casing and runner in contact with water	yes	yes
Greased-for-life roller bearings	yes	yes
Electrical machine	**yes (engine)**	**yes (generator)**
Electrical panels	yes	yes
Medium voltage / high voltage transformer	Yes, if needed	Yes, if needed
Usual building materials of the hydraulic machine	Cast, black steel, stainless steel, bronze	Cast, black steel, stainless steel, bronze
Automatic by pass	**no**	**yes**
Water access	Disassembly necessary	Disassembly necessary

Table 10. Comparison between a pump and a turbine station.

5.3.4 Adaptations to raw wastewater

The main difficulty with raw (untreated) wastewaters is linked with fibrous and filamentous residues that are not caught by the forebay grids (vegetal fibres, strings, threads, etc). Such materials can block on any obstacles in the flow, as for example in the guide vanes of a reaction turbine. Then, some other wastes can cling at them and agglomerate, which can lead to a partial or total clogging of the turbine and of its control systems.

For a **Francis turbine**, the guide vanes and the fixed blades of the runner are obstacles for the wastes. The cleaning of a jammed turbine can imply its whole dismantling, and the replacement of damaged parts, reducing the production of the power plant, and thus, increasing the kWh cost price.

Diagonal and Kaplan turbines face the same set of problems. But it is possible to remove some fibrous wastes by closing regularly the downstream security valve, so as to create a wave back.

On the contrary, **Pelton turbine** geometry is ideal for these applications. Indeed, the simplification of the turbine shapes by choosing progressive flow acceleration reduces waste accumulation. Fig. 6 shows the principle of a 4-nozzle Pelton turbine with such a simplified manifold composed of standard pipes, elbows and tees.

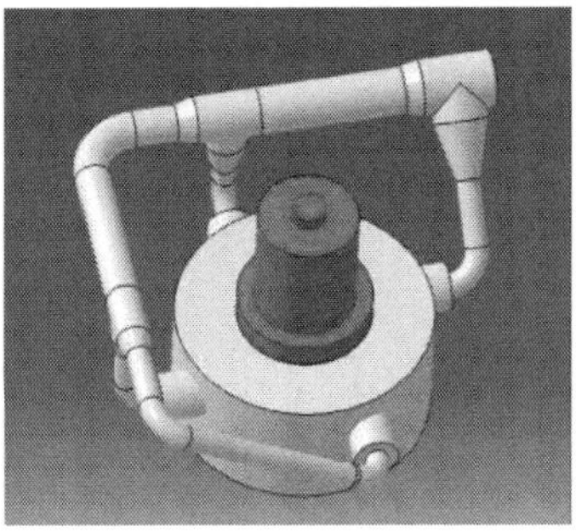

Fig. 6. 4-nozzle Pelton turbine with a progressive flow acceleration to avoid waste accumulation.

Photo 30. The x-cross liner for a nozzle, worn out by limestone.

Photo 31. Achievement of a Pelton runner with mounted buckets (St Jean SHP plant, Switzerland, set in a drinking water network, H = 373 m, Q = 34 l/s, P=102 kW, 2009).

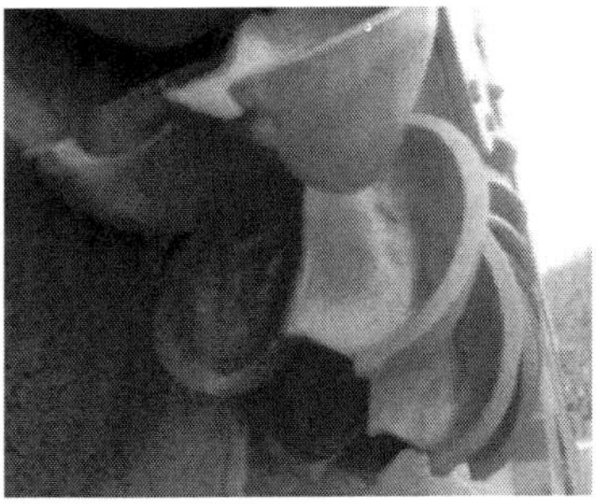

Photo 32. Pelton bucket worn out by sand particles (case study Le Châble-Profray).

Furthermore, it is recommended to avoid:

- the x-cross liners for the nozzles (cf. Photo 30)
- the deflectors (cf. § 5.5), which implies that the turbine and the generator must be able to bear runaway speed for at least the time needed to close the nozzles.

Once these usual design precautions are considered, the only possible clogging risk (but rare) concerns the nozzle tip liner. Finally, compared to a Francis turbine, the cleaning of a Pelton turbine is simple and can be achieved thanks to hand holes to get in the machine without dismantling it.

Regarding wear by abrasion, for **Pelton turbines**, it concerns the needle, the nozzle and, especially, the internal face of the buckets. As far as suitable manufacturing layouts have been achieved, the interchange ability of the needles and the nozzles should not be a problem. On the contrary the replacement and the repair of the buckets are not as simple. One solution is the runner with mounted buckets: the buckets are set together by screwing and pre-stress between two flanges (Photo 31 and 32).

5.3.5 Turbine setting

Whereas section § 3 described turbine setting regarding each multipurpose scheme, this section aims at detailing the possible positions of turbine regarding their types.

- **Pelton turbines and counter pressure turbines**

As a Pelton runner operates in the air, at atmospheric pressure, the reservoir which received the turbine outlet will be set high enough from the consumers to guarantee them a sufficient pressure.

When a higher outlet pressure is required for the existing infrastructure, a counter pressure turbine can be set. For this turbine type, the runner rotates in an air volume maintained at the requested downstream pressure (Photo 33).

Photo 33. Counter pressure Pelton turbine (Fällanden SHP plant, Switzerland, set in a drinking water network, H = 140 m, Q = 16 l/s, P= 17 kW, 2008).

- **Diagonal, Kaplan, Francis turbines and cavitation**

Diagonal, Francis and Kaplan turbines can be directly set as a bypass of the pressure breaker or of a valve. Fig. 7 and Photo 34 shows a turbine directly set as a bypass of the initial regulating valve.

However, the setting of diagonal, Francis and Kaplan turbines is limited by cavitation (Cottin et al., 2011). Such phenomenon can appear for any turbine, but especially for Kaplan turbines (Photo 35 and 36). Cavitation is the transformation of liquid water into steam, through a pressure decrease (Franc et al., 2000). The phenomenon is usually noisy, and always fluctuates strongly. The vapour bubble implosion close to the blade is responsible for its erosion, and for the deterioration of the turbine performances. And the erosion will keep on growing, while the production will keep on decreasing.

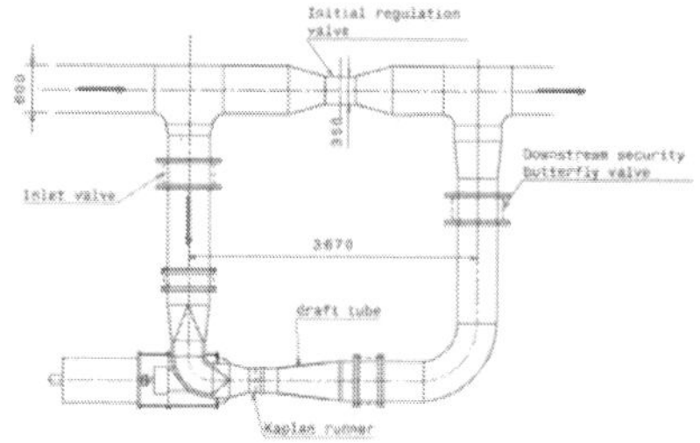

Fig. 7 and Photo 34. Setting of a Kaplan turbine as a bypass of an existing valve in a drinking water network (case study Poggio Cuculo).

Photo 35. Cavitation on blades for a Kaplan runner.

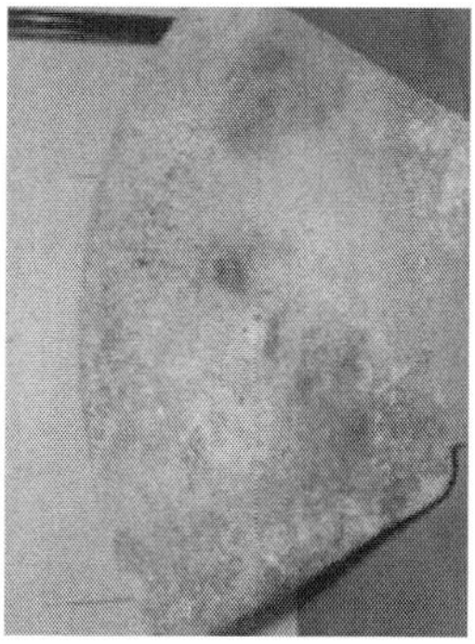

Photo 36. A Kaplan turbine blade, manufactured without hydraulic laboratory techniques, eroded by cavitation after a few months operation.

But cavitation is not a fatality. Laboratory tests permit to identify turbine cavitation behaviour, and to improve it by an appropriate design. Then manufacturers of laboratory-developed turbines can define with accuracy the maximal height regarding the downstream water level at which the runner can be set without cavitation damages.

For the case study Poggio Cuculo, with a head of 28 meters, cavitation could have been a strong constraint. But thanks to the water treatment configuration, the turbine could be set 2 meters under the downstream water level in the reservoir.

5.4 Regulation

Generally, the turbine is regulated according to the upstream water level in the forebay tank, in order to keep it steady.

- When the upstream level tends to rise, the turbine opens up to increase its discharge up to the nominal one. If the upstream level keeps on rising, the surplus can pass through the by-pass.
- When the upstream level tends to go down, the turbine closes itself to take less discharge. If the upstream level keeps on going down, the turbine is shut down.

By controlling the needle stroke for Pelton turbines, the vanes or blades opening for Francis, Diagonal and Kaplan turbines, the turbine can turn to be an efficient and convenient device to regulate discharges.

5.5 Security system

In case of load rejection (due to a storm for example) resulting in disconnection of the turbine from the grid, the machine has to stop automatically. Such shut down must be achieved so as to limit water hammer in the penstock and avoid runaway speed. Indeed, these phenomena could lead to important equipment damage.

The first requirement is that the SHP plant has to be equipped with an emergency power supply. The second depends on the type of turbines.

Francis turbine shut down is achieved by closing the guide vanes and the upstream valve with adapted speeds.

Photo 37. A deflector before the commissioning (case study le Châble Profray).

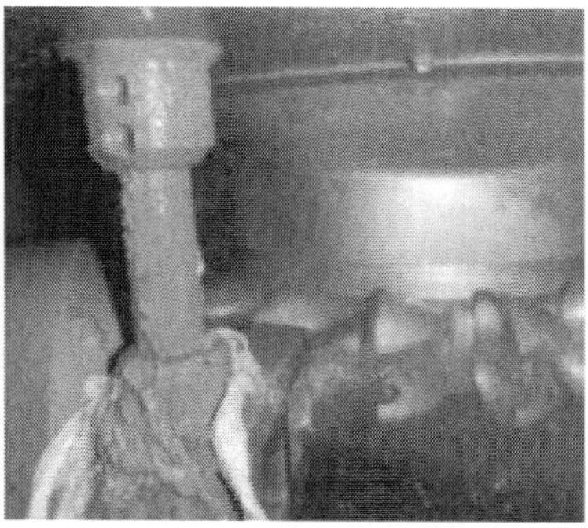

Photo 38. A deflector in the raw wastewater (case study le Châble Profray).

Kaplan or Diagonal turbine shut down is achieved by closing the adjustable guide vanes, the runner blades and the downstream valve with adapted speeds.

Reverse pump shut down is achieved by closing the upstream or downstream valve with adapted speeds.

For **Pelton turbines**, deflectors are a simple and secure solution. Nevertheless, they are not recommended for raw wastewater, as they may be clogged by wastes. In such cases, the turbine will be designed to resist runaway speed, and a special monitoring will be achieved to regulate the valves closures (Photos 37 and 38).

5.6 Maintenance

The maintenance and its cost depend on the water quality and on how the whole design of the SHP plant has been adapted to it, as described in Table 5.

For **drinking water networks**, the maintenance is limited, whereas it can be important for non-adapted SHP plants using **raw wastewater.** To make this maintenance easier, the machine design will integrate hand holes for a direct access to wastes.

It can be noted that most of time, the wastewater treatment plant staff will be in charge of the maintenance.

For the case study Le Châble Profray, in operation since 1993 on raw wastewater, the average usual maintenance amounts to about 40 hours per year. The interventions are linked to the electrical output. Indeed, when the output is lower than the foreseen one for the available discharge, it means that the waste accumulation is not acceptable anymore and the turbine has to be cleaned.

5.7 Bypass

A bypass of the turbine may be required to guarantee the primary function of the existing infrastructure at any time. For water networks for example, it has to be systematically set. It can be used when the turbine is not operating due, for example, to a too low discharge or to maintenance needs. It can also be used when the discharge needed for the existing scheme is higher than the turbine nominal one. In such situation, the turbine uses its maximal discharge, whereas the surplus flows through the bypass (if the head losses are still acceptable for the turbine).

As it replaces the turbine, the bypass has different functions: to regulate the discharges and/or the water levels, to reduce the pressure.

Different instruments exist for pressure reduction in a pipe. They have to be suitable for a continuous operation, and automatically and manually controllable.

For high heads, a Carnot pressure breaker may be the best tool (cf. Fig. 8). It is composed of an adjustable nozzle placed into a long tube immersed in a reservoir. Such device permits to maintain the upstream water level, to regulate the bypassed discharge, while wasting the excess pressure. The nozzle control system is integrated in the process control system of the existing infrastructure and the SHP plant.

When the SHP plant is equipped with a single jet Pelton turbine, the Carnot pressure breaker can be equipped with a similar nozzle, leading to regulation simplification and cost reduction.

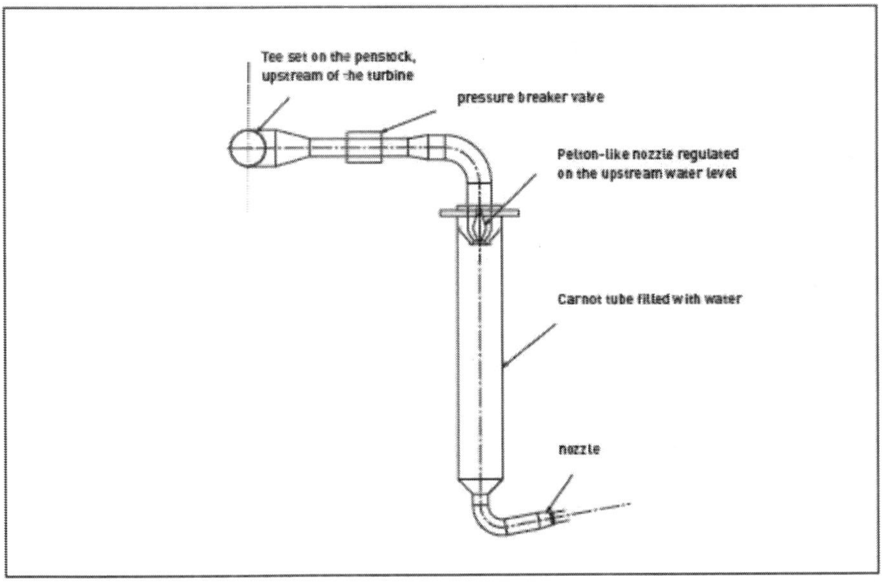

Fig. 8. Carnot pressure breaker.

6. Conclusions

The equipment used for multipurpose schemes does not differ much from the traditional ones used for water streams, apart from the specific conditions of each infrastructure that have to be considered all along the projects' steps.

Regarding environment, as the hydropower plant has to be integrated to the existing infrastructure, the impacts are mainly due to its primary function. One can even mention that the environmental impact is positive as the SHP plant implies an energy recovery.

However multipurpose schemes development is just at the beginning. This is mainly due to the **lack of information** on the possibility to recover energy. Moreover, in some countries, one second obstacle would be **the lack of administrative procedures** adapted to SHP. On the contrary, the procedure in Switzerland is simple. The water network is generally owned by the water office of the commune or city that will often be the plant operator, while the project has to be announced to the authority dealing with the sanitary field.

Small hydropower plants integrated into existing infrastructures is thus a promising environment-friendly market to develop.

7. Acknowledgements

This paper is based on the findings of the EU partially funded projects, Small Hydro Action for the Promotion of efficient solutions (SHAPES; 2007-2010; No TREN/07/FP6EN/S07.74894/038539) and Stream Map for Small Hydropower in the EU (SHP STREAMMAP; 2009-2012; No IEE/08/697/ SI2.52932), for which the coordinator was the European Small Hydropower Association (ESHA). The authors would like to thank Pamela Blome (Wasser Tirol, Austria), Raymond Chenal (Mhylab, Switzerland), Cédric Cottin (Mhylab, Switzerland), Jonathan Cox (Dulas, United Kingdom), Pierre-André Gard (Services Industriels de Bagnes, Switzerland), Jean-Bernard Gay (Mhylab, Switzerland), Marc Leclerc (MJ2 Technologies, France), Tony Leggett, Carmen Llansana Arnalot (Acciona, Spain), Jean Nydegger, Luigi Papetti (Studio Frosio, Italy), Prof. Bernhard Pelikan (BOKU, Austria; ESHA), Siegfried Ploner (Wasser Tirol, Austria), Bruno Reul (Mhylab), dr. Janusz Steller (IMP PAN, Poland), Franscisco de Vicente (Acciona Energy, Spain), Andreas Walner (Wasser Tirol, Austria). The authors would like also to thank the reviewers for their valuable comments.

8. Illustrations copyright

Acciona Agua, Tordera: case study Tordera
Acciona Energy, Pamplona, Spain: case study Sangüesa
Blue Water Power, Schafisheim, Switzerland: Photo 33
Betriebsgesellschaft Marchfeldkanal, Austria: case study Marchfeldkanal
Commune de Savièse, Savièse, Switzerland: case study La Zour
Dulas, Wales, United Kingdom: case studies Llys y Fran
France Hydro Electricité: case study L'Ame
Gasa SA, Lausanne, Switzerland: case study La Zour, le Châble-Profray
IKB, Innsbruck, Austria: case study Mülhau
IMP PAN, Gdansk, Poland: case study Skawina
MHyLab, Montcherand, Switzerland: Photo 11, Photo 12, Photo 13, Photo 14, Photo 27, Photo 31, Fig. 5, Fig. 6, Fig. 7, Fig. 8, case studies Poggio Cuculo, Armary, Le Châble-Profray, Seefeld, Nyon, Le Day
MJ2 Technologies, Millau, France: case study L'Ame
Nuove Aquae, Arezzo, Italy: case study Poggio Cuculo
Services Industriels de Bagnes, Bagnes, Switzerland: case study Le Châble-Profray
Services Industriels de Genève, SIG, Geneva, Switzerland: case study Aire-La-Ville
Shema, France: case study L'Ame
Stanislaw Lewandoski & Emil Ostajewski, Poland: case study Skawina
Studio Frosio, Brescia, Italy: case study Rino
Wasser Tirol, Innsbruck, Austria: case studies Mülhau, Seefeld

9. References

Akash, B.A. &.Mohsen, M.S. (1998). Potentials for Development of Hydro-Powered Water Desalination in Jordan. *Renewable Energy*, Vol. 13, No. 4, pp. 537-542, ISSN 0960-1481, 1998

Bansal, P. & Marshalla, N. (2010). Feasibility of Hydraulic Power Recovery from Waste Energy in Bio-Gas Scrubbing Processes. *Applied Energy*. Vol. 87, Issue 3, pp. 1048-105, ISSN: 0306-2619, March 2010

Bischoff, V. & Salamin, J. (2010). Combining Drinking Water Supply of Two Municipalities to Generate more than 1.4 MW. *Proceedings of the International Conference HIDROENERGIA 2010*, Lausanne, Switzerland, June 16-19, 2010.

Bodin, M. (2008). U.S. Looks to Rediscover Hydropower as Untapped Energy Source. In: *Popular Mechanics*. 29.08.2011, Available from http://www.popularmechanics.com/science/energy/hydropower-geothermal/4281705

Boillat, J.L; Bieri, M. & Dubois J. (2010). Economic Evaluation of Turbining Potential in Drinking Water Supply Networks. *Proceedings of the International Conference HIDROENERGIA 2010*, p.8, Lausanne, Switzerland, June 16-19, 2010.

Budris, A.R. (2011). Using Pumps as Power Recovery Turbines. In: *Water World*, 29.08.2011, Available from http://www.waterworld.com/index/display/article-display/366823/articles/waterworld/volume-25/issue-8/departments/pump-tips-techniques/using-pumps-as-power-recovery-turbines.html

Chenal, R.; Vuillerat, C.A. & Roduit, J. (1994). *L'eau usée génératrice d'électricité.* : concept, réalisation, potential. Office Fédéral de l'Energie, p.75, Berne (English and French)

Chenal, R.; Choulot, A.; Denis, V. & Tissot N. (2009). Small Hydropower. In: *Renewable Energy Technologies*, Section 8, Ed. Sabonnadière C., ISBN: 978-1-84821-135-3, Iste, Wiley

Cottin, C.; Reul, B. & Choulot, A. (2011). Laboratory results of the DIAGONAL project: a step towards an optimal small hydro turbine for medium head sites (25-100 m), *Proceedings of the International Conference HYDRO 2011*, Prague, Czech Republic, October 17-19, 2011.

Denis, V. (2007). Wastewater Turbining Before and After Treatment: the Example of Amman City - Hashemite Kingdom of Jordan. *Proceedings of the International Conference "Hydro 2007"*. p.9, Granada, Spain, October15-17, 2007

Denis, V. (2008). Wastewater Turbining Before and After Treatment. An optimal use of existing infrastructures. *Proceedings of the International Conference HIDROENERGIA 2008*, p.10, Bled, Slovenia, June 12-13, 2008.

European Small Hydropower Association [ESHA], Ademe; Mhylab; ISET;EPFL; LCH; IT power; Austrian Small Hydropower Association; SCTPH; SERO; Studio Frosio. (2005). Guide on How to Develop a Small Hydropower Plant. Thematic Network on Small Hydropower. In *The European Small Hydropower Association*. 29.08.2011, Available from http://www.esha.be/

ESHA ;Mhylab; Acciona Energy;Dulas; IMP PAN;IWHW; Studio Frozio; WMF; Wasser Tirol (2010). Energy Recovery in Existing Infrastructures with Small Hydropower Plants. Multipurpose Schemes - Overview and Examples. FP6, SHAPES, p. 53. In: *The European Small Hydropower Association*. 29.08.2011, Available from http://www.esha.be/

Franc, J.P.; Avellan, F.;Karimi, A.; Michel, J.M.; Billard, J.Y.; Briançon-Marjollet, L.; Fréchou, D.; Fruman, D.H.; Kueny, J.L. (2000). *La Cavitation, Mécanismes physiques et aspects industriels,* p. 582, ISBN978 2 868834515. Collection Grenoble Sciences (in French).

Frederiksen, S.; Wollerstrand, J. & Ljunggren, P. (2008). Un-interrupted District Heating Supply in the Event of an Electric Power Failure. *The 11th International Symposium on District Heating and Cooling,* Reykjavik, Iceland, August 31 - September 2, 2008.

García, J.P., Marco, A.C & Santos, SN. (2010). Use of Centrifugal Pumps Operating as Turbines for Energy Recovery in Water Distribution Networks. Two Case Studies. *Advanced Materials Research* ,Vol.107, No.87. pp.87-92, ISSN: 1022-6680, 2010

Giacopelli, P. & Mazzoleni S. (2009). Lonato SHPP: a successful exploitation of an existing head on an irrigation canal in northern Italy pp.4. *Proceedings of the International Conference HYDRO 2009,* Lyon, France, October 26-28, 2009.

Hahn, C. (2009). Vienna Wastewater Treatment Plant Receives SHP Unit. *Aqua press international.* No1. pp.40-41. (In German and English)

Krasteva, M. (2010). Blagoevgradska Bistritsa: Public-Private Partnership for Turbining of Drinking Water. *Proceedings of the International Conference HIDROENERGIA 2010,* p.8, Lausanne, Switzerland, June 16-19, 2010.

Krouse, W. (2009) Putting Hydrokinetic Power to Work for You. *Proceedings of the Small Hydro Conference 2009,* p.27, Vancouver, Canada, April 28-29, 2009

Low Impact Hydropower Institute (LIHI) (2009). LIHI Certificate No 43. Deer Island Hydropower Project. Deer Island Wastewater Treatment Plant, Boston Harbor Boston, Massachusetts. In: *The Low Impact Hydropower Institute,* 29.08.2011, Available from http://www.lowimpacthydro.org/lihi-certificate-43-deer-island-hydropower-project-deer-island-wastewater-treatment-plant-boston-harbor-boston-massachusetts.html

Orchard, B. & Klos S. (2009). Pumps as Turbines for Water Industry, *World Pumps,* Elsevier, Volume 2009, Issue 8, pp. 22-23, ISSN 0262-1762, August 2009

Papetti, L. & Frosio, G. (2010). Use of Reserved Flow for Hydro Production: Recent Achievements. *16th International seminar on hydropower plants. Reliable Hydropower for a Safe and Sustainable Power Production,* p.6, University of Technology, Vienna, Austria, November 24-26, 2010

Patel, S. (2010). Australia Gets Hydropower from Wastewater. In: *Power,* July 1, 2010, 29.08.2011, Available from http://www.powermag.com/renewables/hydro/Australia-Gets-Hydropower-from-Wastewater_2789.html

Pelikan, B. (2005). SHP Engineering: a New Approach and a Key for the Future. *The International Journal on Hydropower & Dams,* 2005, 12(3), pp.57–60, ISSN1352-2523

Raluy, R.G. ; L. Serra, L. & Uche, J. (2005). Life Cycle Assessment of Desalination Technologies Integrated with Renewable Energies. *Desalination,* 183, pp.81–93. ISSN 0011-9164

Ramos, H.M.; Mello, M. & De, P.K. (2010) Clean Power in Water Supply Systems as a Sustainable Solution: from Planning to Practical Implementation. *Water Science & Technology: Water Supply – WSTWS*. Vol.10. No1, pp. 39-49. ISSN 1606-9749, 2010

Rizzi, S.; Papetti, L. & Frosio, G. (2010). Chievo Small Hydro Plant - from the Obligation of Environmental Flow to the Opportunity of Renewable Energy Production. *Proceedings of the International Conference HYDRO 2010,* p.5 Lisbon, Portugal, September 27-29, 2010

Schlemmer, E.; Ramsauer, F.; Cui, X. & Binder, A. (2007). HYDROMATRIX® and StrafloMatrix. Electric Energy from Low Head Hydro Potential. *International Conference on Clean Electrical Power, ICCEP '07,* pp. 329 – 334, ISBN: 1-4244-0632-3, May 21-23, 2007.

Steller, J.; Adamkowski, A.; Stankiewicz, Z.; Lojek, A.; Rduch, J. & Zarzycki, M. (2008). Pumps as Turbines for Hydraulic Energy Recovery and Small Hydropower Purposes in Poland. *Proceedings of the International Conference HIDROENERGIA,* p. 10, Bled, Slovenia, June 12-13, 2008.

Sulzer Pumps (2011). Reducing Pressure - Increasing Efficiency. Sulzer Technical Review. No 1. pp. 26-29

Swiss Federal Office of Energy [SFOE]. (1995). *L'eau usée génératrice d'électricité, Dossier technique et étude du potentiel,* Diane 10, Petites centrales hydrauliques,1995. In: Swiss *Federal Office of Energy,* 29.08.2011, Available from http://www.bfe.admin.ch/kleinwasserkraft/03834/04171/index.html?lang=en (only in French).

Swiss Federal Office of Energy [SFOE].(1996). Small Hydropower Programme *Trinkwasserkraftwerke, Technische Anlagendokumentation / Petites centrales hydrauliques sur l'eau potable, Documentation technique,* Diane 10, Petites centrales hydrauliques (only in French and German). In: Swiss *Federal Office of Energy,* 29.08.2011, Available from http://www.bfe.admin.ch/kleinwasserkraft/03834/04171/index.html?lang=en

Toader, S; Praisach, Z. & Pop, F. (2010) Energy Recovery System on Drinking Water Pipe at Cluj-Napoca, Romania. *Proceedings of the International Conference HIDROENERGIA 2010,* p.8, Lausanne, Switzerland, June 16-19, 2010.

Wedam, G.M; Materazzi-Wagner, C. & Winkler S. (1999). Implementation of a Matrix Turbine at Freudenau Ship Lock. *Proceedings of the International Conference Hydropower into the next century-III* pp.117-124, ISBN 0952264293. Gmunden, Austria, October 18-20, 1999.

Wedam, G.M.; Kellner, R. & Braunshofer R. (2004). Innovative Hydropower Development in an Urban Environment. *19th World Energy Congress. Congress Papers.* 19p. Sydney, Australia, September 5-9, 2004

White, J. (2011). Recovering energy from an existing conduit. *International Water Power & Dam Construction.* pp.18-20, 0306-400X. May, 2011

Williams A. (2003). Pumps as Turbines – A User´s Guide. ITDG Publishing, London, 2nd Edition, ISBN 9781853395673, 2003.

Williams, A. (2010) Centrifugal Pumps as Turbines: A Review of Technology and Applications. *Proceedings of the International Conference HIDROENERGIA 2010*, p.6, Lausanne, Switzerland, June 16-19, 2010.

Wollerstrand, J.; Lauenburg, P. & Frederiksen, S. (2009). *A Turbine-Driven Circulation Pump in a District Heating Substation. Conference proceedings ECOS 2009, 22nd International Conference on Efficiency, Cost, Optimization, Simulation and Environmental Impact of Energy Systems*, p.9, Foz do Iguaçu, Paraná, Brazil, August 31 - September 3, 2009.

Project Design Management for a Large Hydropower Station

Xuanhua Xu, Yanju Zhou and Xiaohong Chen
School of Business, Central South University, People's Republic of China

1. Introduction

"Hydropower Engineering Design" refers to a whole process, which is undertaken by the design company commissioned by the project owner and includes the following:

- The initial stages covering mapping, survey, exploration and experiment.
- The use of high technology knowledge, taking into consideration the fluvial landform, geological conditions and social economics of the project location.
- Being subject to repeated research and scrutiny by particular engineering and technical personnel.
- Preparing and presenting the design scheme for the project phases.
- Preparing written reports and blueprints.
- The revision and improvement of the design scheme to the point of complete approval or ratification by the owner, adjudication units and various departments.

The typical characteristics of a hydropower design enterprise include a very large management organization, numerous personnel, low efficiency, heavy social burden, and slow technical progress. Consequently a hydropower project design method of production management with a supporting information system is needed which is suitable both for the operation of the design enterprise and competing internationally. The management approach should improve efficiency to face market challenges and ensure survival.

Based on the principles of system engineering and information technology, according to the actual needs of engineering design and project production management, changing and complex hydropower project design can be divided into relatively simple production processes. These processes can be integrated into an overall project production process control system by the use of information technology. The goal of the set of production processes in management control of technology is to simplify complex issues, to standardize simple problems, to sequence standard processes and then to apply information technology to the sequencing procedure. In actual production processes, project decision makers modify the process network parameters and adjust the critical path of process control, according to the production plan and actual progress. This enables the achievement of dynamic management, rapid decision-making and rapid adaptation to changes in the market.

A large hydropower project always faces the following problems: enormous scale of construction, long period of construction, huge investment and high risk, and numerous participating units and personnel. These problems also mean higher requirements for project management for which information technology becomes a more critical factor in project design management methodology. Thus, there is a need to establish a hydropower station design project production management system which has project design as its core business. At present, the widely used P3 platform application focuses mainly on the project planning and control of the engineering construction of large hydropower stations [1-3], on early systems of management information data exchange and integration [4-5], and on petroleum engineering applications [6]. This book presents a method based on the P3E/C platform with the design of the construction of a large hydropower station project as the core of the project management system and its implementation methodology. This enables support for the improvement of large hydropower station project design and production efficiency. At the same time, it also provides references for applying information technology to the production management of large hydropower station design projects.

Aiming at the complexity and therefore increased understanding of design projects for hydropower stations, this chapter analyzes in depth the design processes and characteristics of projects for large hydropower station engineering and puts forward a method of breaking down complex & cyclically changing design processes into a series of relatively simple work procedures or packages of work procedures. The application of information technology to documenting the work procedures and outputs, can realize the control of the design production process, quality control of design processes and products, management of staff performance, product statistics and archiving. The establishment of a production information management system should improve and optimize the design process, and continuously improve the management methods of the design project and the operational efficiency of the company. Therefore, based on the project of a large hydropower station designed by Hydrochina Zhongnan Engineering Corporation, we demonstrate the production management system of a large hydropower station design project using the P3E/C. This approach frames the network structure, and the system running platform puts forward a methodology for the following four technologies key to realizing the system: decomposition of the design process and work-procedure structure, circulation of project work-procedure sheets, generation of a dynamic matrix network diagram, and project production safety processes. Furthermore, also based on P3E/C, it proposes the methodology for realizing the three core sub-systems of project schedule management, activity process management, and working staff performance management.

1.1 Hydropower engineering design characteristics

The production of a hydropower station construction design project is essentially different from that of common goods production. The design project is more involved in intellectual work, while the project outcomes or products result in various drawings and paper documents. This invariably involves many complex design processes and numerous technical staff. Due to high professional demands and the knowledge required, it is difficult to control the production process and estimate personnel performance. This situation directly affects the quality, progress and efficiency of the construction design project. For example, the Mid-South Design & Research Institute of China (MSDRI) is a professional research institute engaged in Hydropower Station Construction Project Design. MSDRI has successfully designed the Hydropower Stations of Hunan Wuqiangxi, Xiangjiaba, Longtan, Fengtan, Dongjiang, and Youchou as well as other large to medium-sized hydropower stations. The MSDRI has established its own local area computer network (LAN) which is equipped with advanced servers and associated equipment. It has also developed a web-site and comprehensive management information system. However, the engineering design business management of hydropower station design in MSDRI is still a combination of manual process control with computer software design document processing. Specifically, the hydropower engineering design characteristics are outlined below.

1.1.1 Knowledge-intensive intellectual work involved with multiple technologies

Hydropower engineering design is a type of knowledge-intensive intellectual work. Inside the design company, the specialties include geology, exploration, hydrology, planning, water and construction engineering, mechanical and electrical engineering, reservoir resettlement and environmental protection, scientific research and experimental work, construction, operations, HVAC, fire-fighting, safety evaluation, soil and water conservation, engineering quantity surveyors and cost accountants and other professional staff. According to the needs of their professional work, qualified designers should have a good understanding or be familiar with the relevant knowledge of economics and management. In addition they need to be expert or knowledgeable in one of the following professions: hydrology and water resources, electric power system, engineering design, engineering geology and hydrological geology, land management, asset valuation, city planning, traffic engineering, power engineering, regional planning, agriculture, forestry, environmental ecology, biology (aquatic and terrestrial organisms), soil, hydraulics, aerographics, river kinematics, risk assessment theory, structural mechanics, material mechanics, elastic mechanics, hydraulic machinery and engineering machinery, electronic science, simulation theory, and the theory and method of general budget estimating, etc.

1.1.2 Work content, multiple agencies involved and wide social influences

Survey, design and research are the three main components of Hydropower project design, and the detailed work includes field surveying, investigation and research, in-door design, and consultation, assessment, review, on-site services, etc. Investigation and research content relates to the various departments of social, economic, environmental, nature, national economy and so on. It also needs to mobilize large numbers (possibly hundreds) of people to work for months or years to complete the task. The whole process of design also has very wide social influence.

1.1.3 Obvious division of work stages and governmental dominance

Several problems arise during the construction of hydropower stations, such as the occupation of large amounts of land (especially cultivated land and gardens) and migrant resettlement. Some problems involve cross-region resource development, immigration, taxes and reservoir operation management. The capacity of a hydropower station can range from millions of cubic-meters to tens of billions of cubic-meters. In this environment, any major accident could cause huge loss of life and property downstream. Therefore, the government has set up strict regulations for the necessary working stages of hydropower station construction including: river planning, pre-feasibility study, feasibility study, design bidding, construction blueprint design, completion and final acceptance. The later stages cannot be started until previous stages have been approved.

1.1.4 Quality products are based on collective wisdom & team spirit

Subject to the complex professional structure, cyclic and changing work-procedures, and based on the proposed scheme, professional personnel participate in the design of the hydropower station project in a coherent, and generally co-ordinated manner under the guidance of a chief engineer. Chief design officers and chief engineers, finding problems in completing their individual professional work, then modify and optimize the design scheme, and finally summarize their recommended scheme after repeated comparison and scrutiny. A good scheme and quality product can only be born out of the wisdom of the professional experts combined with the close cooperation of a large number of designers.

1.1.5 Products appear as information products such as drawings or electronic documents

Different from industrial products in physical form, design products appear as information detailed in written reports, drawings or electronic documents. In order to realize their functional requirements or value, industrial products can be directly used by the consumer while the design product for a hydropower station project needs to be carried out by processes of construction and resettlement of people affected.

1.1.6 Each of the various types of projects requires individual effort of survey and design

Hydropower stations can be located on different rivers and on different locations on the same river. Each of them confronts different terrain and geological conditions. The amount and difficulty of construction land requisition and resettlement for each station is typically not the same. Differences are also apparent in the range of power supplies, the transmission distances, the scale of engineering construction and the layout and form of engineering buildings. Accordingly, survey and design research work must be separately aimed at each project in order to make the design products suitable for the requirements of the geological and topographical conditions, local social and economic conditions, and actual power supply demands.

2. Key technologies of production process control

To efficiently complete the project design production within given time frames, four key factors need to be dealt with: A. Decomposition of the design process and process hierarchy. This solves problems for design project identification and feasibility. B. Communication of the implementation of the project process. This deals with responding to questions on the process control of the project design and of the design personnel. C. Realization of a dynamic matrix network graph. This is concerned with the schedule control of the project production. D. Realization of project product safety management, which includes the settlement of project product copyright issues.

2.1 Decomposition of design process and working-procedure hierarchy

2.1.1 Hydropower station project design processes and their internal structure

The whole process of hydropower station design can be divided into the following main production procedures: selection of the location, waterline and type of the dam, hydro-junction layout research, project scale arguments (concerns the normal water level, limit water level for flood periods and flood demonstrations, dead storage level choice, and the selection of installed capacity), design work for the selected scheme, report generation, evaluation and consultation, reviews and revisions, work completion, etc. Within each of the main production processes above, there are unique design professions involved, various professional force inputs, different focus and information accuracy. Each professional work activity has its own sequences, provides mutual information and contributes to mutual authentication.

The project design process is complex, and the multilayer production plan based on P3E/C can solve and coordinate this complex problem. Through the work breakdown structure (WBS) we define:

- Activities (operations, activities or tasks can be organized into package of work procedures)
- Steps (steps within operations)

The senior management layer starts top-down planning, while the base-level of the project team carries out bottom-up decomposition for the project plan. A harmonious and unified plan is then formed with milestones suitable for the needs of different management levels (P3E/C can recognize the milestones layers). Different levels of the program reflect different focuses of attention and measures that need to be taken to ensure communication consistency and validity between all levels of management and/or between the project team and external related units. This enables the plan to guide and control the project implementation process.

According to the project design characteristics, the design company's WBS system is formed as a tree structure. This system, using a set of coordinated process codes, integrates complicated production processes into an organic production process control model which assumes the project plan controls as the main activities, and then adds on quality control, product transfer and management, performance appraisal in the program execution process, and finally forms the production management information system.

2.1.2 Hierarchical structure of work-procedures for hydropower station project design

A hydropower station is designed in terms of project management, professional division, and professional offices which progressively extend to enterprise level. It can be divided into three levels.

For the first level, the main production processes must be set up. These processes, which include various professional offices providing product exchange links outside the office, are the focus of project management at the enterprise level (including the leader in charge, administrative department, project manager, chief design officer etc.).

The second level is to set up internal production processes in professional divisions. These processes, which include controlling product exchange links internal to the professional division, are the focus of project management at the division level (including the leader in charge, specifically general managers, and main designers).

At the last level, the product submission links in various professional offices are designed. They are the key point of control for the main designer and leaders in the professional office.

Certain feedback work-procedures (such as project selection) in various professional divisions can follow the work-procedure approach of the management package.

Figure 1 roughly describes the process of the above hierarchical structure. At the same time, the C13 professional internal structure and its subsidiary specialized work-procedures are also presented in the diagram. Different professionals may launch the work-procedure structure according to their own actual situations.

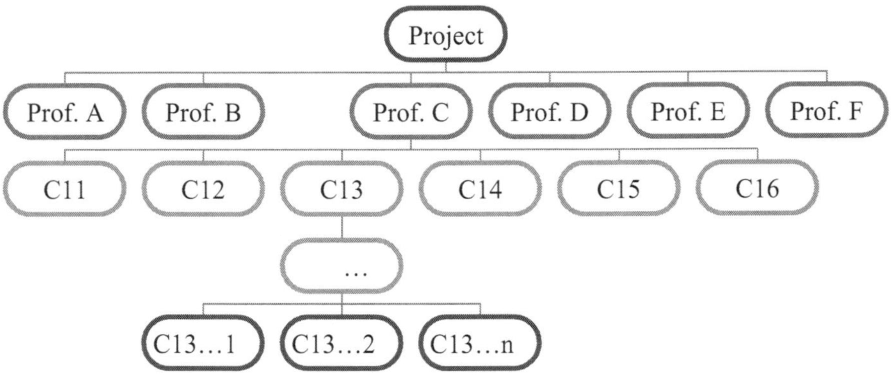

Fig. 1. Working-procedure hierarchical decomposition structure diagram of hydropower station project design (Note：Prof.=Profession).

2.1.3 Work-procedure structure

Work-procedure structure must be standardized, otherwise, neither are the input/output parameters measured nor is the quality of the results in the work-procedures controlled. The basic model of the process structure is as shown in Table 1. The work-procedure structure pattern includes the following main contents.

1. The application of the design process

Some work-procedures occur only in certain or a few design processes. For some professionals, some processes are not relevant, i.e. there is no process. Most of the professionals in various design processes have work and design processes. However, for different professionals at different stages of the design process, these design processes differ from each other in terms of the input workload, requirements for precision in the results which is built into the procedure as a basis for the work, and the focus of the different professionals.

The work-procedure structure requires that various professionals are divided by unified design processes in accordance with their own situation, and break down the work in different design phases into various design processes. The final product of a previous design process provides the basis for the work and input conditions of the next design process. With the gradual deepening of the design process, the design process focus, work depth and product quality requirements become clearer. The end of a design process represents completion of the design work in a certain phase.

2. The encoding work-procedure and its significance

Only the complex work-procedure work in coded representation can be described in mathematical terms for use by the main network in information system control. Each work-procedure corresponds only to its corresponding coding which has a specific meaning.

Application of design process	Encoding work-procedure	Work-procedure name	Information input	Rules	work-procedure parameters	Work time	Output QC	Output orient-ation
....								

Table 1. The working procedure structure of professional design stage.

3. The work-procedure name and output products

The work-procedure name is an abbreviation of the work content in a process, which should align with the output product in the process. Each procedure corresponds to a product (process) or a variety of products (process package), although the product may be repeated a few times (such as when comparing the results of a scheme). However, its occurrence should be used as product to be managed every time. There are two reasons for this. One is to make it easy to track back, the other is to facilitate measurement.

4. The information input in the work of the work-procedures

The working conditions are input conditions in the process. These conditions include the basis of the work which is necessary for work completion. The basis of the work can be either created by organizers, such as stipulating a design principles guide book, outline design, professional design rules or proposed design, or the output product or result provided by the previous step. The operator cannot complete the process successfully without clear information input.

5. The work content and rules in the process

In the process of decomposition, the effective links in the whole design process must be taken fully into consideration. A single work process, under certain conditions for one person, is classified as one step in the process, so as to reduce disputes over trifles, clarify responsibility and set up convenient measurement and assessment metrics. The rules of work mean principles and methods which comply with prescribed requirements in the work content, such as design procedures, standards, design handbook, reference books or relevant documents. Designers should master the work content and apply it in a flexible way, to complete the job quickly and accurately. The process structure table must clearly indicate or establish the principles and methods on which the work procedure is based, to identify the basis for proofreaders to complete their work.

6. The work-procedure parameters

Work-procedure parameters can be set according to the management level of the enterprise, the characteristics pertinent to the engineering project and the allocation system of the enterprise. In general, they should include the basic work time parameters which are involved in the work content, the job complexity parameters which are involved in the same work content of a different project, and any repetition or rework parameters, etc.

7. Work time

The work-procedure operation-time parameters (converted into operation time on a regular basis) are set down and chosen according to the production level of each professional office. However, specific projects are provided for in the total work time for the contract, therefore according to the specific circumstances of the project design, uniform adjustment of all professional work time parameters is required to meet the owner and contract requirements.

8. Quality control

The functions performed by different process products will always vary either in different stages of the same project or in different design processes in the same stage, so a control methodology for the quality of outputs may be different in each case. The control methodology for process products can be established according to its required quality management characteristics and specific process classification.

9. Output orientation

In the work-procedure structure, the output point must be made clear in every process. It is convenient for the system to conform to the specified work-procedure which follows the given direction, as the work input condition.

2.2.1 Transfer process for work-procedure sheet

The work-procedure sheet is the core data structure of the system, which represents not only the work-procedure content and logical relationships, but also the important basis for calculating project duration and personnel performance. Table 1, the work-procedure structure table according to its circulation needs refinement, as shown in Table 2. One of the key issues in design project management is to realize the circulation of the work-procedure table in the system.

Application of design process	Encoding work-procedure	work-procedure name	Input (Work Condition)			work-procedure Work		Quota Work-day					
			Input Profession	Code	Input Content	Content	Basis (Standard & Fomula)	Basic Work-day	Difficulty Level	Repeated Coefficient		Conversion Coefficient K	Counted Work-day
										Number of Times	Coefficient		
Hydrogeological investigation test 111	11101	Water quantity investigation				Water quantity investigation			2	1	0.8		

Table 2. a) The working-procedure table of professional stage.

Design		Proofread Scale				Output(parameter or achievement)			
No.	Coefficient	Proofreader	Chief in Profession	Chief Design Officer	General Chief	Content	Reciever		Code
							Profession	No.	
		365	178			Achievements of water quantity investigation (investigation files, Calculation & analysis draft, Drawing, similarly hereinafter)			11055

Table 2. b) The working-procedure table of professional stage.

The work-procedure table contains many data fields: code, input / output, working documents, difficulty coefficient, overlap coefficient, conversion coefficient. The process table is set up in a framework for circulation in the system to the completion of production for the design project. However, P3E/C does not correspond to the table directly, but this can be achieved through customization and matching the process table with the project operation. Because P3E/C operations include many attributes, some correspond to part of the data field in the process table, such as code, input/output, and working documents. However, a certain number of data fields in the process table, such as the difficulty coefficient, overlap coefficient, conversion coefficient and etc., have no direct corresponding attributes in the P3E/C operation, which was achieved through custom fields in P3E/C. Custom fields can be either global or constrained. By adding custom fields into the operation, all the data in the process table are mapped into the attributes of P3E/C, thereby realizing the successful circulation of the work-procedure table in the system.

Circulation control of the work-procedure table is realized through the P3E/C program, which contains project management (PM) and project reporting (PR) components. The PM implementation is used to make work-procedure table creation, distribution, and feedback on supply and demand indices a reality. P3E/C provides a variety of ways to achieve this. One is to log onto the PM main module which feeds back the completion time, the percentage of completion of the task and resource quantities in a given period. The other way is to log onto the Progress Reporter. The latter method is recommended, because it consists of a B/S structure which is very suitable for feedback on remote design work in progress. On the feedback interface of the P3E/C work-procedure sheet, the task for user involvement is listed automatically in this period. Users can: a) record the actual opening and completion time; b) obtain information from the project manager, while also giving feedback to the project manager; c) record the task log on a note book; d) articulate task related documents; e) give detailed feedback on accomplishments; f) fill in the form describing a single task, actual work hours for each day or each week. Once finished and submitted, this form cannot be edited further, and waits for approval.

2.2.2 Realization process for the dynamic matrix network graph

The dynamic matrix network graph is the core of project progress control. The single-code network diagram (PDM) is used to represents the logical relationships of the various

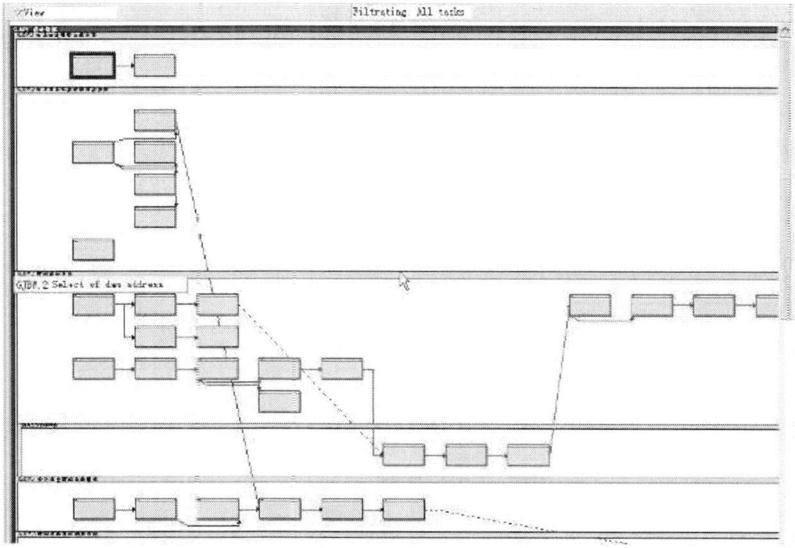

Fig. 2. Schematic diagram of an activity dynamic matrix network.

processes in the P3E/C. Data in the network graph is from WBS (project decomposition) and the definition of operation. Not only the time limits for a project require to be defined but so do the logical relationships in operation. In the P3E/C operation, the logical relationship in operation can be defined as four types: finish to start (FS), start to start (SS), finish to finish (FF), and start to finish (SF). These relationships are set in the job attributes. With WBS, operation and operation logic, P3E/C generates the project network diagram automatically and will use a direct or adverse method to calculate the critical path. When the job attributes (such as the time limit for a project, the modified logical relationships) are set, the network will dynamically update. The PDM of P3E/C is shown in Figure 2 (the red line represents the critical path).

2.2.3 Realization process for the safety management of a design project product

The main products of the design project are various types of documents or computational drafts. This kind of product is easy to copy, so product safety management is a very important component of design project management. The process is achieved by using role safety management strategies provided by P3E/C, which the user directly associates with Enterprise project structure (EPS), organization breakdown structure (OBS) and WBS. The safety management strategy for a design product is shown in Figure 3.

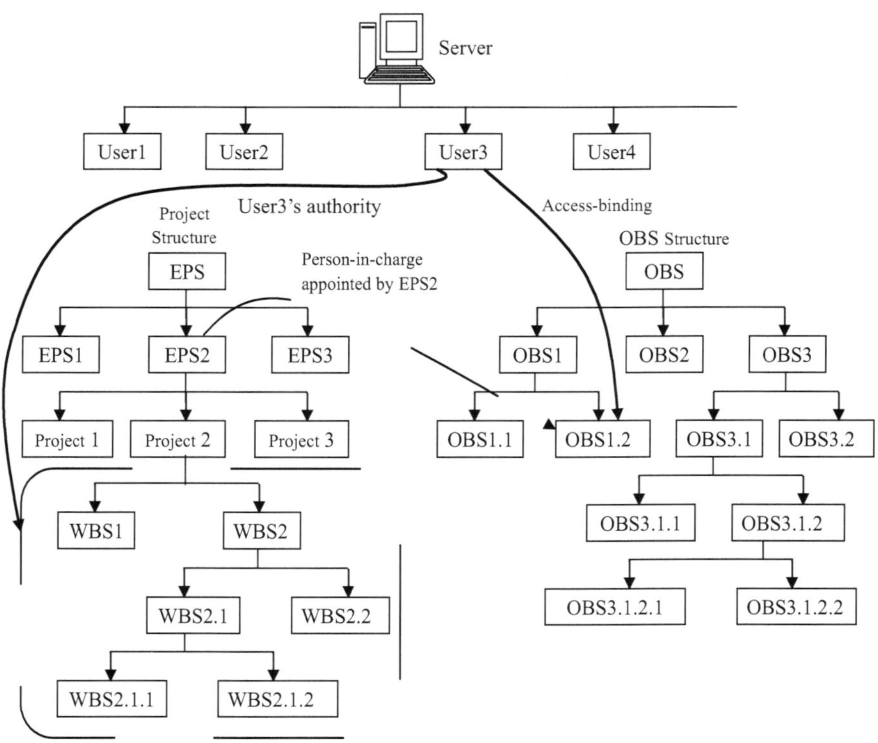

Fig. 3. Schematic diagram of product safety system for design project.

The user's rights depend on his level in the OBS, and the projects (or WBS) assigned to related organization level (such as a Department) in the project structure. Therefore, the privileged management of a project makes its definition simultaneous with EPS, OBS, and WBS, which maintains communication with the entire project structure.

In addition, using document management in an individual work center of P3E/C, can easily deal with the approval process of the design product, and the process documentation and management needs of the document version, view, download, and security audits.

2.3 System platform construction

2.3.1 System network structure

The project design staff can vary, therefore remote design requirements should be further supported based on the local area network. A Virtual Private Network (VPN) is used to communicate with the design unit headquarters and a remote designer. The VPN establishes a temporary secure connection through a public network (usually the Internet), which is a safe, stable tunnel going through the complicated public network (VPN is an expansion of enterprise networks), and forms a wide area communication network for the project design. The network communication structure is used for the whole system of the project design, as shown in Figure 4.

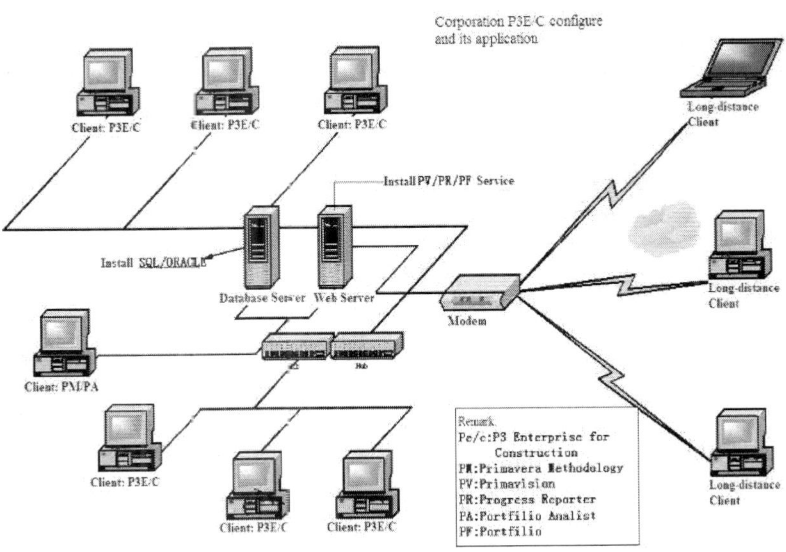

Fig. 4. Structure diagram of a project design management system network for a large hydropower station.

2.3.2 System operation and development platform

On the basis of investigation and comparison, the realization of this system uses P3E/C as the platform for system operation and development. The P3E/C software platform, with perfect function capability, is a product of the America Primavera Company, which integrates with advanced project management thinking and methodology. Based on the structural design, using the P3E/C platform is the best choice for realizing the core function of a production management system for a design project. P3E/C can realize many functions, such as scope management, plan/schedule management, quality control, personnel evaluation, portals, statistical analysis (such as Earned Value Analysis), and security/access control. Combining the use of P3E/C with the case study from the hydropower station engineering design project in the Survey and Design Institute of Central South China, resulted in no less than 20 concurrent users of P3E/C for the project design and management staff. The further development of the system portal, partly for statistical reports, and other information system interfaces in the system, will use Microsoft ASP.NET platform in its realization.

Application server installation included Weblogic and P3E/C components: MyPrimavera and TeamMember server (including Collaborator and TimeSheet). There are two types of client: the LAN client (C/S) and Internet client (B/S). The LAN client needs to install the C/S client and for a manager, project manager or planning personnel, the PM/PA and MM components need to be installed. The other client uses the browser client.

2.4 System function realization method based on P3E/C

The production management information system for project design cannot be called a complete project management information system, because it focuses on the production process control of the design project, integration of quality management, staff performance appraisal, and production management. Unlike the latter system, it does not include the management of project funds and complete human resource management, and contains only some of the requirements of project information management. The system is a kind of management system which has a professional basis and adapts to the changes in an enterprise organization's form over a considerable range. It is not only suitable for the existing professional situations, but can also adapt to the project management organization form, and even fit with the organizational structure in existing professional design divisions. Its purpose is as follows:

- To simplify the production links.
- To devolve the implementation of the production organization to an individual level.
- To decompose the distribution and distribution examination at different design stages in existing professional divisions into processes or process packages.
- To implement the job performance appraisal at an individual level.
- To turn the implementation of quality management into processes or process packages.
- To integrate planning control, quality management, employee performance and product filing during production process into an organic whole.
- To form a standard information management system from the production organization system.
- And finally, to help the hydropower design company save energy in its activities of exploring markets, gradual transitions in business orientation, and encourage research and development in science and technology in order to improve the company's competiveness.

Project management will be facilitated by the production management information system, which is critical to the reorganization of production processes, so that the production process of the design project can be organized in accordance with the requirements of the procedure. Thus, it can not only inherit the advantages of general professional management, but also involve the subtle links of process management. The management focus has been shifted to product demands, while the staff's attention is diverted from pure obedience on distribution to paying active attention to the demands of related procedures.

According to the requirements and objectives, the corresponding functional subsystems are divided with reference to the knowledge system of project management. The logical relationships are as shown in Figure 5 below.

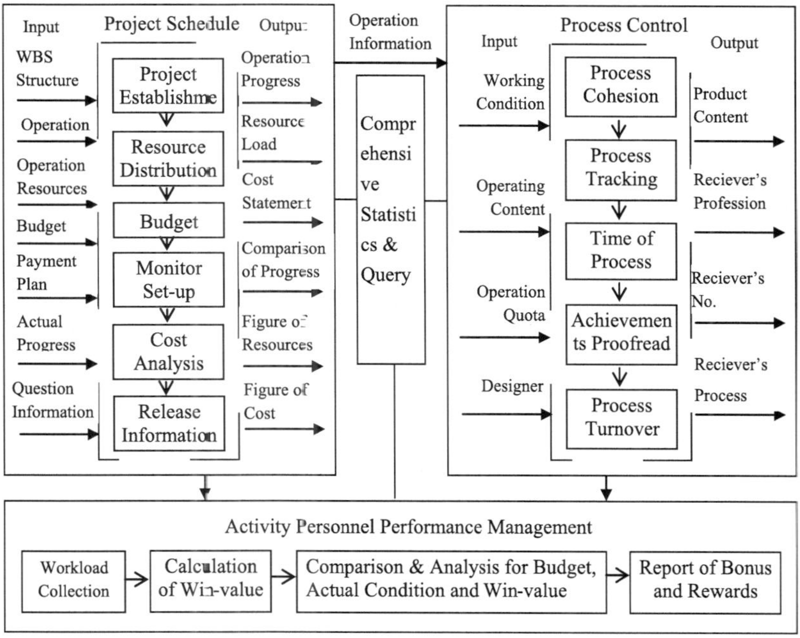

Fig. 5. System function structure diagram based on the P3E/C.

2.4.1 Project schedule

The project schedule includes basic information on the project, such as EPS, OBS, WBS, staffing and resource plan, schedule distribution and release etc.

The project schedule is realized through the function of WBS in the PM component of P3E/C. The WBS is a kind of hierarchical structure coding which decomposes the project scope step by step. WBS is a mesh elements grouping, designed for delivering the final product, which organizes and defines the scope of the project. Unlisted WBS work is excluded from the scope of the project. Therefore, setting up the WBS is not only the main content for making a plan, but the important basis for defining later operations (processes).

In P3E/C the WBS function design sets up a special WBS window to prepare the WBS for the project.

WBS code uses: project code + design phase code + the structure of process code. It is a three layer code, which uses the P3E/C WBS to realize it directly. The timing plan of the project will be realized by the following steps: 1. Establish the WBS work breakdown structure; 2. List the name of the operation; 3. Estimate operation time and arrange the work plan; 4. Optimize the work plan.

2.4.2 Activity process control

The process control realizes process tracking, control, decision-making, statistical analysis in process operation, manages or controls operation status such as time usage, the procedure connections, work conditions, product or achievement checking (including who designs, reviews, and provides the input conditions), and to whom the results are disseminated. The dynamics of the processes are reflected through an electronic board (network view).

The operation (TASK) is a basic design unit in P3E/C planning. It represents tasks required to be undertaken to complete the project. In general, many operations consist of a WBS, several WBS from a higher WBS until the project is fully configured.

The P3E/C operation has many attributes, corresponding to the data fields of the process table in the system. These include:, procedure code, name corresponding to P3E/C operation code and description, input/output conditions corresponding to a logical relationship in operation and precedence/successor operation; the output results corresponding to products and documents in operation, quota work corresponding to period types, operation calendar and the percentage of completion, designs/reviews corresponding to working documents.

The circulation of a process table in the system is achieved by distributing tasks automatically and periodically using the PR component in the P3E/C. Operators use P3E/C to realize timesheet approval through the PR timesheets feedback. Only an approved timesheet can be used to renew the content in P3E/C database.

2.4.3 Performance management for activity personnel

The system administrator establishes individual accounts for operation personnel. The system will automatically unify the work days for the process operators and the work days for auditors into their personal and statistical accounts. At the same time, the system calculates payments such as bonuses, etc. according to the settlement rules and provides a mechanism for various managers to evaluate personnel performance. Furthermore, it is easy for an employee to check information in their own accounts.

Staff performance in this system is a collection of various types of staff work hours. This part of the implementation is based on the correct definition of EPS, OBS, WBS and operations. The P3E/C earned value function is used to calculate real work hours and evaluate work performance. Through analyzing the project budget, actual and earned values, and progress delays can be determined. Comparing artificial time consumption and budgets, it is possible to determine whether savings have been made or overdrafts required etc. In this way, causes are easily found and remedies applied.

In addition, the use of the P3E/C custom field functions can greatly expand its applications. For example, through increasing the custom fields of operation, performance quality is evaluated by assessment and scoring, as a reference for performance evaluation. Definition of the evaluation parameters can be as follows: points, work discipline, difficulty of the work, work attitude, work intensity, coefficient of difficulty and so on. Composite evaluation marks can be obtained through an appropriate calculation formula.

3. Conclusions

There are two main types of platforms used at present for project design management (both at home and abroad): Microsoft's *Project* platform, and the American company Primavera P3E/C platform. The *Project* platform is focused on progress control. It is convenient to use, but can only be used on a single machine and hardly be applied to large hydropower complex project design. P3E/C provides a platform which can widely realize and develop functions, and be able to complete the design and management of large complex engineering projects via a network.

It is a new attempt to apply information technology in managing the design process of a hydropower station construction project. The main contribution of this chapter is to describe the use of the P3E/C platform for the realization of this kind of management requirement based on the large hydropower station project design case study of Hydrochina Zhongnan Engineering Corporation. Based on the systematic analysis of hydropower engineering design characteristics and research on key technologies of project design production process control, we designed a network communication mode and network structure for the operation of the system, and constructed a system operation platform. The advanced management concepts, based on the P3E/C platform, resulted in a system function structure for three core sub-systems: planning a project's progress, operation process control, and performance management for design project personnel. In this way, it provides good support for the successful production of a large hydropower station project.

4. Acknowledgment

The chapter is supported by the Natural Science Foundation of China (70871121, 71171202, 70921001, 71171201).

5. References

[1] Liu Shaoquan. The Application of P3 Progress Plan Software in the Project Management of Ertan Hydropower Engineering Project. Sichuan Water Powe,. 1998.17(4):25-27.

[2] Zhou Hougui. The Application of Primavera Software Package in the construction of Three Gorges Engineering Project. Project Management, 2003.10:21-23.

[3] Chen Zhou. The Application and Practice of P3 Project Management Software. Power Information, 2006.4(4):72-75.

[4] Hou Xin, Chang Qianduan. Research and Implementation of Data Conversion of P3 and Enterprise Information System. Journal of Industrial Engineering/Engineering Management, 2005. Vol.19:58-61(Supplement).
[5] T Devogele, C Parent , S Spaccapietra.On spatial Database integration [J].International Journal ofGeograhica Information Science , 1998,12 (4):335-352.
[6] Tang Min. The Application of P3 Project Management Software in the Petroleum Engineering Design. China Science and Technology Information, 2005(13):23-23.

14

Damming China's and India's Periphery: An Overview over the Region's Rapid Hydropower Development

Thomas Hennig
Philipps-University of Marburg, Germany

1. Introduction

To sustain their impressive economic growth and to provide electricity for the two most populated states, PR China and India are aggressively developing their energy sector. Regenerative energy, mainly hydropower, plays a key role in their present and future energy sector strategy. The speed of hydropower development shows a dimension which is unique worldwide. This extreme rapid development is mainly based on controversial large dams, but also on the fast development of small scaled hydropower projects.

The following chapter uses a geographical perspective to compare and to analyse the two states in their past, present and future hydropower development. Based on a political-ecological approach the chapter studies the institutional, spatial, environmental and socio-economic challenges and problems related to hydropower development in China and India. Although the focus is on large hydro schemes, and often they are so large that they can even referred to as mega-projects, the chapter includes also the development of the small hydropower sector.

The chapter compares and discusses similarities and discrepancies of the hydropower development in China and India. First it studies the relevance of the hydropower sector in the wider field of power development and power mix and further it discusses spatial as well as institutional challenges related to the fast developing hydropower sector. Second the chapter analyses in a more detailed scale the massive hydropower development in Asia's water tower, which is the transition area between Northeast-India and Southwest-China. This region has worldwide the highest potential in hydropower development and is therefore tagged as one of the world's future power houses or batteries. It shows the spatial consequences from that development and discusses it in its institutional, environmental and socio-economic implications. The last paragraph goes beyond that region and studies the conflicts and trends of internationalising hydropower business in Asia, mainly in the adjacent transboundary watersheds.

The chapter is based on extensive field work in China, mainly in Yunnan and on minor field work in India. The author visited many existing hydropower stations and potential hydropower locations and conducted semi-structured interviews during several field visits between 2009 and 2011 with researchers, members of hydropower companies, government officials, and policy advisors. Further interviews were conducted with representatives of foreign non-governmental organisations and consultants doing services in China and India as well as with other researchers working on the topic in China. Given the delicate political nature of the issue, all interviewees from China were anonymity assured.

2. Overview of China's and India's power sector

2.1 Present state of the power sector in the two countries

China and India are the two most populated countries in the world. Additionally the economic growth rate of both countries belongs to the highest in the world and the GDP-growth has also maintained high over a long period. The national economic growth in both nations is inevitably accompanied with an increasing demand for energy in general and electric power in particular. To sustain their impressive economic growth both states have been developing aggressively their energy sector. The challenge of their rapidly growing energy and power sector in the 21st century is unparalleled in the world. Finding a proper solution for this energy bottleneck will determine the economic, social and sustainable future of China as well as of India.

A key issue within the present and future energy sector strategy is to solve the generation and supply of electrical energy. On one side the power sector hast to grow slightly faster than the GDP-growth and on the other side is the present need for power for an average Chinese and Indian merely 1 kW/person, which is far inadequate compared to 11 kW/person for an average US person and ~ 3.5 to 5.5 kW/person for a Western European (Cahen & Lubimorsky, 2008, as cited in Chang et al. 2010).

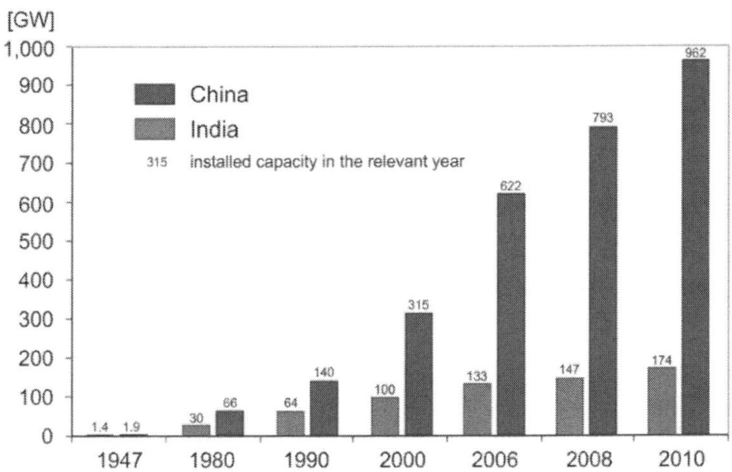

Fig. 1. Comparing China's and India's total installed power capacity for selected years

Despite the fact that the challenges of the power sector are in many ways similar, the present state of the sector varies considerably between the two nations [cp. Fig. 1]. Compared to India, China had from the beginning a higher installed capacity as well as a higher energy production output. Over the years the gap between the two nations grew steadily. In the beginning of both nations (late 1940s) the installed capacity was with 1.8 GW (China) and 1.4 GW (India) quite low. Today (Dec 2010) China's installed capacity is 962 GW, while India's part grew to 171 GW. In mid-2011 China exceeded as the second country the 1,000 GW of installed capacity and soon it will lap the USA as the world's country with the highest installed capacity

The growth of the Chinese energy sector is so impressive, that in less than two years China adds a new installed capacity which is similar to the total installed capacity of a strong West European economy like Germany or France. This fact is even more impressive, when it gets considered that since 2007 China closed about 77 GW of old, small and inefficient thermal power plants.

Despite the strong growth China's power sector faces cyclic shortages, the latest were in 2002-05 and in 2010. Contrary to the temporary Chinese power problems, is the Indian power sector characterized by a chronicle deficit, which is still about 7%. India's actual shortage would be much higher when the peak demand is considered as well as the unscheduled power cuts. The ambitious goal of the Indian government solving that problem by 2012 seems rather unrealistic.

2.1.1 Thermal power

It is well known for both countries that the primary energy resource is coal. About two third of the installed capacity is thermal power (mainly coal, but also gas and petroleum); its share in the energy production output is even higher. Similar in both countries is also that the coalfields are quite far away from the load centres. Additionally India lacks on sufficient high quality coal, China only to some extent. For that reasons coal either has to get transported over long distances which affects the efficiency of the railways or they have to get imported from other nations (e.g. from Indonesia, South Africa, Australia, etc.). In particular for China's booming South Coast it is cheaper to import the coal from abroad, rather than transport it from Northern China.

To solve India's serious power shortage the government pursues the construction of huge 'hubs' of thermal power stations around selected sea ports. Mainly based on imported coal, state-owned as well as private power companies construct here large thermal power stations. In those hubs the Indian government allows also smaller power plants which often cause serious environmental impacts.

While China's power generation is mostly in the hands of a few state-owned and to some extend also provincial owned generation companies, the share of India's private sector in capacity expansion has gone up substantially over the past decade. It is expected that its share is growing from presently one third to 50 per cent of the total incremental capacity over the next Five Year Plan (2012-17).

2.1.2 Nuclear power and renewables

Beside the massive development for thermal power stations both countries are seriously working in diversifying their power mix and reducing the ratio of carbon emissions. One option has been massive investments in nuclear power development, which plays so far in both states only a minor role. Despite its scheduled impressive absolute growth rate, relatively nuclear power plays also in future only a minor role. Due to the Fukushima accident in early 2011, China put some of its nuclear plans on hold, while India is proceeding with its nuclear ambitions.

Another option for reducing carbon emissions and diversifying the power portfolio is the development of renewables. In both countries are the absolute growth rates impressive, also they started at a relative low level. India is here relative more successful than China. Massive investments will boost the installed capacity of Wind, PV and Solar energy. Both countries have large territories and one challenge is that the generation areas are far away from the load centres. Massive investments in the relevant transmission system are due to the unstable generation of renewables often not justified.

Under the above premise solely hydropower is the only real alternative to thermal power, for it is considered as a renewable, clean and cost-effective resource. Its relevance has also to be seen in the context of reducing carbon emissions and dependency on fossil energy resources as well as of optimizing the power portfolio (base vs. peak load).

Following is explained the relevance of the hydropower sector in the present and future power scenarios, but it should get distinguished between the small and large hydropower sector.

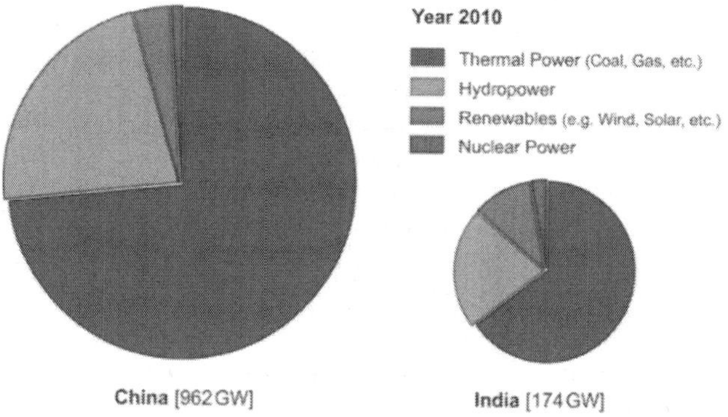

Fig. 2. Comparing the power mix of China's and India's total installed capacity in 2010

2.2 Role of the hydropower sector in the present and future power scenarios

Parallel to the rapid development of China's and India's power sector in general increased also their hydropower sector.

In 2010 out of the 962 GW of Chinese installed capacity, immense 213 GW or 22% were generated through hydropower. The same 22% is also India's hydropower generation, but this is absolute only 38 GW.

In the global context China ranks first and has by far the highest installed hydropower capacity (cp. Tab. 1). The 213 GW (2010) are so high, that it almost exceeds the cumulative installed capacity of the USA, Canada and Brazil which rank second, third and fourth. India ranks regarding the installed capacity after Russia on place six.

2008	Installed capacity [GW]	% of total capacity	Annual hydroelectric production [TWh]	Economic exploitable hydroelectric production [TWh]
PR China (2010)	*213,4*	*22,2*	*721,0*	*1.753*
PR China	*171,0*	*21,5*	*580,0*	*1.753*
United States	79,5	5,8	254,8	376
Canada	73,4	61,1	377,4	536
Brazil	77,5	85,6	365,1	818
Russia	49,7	17,6	180,0	852
India	36,9	15,8	115,6	442
Norway	28,3	98,2	140,1	206

Table 1. Comparison of the world's largest hydropower producers regarding installed capacity and annual hydroelectric production for 2008. Notice: Except China there are no substantial changes in other countries between 2008 and 2010

(Data compiled by author from different sources; 2008 data: World Energy Council. 2010 Survey of Energy Sources)

2.2.1 Brief overview about historical hydropower development

Before the present and future role of the Chinese and Indian hydropower sector is discussed, a brief historical overview is given. The Sidrapong hydropower station near Darjeeling (2*65 kW; later renewed and increased), finished in 1897, was India's first hydropower station; while China's first one was built in 1912 in Shilongba (2*240 kW), near Yunnan's provincial capital of Kunming.

With the founding of their nations in the late 1940s China and India had only a specific number of smaller and a very limited number of larger hydropower stations. This situation was very different from Europe and the United States, which had at that time not only a bigger number of dams but also larger hydropower stations constructed. The US Hoover's Dam initial 1,345 MW power plant constructed in the early 1930s on the Colorado River, was at that time by far the world's largest hydro dam. Only a decade later it was already eclipsed by the 6,809 MW of the Grand Coulee Dam on the river Columbia.

The situation began changing in the 1950s. In particular in the context of the early industrialization of that period the hydropower sector developed fast in China and India. At that time both states were supported and assisted in their hydropower programmes by the Soviet Union. In particular larger projects incl. dams were very prestigious for the young nations, although they required large initial capital outlays. The large dams of that time were mostly multipurpose projects and hydropower generation was only of minor relevance and often justified the large irrigation investments related to that projects.

In both countries the relative share of the hydropower sector grew fast and reached its peak about 30 years ago. In India it was in the late 1970s as the installed capacity was almost 40% and in the energy output the relative share was 45%with even higher even. In China the peak of the highest hydropower share in the energy portfolio was almost a decade later and, compared to India, the peak was also less (around 30%). Today in both countries the share of the hydropower sector is declining rapidly although the absolute growth is impressive. The relative energy output is declining even faster.

2.2.2 Hydropower potential

China and India combine together about two third of the world's 45,000 large dams. The figure includes dams over 15 m in height or, if less, the reservoir should have a capacity of more than three million cubicmeter. Despite the fact that China and India have combined more than one third of the global gross installed hydropower capacity, is their development level still quite low.

The exploitable hydropower potential has been analysed in both states by different national large-scaled general surveys. China's latest and fourth survey was published in 2005, whereas India's last re-assessment studies were completed in 1987. In consequence from these surveys both countries identified major hydropower bases; India six and China fourteen, whereas the Yarlung Tsangpo is officially not recognized as a base so far (cp. Fig 3).

The stage of development can vary considerably between these hydropower bases. In those surveys it gets distinguished between the technically exploitable and the economic exploitable capacity, whereas the first is naturally far higher. For China the economic exploitable hydropower potential is assessed to 402 GW (1,750 TWh/a), while that for India is 149 GW (660 TWh/a).

The core of that potential is located in the southern adjusting mountain ranges of the Tibetan plateau, which is ranging from the Karakorum over the Himalaya up to the Hengduan mountains. This region has globally the highest hydropower potential and is therefore called Asia's water tower. Within that water tower the key region for the present hydropower development is the transition area between the states of Northeastern India (mainly Arunachal Pradesh) via Myanmar to the Southwest Chinese provinces (mainly Yunnan and Sichuan). The present state of hydropower development in that region as well as potential future projects are discussed in chapter 3.2.

2.3 Spatial and institutional challenges of hydropower development

Neither in China nor in India hydraulic resources with their extremely uneven distribution do match with regional economic development (Chang et al. 2010). The economic hubs with their often high energy demand, which are in China the Southern and Eastern coastal regions and in India selected metropoles of South- and West-India, are mostly far away from the hydropower generation areas.

To get the electric energy to these far away load centres a robust, efficient and expensive high-voltage transmission system is required. In that context it is not amazing that in particular China and to some extend also India are increasingly using UHVDC (ultra high voltage DC) for long distances transmission. This is presently the technically most advanced technology and has only minor losses, although it is far more expensive than the common high voltage AC lines. Quite a few countries are using worldwide this technique, but often only over short distances. China developed over the past few years the world's most powerful DC lines. It should be already here indicated that the development of hydropower is a major driver for the development of the globally most advanced transmission technologies.

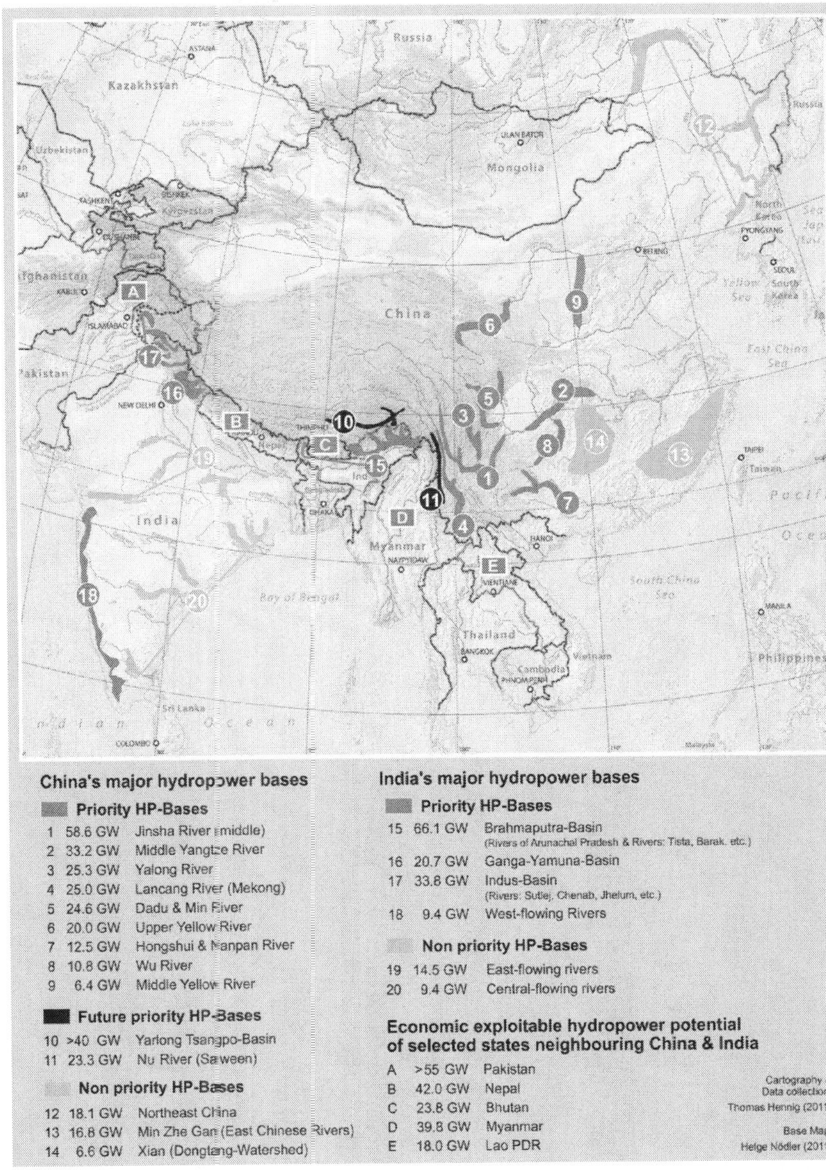

Fig. 3. Economic exploitable hydropower potential of China's and India's hydropower bases as well as of selected neighbouring states.

Another challenge of hydropower development is the fact that most of the key hydropower bases are affected by the monsoon climate with its uneven intra- and inter-annual runoff distribution which causes great differences between rainy and dry seasons. This results in a large quantity of non-beneficial spillage during the rainy season and a deficient generation in the dry season. Then hydropower has to get compensated mainly by support from thermal power. To reduce this discrepancy both countries are also investing heavy in the construction of pumped storage systems. Compared to the large scaled hydropower stations, the large scaled pumped storage projects are mainly located near the load centres. In 2010 China had a hydropower storage capacity of 14.6 GW, which ranks third in the world. But in an ambitious programme it should almost triple to 41 GW over till 2020 (Reuters, 9 June 2010).

India's hydropower storage capacity is presently merely 5 GW, but is also growing slowly.

Most of the hydro projects in China and even more in India are not developed solely for power generation and are therefore multipurpose in nature. Due to the large initial capital outlays for these projects, hydropower provides often the means to pay for required irrigation and related food security as well as for water supply or flood control. In particular the early prestigious projects were multipurpose dams, in which the primary purpose was mostly irrigation. The problem is that the timing for irrigation often competes with the main demand for power generation.

Only over the past decade many projects have been constructed in more suitable humid and mountainous areas, where hydropower became either the solely purpose or it has been the major one. Irrigation plays in most of the recent projects a less crucial role and therefore flood control and improved navigation become beside power generation a primary purpose.

Large hydropower projects require also large capital investments which are often, except for limited prestigious projects, even for governments difficult to arrange. China and India were for long no exceptions, but for China the situation changed over the past decade.

China became due to its growing economic and financial strength less dependent on foreign sources for the financing of its dams. Only at certain projects the Worldbank (WB) or Asian Development Bank (ADB) has been get involved; the 3,300 MW Ertan hydropower station in Szechuan province is a famous example. Built in the late 1990s it was largest hydropower project before the Three Gorges Dam China's. Finished before the World Commission on Dams published its strategic priorities for improved decision-making, management and planning of dams, Ertan received one of the largest single project credits from the Worldbank and it was China's first hydropower plant to be built through international bidding as China's first 200m dam.

China's provinces have been funding only a few large hydropower stations, mainly relevant projects to sustain their own energy needs; hence the majority of the recent and future large and prestigious hydropower stations are funded by the central government. This situation of financing so many large hydropower projects by a state government is worldwide unique. China's main challenges of ongoing large hydropower projects are the resettlement issues including related costs. There are currently about 23 million registered relocates in China (Hensengerth, 2010), if the number of former past projects is included, the figure is much higher. The consequences caused by resettlements (social implications, large costs, etc.) are a key reason that there are hardly any new large and present hydropower projects in the densely populated Eastern part of China.

Compared to China's financial strength, India still has great difficulties to mobilize finances for the implementation of its planned large hydropower schemes. Therefore it can presently only fund a few large, public projects that are considered of national importance. Alternative funding is for India quite important. It still relies on financial institutes (e.g. WB, ADB or International Finance Corporation) and on cooperations with international agencies (e.g. Germany's KfW) or private partners (e.g. Norway's Statkraft). Parallel is India seriously looking for tapping into the BOT market while allowing private hydro developers (Ramunathan & Abeygunawardena, 2007).

In the context of the challenging power and energy sector incl. the large financial constraints related to it, both countries initiated various reforms of the energy market inclusive the hydropower sector. Following is given a short overview over these reforms.

China	India
Generation companies	
public	
Datang -China Datang Corporation Huaneng - China Huaneng Corporation CPI - China Power Investment Corporation Huadian - China Huadian Corporation Guodian - China Guodian Corporation CTGPC - China Three Gorges Project Corp.	NHPC - National Hydropower Corp. of India NEEPCO - North-Eastern Power Corp. Ltd. SJVNL - Sutluj Jal Vidyut Nigam Ltd. (JV between the Gvt. of India & Gvt. of HP) NTPC - National Thermal Power Corp. THDC - Tehri Hydro Development Corp. (JV between Gvt. of India & Gvt. of UP) NHDC - Narmada Hydroelectric Devel. Corp. (JV between NHPC & Gvt. of MP)
private	
Hanergy - formerly: Farsighted Investment Group CHC - China Hydroelectric Corp.	Reliance Power Athena Energy Jindal Steel and Power Ltd. Jaiprakash Associates Ltd. KSK Electricity & KSK Energy DSC Ltd.
Other companies (transmission & distribution; financing; design; consultancy, etc.)	
SGC - State Grid of China CSPG - China Southern Power Grid Sinohydro Corp. Hydrochina Corp.	PGCI - Power Grid Corp. of India PFC - Power Finance Corp. WAPCOS - Water & Power Consultancy Services

Fig. 4. Major drivers and actors of China's as well as India's hydropower development.

2.3.1 Reforms of India's electricity and power market

India's power market was already opened in 1991 and various policy initiatives were taken for increasing the hydro capacity and the participation of private entrepreneurs. Despite these efforts in 2007 the public sector had still a predominant share of 97%. In one of the initiatives the Central Electricity Authority (CEA) prepared in 2001 a vision document giving a road map for expediting hydropower. In that report about 400 schemes totaling about 107 GW have been ranked from the point of view of attractiveness. In a further step the status of a Mega Hydropower Projects were introduced with the goal for bringing substantially down tariffs due to reduced levies, taxes, etc. The threshold for a Hydro-Mega-

Project is 500 MW and 350 MW for peripheral place like Northeast India, Sikkim and Jammu & Kashmir. In another move a three stage clearance procedure should encourage private entrepreneurs to enter into the otherwise high risk hydropower investments. Stage I results in a pre-feasibility report; stage II in a detailed project report (incl. also pre-construction activities, infrastructure development and related land acquisition) and under stage III the investment decision will be given after approval of all the documents.

Despite these initiatives almost all the Mega Hydropower Projects are owned and invested either by the central- or by state owned power generation companies or corporations (cp. Fig. 4). So far the 1,000 MW Karcham Wangtoo hydro scheme in Himachal Pradesh (commissioned in 2011) is the only private funded Mega hydropower project; however in Northeast India a few more are under construction or in planning. An overview over the present and future multiple hydro schemes in Northeast India, which distinguish in certain ways from the rest of India, is given in chapter 3.2.1.

To encourage hydropower development as well as looking for a tool for local development in peripheral regions, the Indian Prime Minister launched in 2003 in a landmark move the so called '50,000 MW hydro initiative'. Under this scheme preliminary feasibility reports for 162 large and medium sized projects with a capacity of 47.9 GW were prepared. Out of these 162 schemes, 133 are in the Indian Himalayan Region (IHR), mainly in Arunachal Pradesh.

Contrary to large hydro schemes, medium to smaller hydropower projects in India are funded and allocated by a competing market of diverse public and private companies. In particular its Ministry of New & Renewable Energy encourages and supports small hydropower projects. Additionally the CDM market attracts hydropower investments, in mid-2011 about 80 hydropower projects are already registered and another 120 projects are at validation (http://cd4cdm.org; 09.Sept.2011).

India's hydropower development has also to be seen in the context of building up a strong and vibrant national grid which so far is quite fragmented and split into five regional grids. Today almost half of India's present power transmission is done by regional grids or private funded transmission lines. The latter permit in particular for high energy consuming enterprises a direct commercial relationship to a generation company; some of the large hydropower stations have such an arrangement.

Avoiding a further fragmentation was 1991 the national Indian Power Grid Company incorporated, which carries now about 51% of the India's generated power. One of its major objectives is creating a strong and vibrant national grid. Therefore it already gained that four of the five Indian regional grids are now operating synchronously. Another objective is connecting the regional grids with that of neighboring states like Sri Lanka, Nepal and Bhutan.

2.3.2 Reforms of China's electricity and power market

With the breakup of the former Chinese Ministry of Energy in 1997 and the subsequent State Power Corporation in 2002 the Chinese power generation was separated from the grid but also from the projection. The state power monopoly was in the context of generation split into five national generation holdings (so called 'Big Five': Huaneng, Datang, Guodian, Huadian and China Power Investment) and later came along the Three Gorges Corporation as the six major hydroelectricity generation company. These six national holdings control a major part of the Chinese power generation and regarding large hydro schemes they control, beside some provincial owned holdings, the market almost exclusively (Hennig, 2007).

Similar to India only a few larger hydro projects are also funded by private companies, whereas they often have an affinity to the political circles in Beijing. The smaller the hydro projects become, that often private entrepreneurs and also local cooperatives invest in hydropower, most of these private investors come from the economic boosting areas of Coastal China. A more detailed overview is given in chapter 3.2.

China's policies designed to support hydropower are formulated at different levels. Chief among them is the Programme for the development of the Western Regions, officially instituted in 2001. Mainly infrastructure development (incl. hydropower) is seen as a key instrument for closing the disparity gap between the rich coastal Chinese provinces and the poor interior western provinces. Under that framework falls also the policy of Send Western Electricity East and Send Yunnan Electricity to Guangdong, another policy is the Rural Electrification Programme which promotes more local and also smaller hydropower development (Magee, 2006; Tullos et al., 2010)

Due to the above indicated facts that most of the hydropower regions are far away from the load centers the various grids and transmission lines play a crucial role in both countries. Until recently the Chinese provinces as well as the Indian states owned mostly their own grids. In the context of the above mentioned Chinese power reforms, two large power grid holdings emerged: China Southern Power Grid Company which is in charge of five South-Chinese provinces and the State Grid Corporation which controls the rest of China. Over the last years all the provinces got interconnected in one of the two grids, except Tibet where it is due to the difficult terrain still under construction. As the first country worldwide China started in 2009 a powerful long-distance high-capacity power transmission technology using direct current. Aiming to bring more power from its remote western and northern regions to the energy-hungry East and South coasts, China is the only country in the world which plans to build large ultra-high voltage (UHV) power-line networks. For that goal four other major lines are under construction or were finished in the last two years.

2.4 Brief introduction of China's and India's small hydropower sector

There is no internationally consensus on the definition of small hydropower (SHP). In Europe is the generally accepted norm 10 MW (European Small Hydropower Association), while in India it refers to a capacity up to 25 MW and in China depending on the context between 25 and even 50 MW.

In India SHP was neglected over a long time. Only after the present and renamed Ministry of New and Renewable Energy was formed in 1992, the development of SHP projects acquired good pace. However the policy and therefore the implementation varies considerably between the Indian states. India is planning to increase the SHP power generation capacity from present 900 MW (2009) to about 7,000 MW by the end of the 12 year plan (2017). From the potential 5,415 identified SHP-sites, merely about 800 have been implemented so far (Nautiyal et al., 2011). Except for remote villages, mainly in the Himalaya, most of India's SHPs are feeding into the regional as well as the national grids. Naturally the largest potential is in the Himalayan states and along the Western Ghats.

Also due to the other definition of SHP, China's potential as well as its implemented capacity is far higher. Already in 2005 more than 40,000 SHPs had an installed capacity of 38 GW and an annual average power generation of 130 TWh, which was about one third of the total hydropower generation in that year (Huang & Yan, 2009). China's major backbone for rural electrification has for long been SHP and about 653 rural counties had achieved preliminary electrification from SHPs. Today, with the rise to about 55 GW installed SHP capacity (in 2009) China's SHP sector has become a major grid feeding player. Despite its strong growth, the development is still far below its estimated potential of 128 GW capacity and 450 TWh/y average generation (Huang & Yan, 2009).

Additionally China has been establishing about 15 bases and Yunnan province another two bases, each having a capacity of about 1,000 MW each, which mainly consists of small hydropower projects. These SHP-bases are a specific and unique characteristic of China's ambitious SHP development. Their development is mainly driven by private entrepreneurs as well as local governments and local power grids.

3. Hydropower development in Asia's watertower

3.1 Common natural and cultural heritage of the case study areas - the key region between SW-China and NE-India

As mentioned above the southern adjusting mountain ranges of the Tibetan plateau, which are ranging from the Karakorum over the Himalaya up to the Hengduan Mountains, have the world's largest hydropower potential and supply water for almost half of the global population and regulate the climate in upland and lowland areas of Asia adjacent to it (Jianchu, 2007). Within that region, the core area for present hydropower development is the transition area between the states of Northeastern India (mainly Arunachal Pradesh) and the Southwest Chinese provinces (mainly Yunnan). This key region for hydropower development features, despite its present political splitting into different states, geographically a similar natural as well as cultural characteristic. It can even be regarded as one cultural landscape.

Following is given a short introduction of that region.

3.1.1 Environmental diversity

The Tibetan Plateau in general and its south-eastern extension (incl. its adjacent mountain ranges) in particular play as part of the 'Asian water tower' a crucial role. Some of the world's largest rivers with their tributaries flow through the region (e.g. Yarlung Tsangpo/Brahmaputra; Jinsha/Yangtze; Lancang/Mekong; Nu/Salween, and Irrawaddy). In that region the rivers change their topography (hydraulic gradient) from plateau via a long transition area with often deep gorge topography or rain-drenched Himalayan slopes to that of lowland topography and landlocked alluvial plains. Combined with its wide range of climatic setting is the region one of the global core areas for hydropower development, both that of larger scale as well as of smaller scale.

The area is a product of the collision and subduction of the Indian subcontinent with Eurasia, resulting in an impressive geodiversity but also in a fragile geological base and active seismic-tectonic instability. This unique physiogeographic setting includes with the

Eastern Himalaya and the Hengduan Mountains some of the highest mountain ranges in the world. But it includes also diverse Karst landscapes and huge plateaus and basins. The territory's geodiversity, combined with a climatic setting that ranges from tropical to temperate, has led to a unique diversity of ecosystems: from tropical rainforests in the south to shrub and grasslands in the mountainous north.

This small region is part of three of the world's major biodiversity 'hotspots' and several important ecoregions (WWF 2001) Alone the province of Yunnan hosts about half of China's biodiversity and boasts the second highest species abundance index in Southeast-Asia (Kwai Wong, 2005) and Northeastern India houses 21% of India's important bird areas, identified as per international criteria (Vagholikar & Das, 2010). Larger areas in that region are still poorly documented and biologists have been discovered in recent years a number of new species including large mammals.

3.1.2 Heritage and ethnic diversity

For more than 2000 years the region and in particular areas of today's Yunnan province were a major trade hub along the southwestern silk road connecting China with the economic centres of Southeast Asia, Tibet and India. The regions long and outstanding role in regional trading history has led to thousands of years of migration in and out of the region. While many of the migrant groups were small and assimilated with the local population, several were big enough to establish independent local and regional empires (e.g. bronze-age Dian-culture, medieval Nanzhao and its successor, the Dali empire and the medieval Ahom empire; Hennig & Linde, 2008).

A result of this history is a remarkable ethnic and cultural diversity, which makes this region the ethnic-richest area in India as well as China. Most of the hydropower projects under construction and / or planned are situated in areas inhabited by these minorities.

3.1.3 Economic potential and regional commitments and initiatives

Northeast India is a collective term for the eight states of that area, which are connected to the subcontinent only via the 22 km narrow Siliguri corridor, which is popularly referred as Chicken's neck. Hence, the region is similar to Yunnan in a peripheral and economic disadvantaged position. In spite of massive infrastructure improvements the economic and political centres of both countries are still far away.

Despite this economic disadvantage the region has the privilege to be a key passage between China and India as well as to Southeast Asia. Therefore it is in a good geostrategic position regarding favourable revenue arrangements and allocating funds for regional infrastructure development, mainly construction and upgrading of roads, railways, waterways, and pipelines. This infrastructure development has to be seen in the context of the rapid economic growth which causes a seeking for additional resources, trade- and market links to sustain it (Hennig & Linde, 2008). Therefore both countries expand their bilateral partnerships - often based on already existing political dependencies and patronages as well as increasingly engaging in multilateral partnerships. Fostering such transboundary economic partnerships are particularly beneficial to both regions due to their proximity to the other side and to the Southeast

Asia. But it has so far not affected a joint hydropower development or even a joint watershed development initiative. Two major examples for the regions engagement in joint development cooperations are the Greater Mekong Subregion (GMS) Economic Cooperation Programme, and the track-II Kunming Initiative.

Among the various rich 'green resources' hydropower is becoming in both regions a key pillar, both for the large as well as for the small hydropower sector.

3.2 Present state of hydropower development (incl. upcoming projects)

Despite the above mentioned similar common cultural and natural heritage, the hydropower development is very much affiliated by its political ties, therefore it will be separately described and only later jointly assessed.

3.2.1 Northeast-India

In mid-2011 India has been developed only 25.6% of its present economic exploitable hydropower potential of 149 GW. Three quarter or 117 GW of India's hydropower potential is located in the Indian Himalayan Region, an area which accounts contrary only for approximately 18% of India's total geographical area. Within the Himalayan Region more than half (=63.3 GW) comes in the Brahmaputra Basin of NE-India (Agrawal et al., 2010).

Beside the main stem of the Brahmaputra, which is known in China and Tibet as the Yarlung Tsangpo and in Arunachal Pradesh as Siang, there are few major tributaries, each having a large hydropower potential. While the Tista is draining Sikkim and West Bengal and the Barak Manipur and Assam, all the other relevant tributaries are in Arunachal Pradesh. Most of those North-Bank tributaries of the Brahmaputra are of Himalayan origin and are fed by glaciers in their upper reaches, e.g. Subansiri, Kameng (Jia Bhareli), Dibang, Lohit, etc.

Due to the fact that more than 40% of India's hydropower potential comes from the Brahmaputra Basin, the region and in particular Arunachal has been proactively tagged as the 'future powerhouse' of India.

The focus for massive hydropower development in the Northeast of India was pushed by the Indian Central Government. With the gradual liberalisation of hydropower policies, the Indian states were allowed to invite private players. Sikkim kick-started this process about ten years ago allowing private companies exploiting the hydropower of the Tista river and its tributaries. Today the tiny Sikkim-State is together with Arunachal Pradesh at the forefront in the initiative to sign multiple Memoranda of Understanding/Agreement (MoU/MoA) with private and public power developers. Alone the state of Arunachal Pradesh allotted hydro schemes with a capacity of more than 40 GW till 2010 and most of these projects involve private Indian players (Vagholikar & Das, 2010). The impressive share of private entrepreneurs is contrary to the national trend. It is also fostered by the above mentioned initiatives by the national government as well as the state policy.

So far only 423 MW of the exploitable hydro resources have been developed in Arunachal, which is merely 1% of its potential. By 2012, the 2,000 MW Lower Subansiri Scheme will be

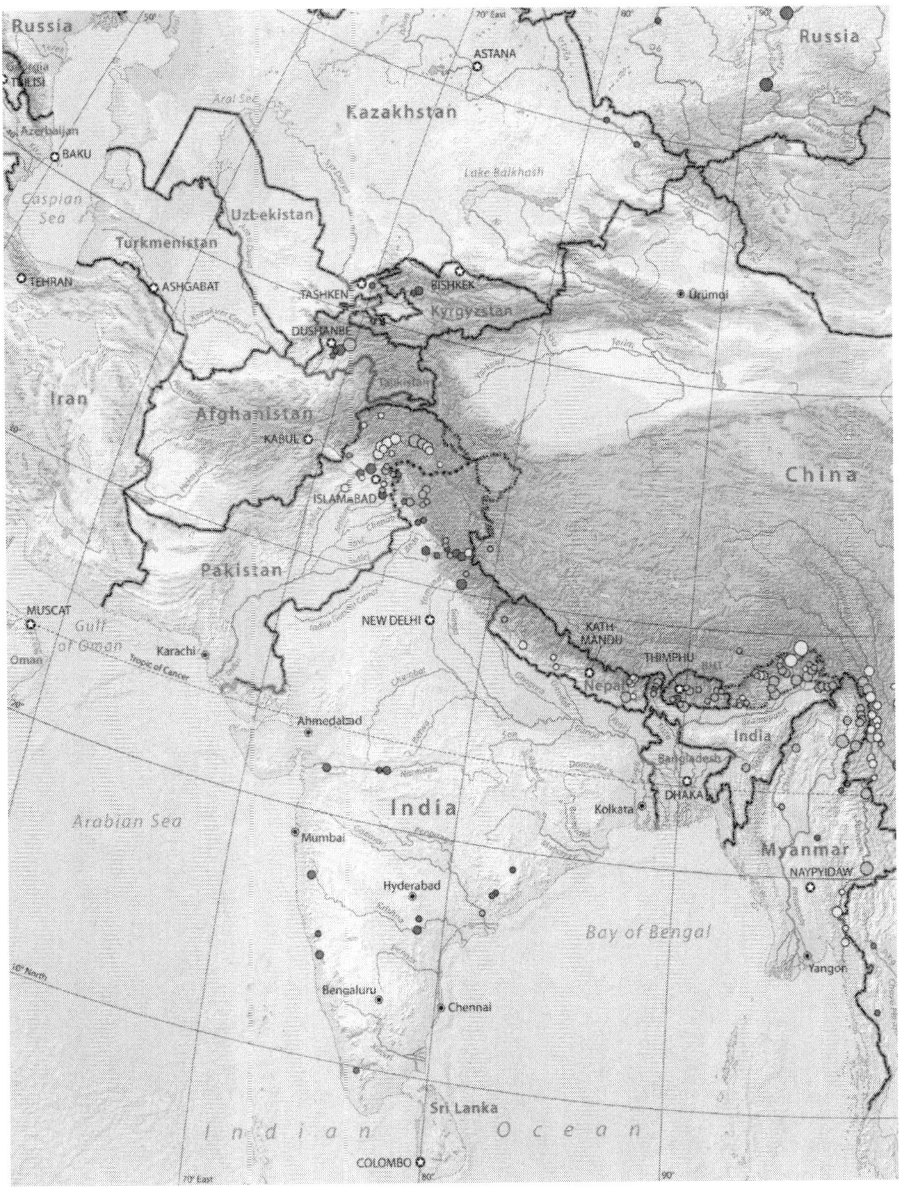
Russia
ASTANA
Kazakhstan
Aral Sea
Azerbaijan
BAKU
Caspian Sea
Lake Balkhash
Uzbekistan
Turkmenistan
TEHRAN
ASHGABAT
TASHKENT
BISHKEK
Kyrgyzstan
DUSHANBE
Tajikistan
Ürümqi
Iran
Afghanistan
KABUL
China
ISLAMABAD
Pakistan
MUSCAT
Gulf of Oman
Oman
Karachi
Tropic of Cancer
NEW DELHI
KATHMANDU
THIMPHU
Nepal
India
Ahmedabad
Bangladesh
DHAKA
Kolkata
Arabian Sea
India
Mumbai
Myanmar
NAYPYIDAW
Hyderabad
Bay of Bengal
Yangon
Bengaluru
Chennai
Sri Lanka
Indian Ocean
COLOMBO
70° East
80°
90°
10° North
20°

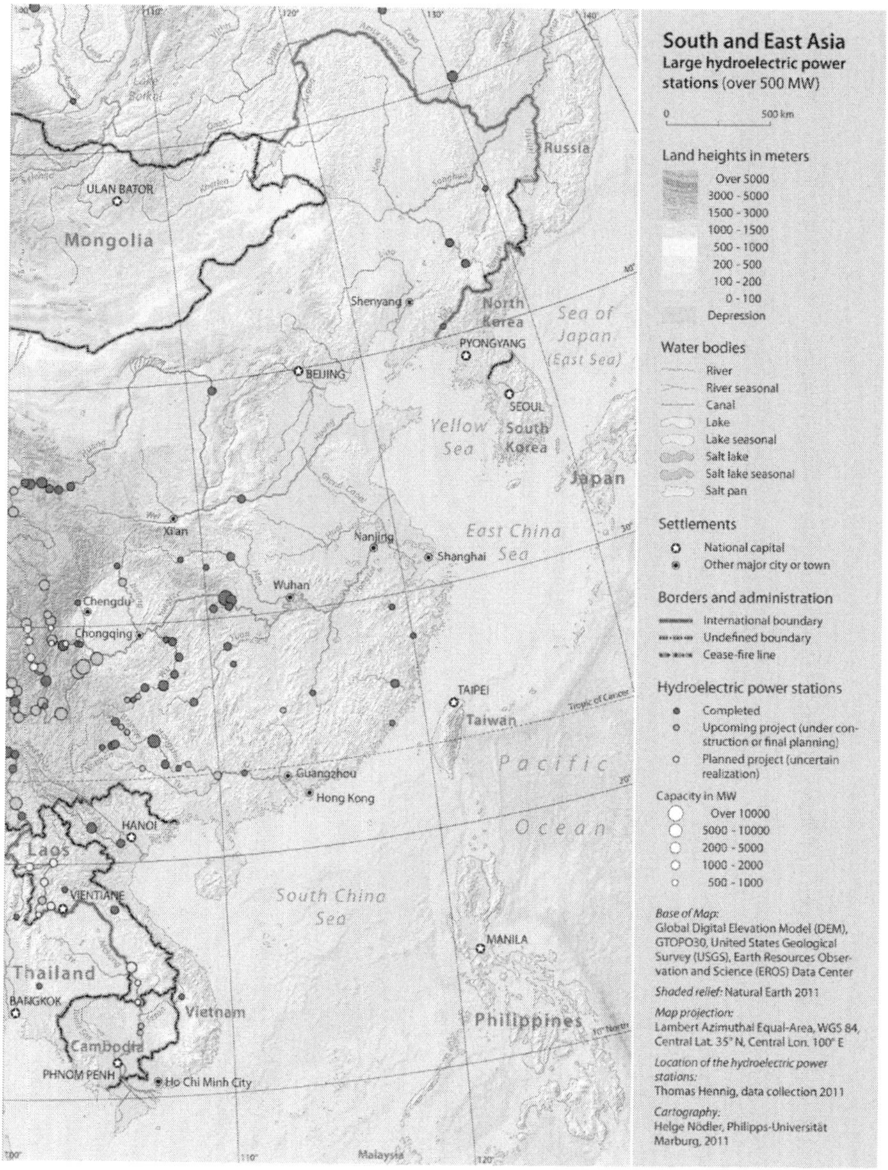

Fig. 5. Map of present and forthcoming large hydropower projects (above 500 MW) in Asia.

commissioned and from then onwards Arunachal Pradesh becomes an exporter of electricity. It is the first Mega-Project in that region and it will be operated by the state owned NHPC.

So far the Northeastern power grid is the most inefficient one in India and in particular the distribution is the weakest link in the power system. With almost 40% the transmission and distribution losses are the highest in India and far above the Indian average. In some regions the losses range up to two third of the generation (Rao, 2006). The construction of an efficient transmission system is a key issue of the regions hydropower development.

The transmission system for power evacuation from the Northeast is of serious geopolitical interest. Actually it should get finished at the end of the present five year plan in 2012 as a 800 kV HVDC bi-pole line and a 400 kV double circuit AC lines in a hybrid system, which would be one of the global most advanced transmission lines. This forthcoming transmission corridor, which requires a stretch of about 800 to 1,200 m, should at a later stage evacuate 25-35 GW of electricity mainly to energy hungry Northern and Western India. It has to pass through the geopolitical sensitive narrow Siliguri corridor and should later also include hydro generated power from Sikkim as well as from Bhutan and Myanmar (Rao, 2006; CEA, 2008).

But due to lack of central funding the prestigious project is seriously delayed. Alternatively the Indian government is seriously looking for a cheaper version passing through Bangladesh. Additionally India tries to increase private investments of such geostrategic important transmission lines as well as to increase the power and capacity of those lines in order to reduce land requirements and transmission losses (Planning Commission of India 2011).

All the project developers of Northeast India need to obtain a Long Term Open Access for transmitting power through the corridor. The first mega hydro projects which will be gradually commissioned over the next years, will be located on the Brahmaputra-tributaries more close to the Indian mainland (rivers Kameng and Subansiri respectively). They will be projects by public funding, either by NHPC or by NEEPCO. India's presently largest privately funded project is the 2,700 MW Lower Siang project on the main stem of the Brahmaputra. It is owned by Jaypee Power Ventures and should be completed by 2017. That project is like other private funded projects on BOT bases, which means the owner is the local state (e.g. Gvt. of Arunachal Pradesh) and the investing company gets over a concession period of 40 years the right of utilisation.

Due to environmental and social concerns various Indian State Governments (including Arunachal) encouraged the development of run-off-river (RoR) type hydro projects. Quite a few of formerly scheduled storage schemes in the Pre-Feasibility Reports were converted into RoR-schemes, either as one big one or as 2 to 3 smaller ones.

The controversial upper Siang project (about 10,000 MW) on the Brahmaputra/ Yarlung Tsangpo would be by far India's largest hydropower station. Due to China's recently published plans for future development of the Yarlung Tsangpo, India revived immediately its plans for the prestigious project. As such a large scaled and scoped hydel project has never been attempted in India, it is presently discussed to hive off the project into two or even three modules, so as to keep better tab on costs and aid implementation (cp. chapter 4.2).

3.2.2 Southwest-China (Yunnan-province)

The Southwest Chinese provinces are the backbone of China's ambitious present and future hydropower development and within that area the province of Yunnan is the focus for China's forthcoming hydropower expansion plans.

The ambitious programme makes Yunnan, which is in its size almost comparable to Germany, in a few years to the world's most powerful hydropower base. It will then have the largest installed hydropower capacity as well as the largest annual hydropower production, more than states like the USA, Canada, Brazil or Norway.

What is the backbone for this development?

The principal item is the construction of large and even mega hydropower projects on the major rivers and along their tributaries (Hennig, 2009 & 2007; Dore & Xiaogang 2004), secondary also the development of smaller hydropower bases. For Yunnan are three rivers special relevant; the Jinsha river, which is the upper part of the Yangtze river; the Lancang or Mekong river and the Nu or Salween river. The proposed power stations on the main stem of these rivers are all planned as cascade systems. Most of the projects will be very large ones, and often they are designed as run-off-river systems, where parts of the river are diverted through huge tunnels. Only a few projects are designed as huge reservoirs and are in combination with a large installed capacity even called mega-projects.

The Jinsha has by far the highest hydropower potential worldwide. Not in Yunnan, but on the Yangtze is also presently the world's largest hydropower station in terms of installed capacity (20,300 MW); the Three Gorges Dam. Its first commercial operation began in 2008. The owner, China Three Gorges Corporation (CTGC), is presently constructing a cascade of four other hydropower stations on the Jinsha river, which have a combined capacity double that of the Three Gorges Dam. Two of the projects (Baihetan and Xiluodu) belong to the largest worldwide. Presently China is one of the few countries which uses 700 MW turbines. But based on Jinsha's Wudongde and Baihetan hydropower stations, China gets as the first country the ability for independently design and manufacture the world's largest hydropower-generating single units (1,000 MW).

In the upper reaches of the Jinsha is a second cascade under construction. Originally it should be a cascade of 8 major dams, but one was cancelled due to environmental and social concerns. A few other dams were stopped for a while due to environmental problems (see next chapter). The cascade is developed by the Jinsha Hydropower Development Corp., but the projects are mostly owned by different state owned companies. The 2,400 MW Jin'anqiao-project is not only the first commissioned project of the cascade (in 2011), it is also China's first large hydropower station invested by a private company, Hanergy.

The Lancang or Mekong river reduces in Yunnan its altitude difference by 1,780 m and has therefore a high potential. The lower cascade of proposed 8 dams is Yunnan's first major hydropower project. It is now developed by Hydrolancang, a subsidiary of Huaneng company. The 1,500 MW Manwan Dam completed in 1995 was the first dam on the Mekong and also Yunnan's first major dam. Its funding had at that time model character, because no international donors were included. With the second dam (Dachaoshan; 1,350 MW; commissioned in 2003) Yunnan became a power exporter. The third dam (Jinghong; 1,750 MW; commissioned in 2009), was China's first hydropower joint venture with a foreign country (Thailand). After completing a 1,070 km long transmission line, Thailand will be also a primary consumer of electricity generated by the dam. With the fourth dam, Xiaowan (4,200 MW; finished in 2010), China exceeded its 200 MW installed hydropower capacity. Presently it is China's second largest dam but in a decade it will be only of a 'mid-ranged' size. On the upper part of the Lancang/Mekong China is presently starting a second hydropower-cascade. But due to concerns of the lower Mekong riparian states its details are still not finalized.

The proposed cascade of 13 dams along the Nu/Salween river raised international attention, because it has been still halted due to environmental concerns. Background is that the Nu is the only undammed major river of China and a few of the project sites are affecting the UNESCO world heritage site 'Three parallel rivers'.

On the other (former) undammed river, the Yarlung Tsangpo, is since 2011 a first hydropower station under construction, the 510 MW Zanwun dam. Subsequently a major cascade is scheduled on the Yarlung Tsangpo.

Beside these mostly large or even mega dams along the main stem of the rivers a number of other hydropower stations are finished, under construction and/or planned within the respective watershed. In the other three relevant waterheds of Yunnan (Irrawaddy, Red River and Nanpan) a large number of moderate to large hydropower stations are presently coming up.

Most of those projects feed into one of the three major transmission lines, which send power to the economic hubs of China's east and south coast. The development of an effective grid is an integrative part of China's hydropower strategy.

(Photo: Hennig, 2006)

Fig. 6. The 108 MW Dayin-1 hydropower station on the Dayin/Tarpein river was finished in 2007. It was the first of a series of 6 cascades in the Chinese-Burmese border region. The dam is located in a protected area. It was constructed by private entrepreneurs from the Chinese east coast and was later purchased by the state-owned Guodian Corp.

Part of that is also the construction of an efficient UHV-DC transmission system. China is presently the only country building such a network. Since 2010 the first line connects Yunnan with Guangdong (5,000 MW) and a second line between Yunnan and Shanghai (6,400 MW) is under construction. Their DC-voltage level of 800 kV is far above the present existing maximum of 500-600 kV.

At present there are in Yunnan 157 hydropower CDM projects approved, another 110 are at validation (http://cd4cdm.org; 09.Sept.2011). Most of those projects are small hydropower stations (SHP). Surprising in that context is also, that the majority of SHPs in Yunnan are grid feeding and an integral part of the power transfer. Beside the omnipresence of SHPs in Yunnan, there are two major SHP hubs existing. This is a cluster of SHPs and some bigger projects within a small area and each has a cumulative installed capacity of about 1,000 MW. Yunnan's two SHP-hubs are in the Fugong-county of Nujiang-prefecture and in Yingjiang-county of Dehong-prefecture. Recent studies in Yingjiang indicate serious environmental cumulative consequences of the recent SHP-development. Cascades of SHPs, which are according to international classifications mid-sized projects, result in the drying up of a number of smaller rivers in the dry season. This is caused by water diversion and the failure of ensuring a riparian distance between the hydro projects and the failure of enforcing a minimum water discharge.

3.3 Environmental and socio-economic challenges of the hydropower development in SW-China and NE-India

China as well as India officially reject the famous and influential 2001 report by the World Commission on Dams (WCD). However, both governments are not averse to international cooperation. Their domestic hydropower and dam legislation policy, which includes resettlement issues as well as Environment Impact Assessment, has been influenced by international debates and by their own characteristic domestic learning policy. Despite not recognizing the WCD-guidelines, hydropower projects which are internationally funded (e.g. from WB, ADB, IPC or from Western donors) have to follow the rules by the World Commission on Dams. This counts also for hydel projects requesting for CDM.

Whether or not hydropower projects are constructed is decided in both countries on the basis of their assessment of their economic development and the related energy needs. Comparing both states over the last two decades it is peculiar that China's hydropower sector is growing faster than predicted, while India's is growing more slowly than expected and needed.

Despite this simple characteristic the hydropower sector in both states belongs to the fastest growing worldwide. This impressive growth is naturally accompanied by certain environmental and socio-economic problems and challenges. Following a few of these challenges are explained and discussed more detailed.

3.3.1 Environmental and resettlement policy in China (Yunnan)

There are currently about 23 million registered relocatees in China (Hensengerth, 2010) and according to the WCD-report between 1949 and 1999 about 12 million people were displaced due to reservoir construction. Millions more were displaced due to the construction of the Three Gorges Dam and other large hydropower schemes. In many cases people were moved to other areas and provinces, which raised assimilation problems and resulted in conflicts with the resident population (Heggelund, 2006). Growing problems related to resettlement were solved by passing a number of new laws and regulations. Now the implementation of newly approved resettlement plans are in the responsibility of county and sometimes even provincial governments, while the monitoring and evaluation are carried out by external agencies.

The immense and steadily growing costs of resettlements are a major objective of the lack of new hydropower projects in the more densely populated areas of eastern and southern China. Only prestigious and important projects like the controversial Danjiangkou reservoir as part of the S-N-water transfer project get acceptance. It is with 350,000 people the largest present resettlement project and one of the very few Chinese projects which are heavy delayed.

Studies by the author of recent hydropower related relocation projects in Yunnan (interviews conducted in 2010 and 2011) indicate that they were done according to the Chinese rules. Relocated villagers (incl. village committees) do not complain about the relocation itself, but about the intransparent decision making and the lack of participating in relevant decision making and co-determination.

The development of the Chinese Environmental Impact Assessment (EIA) since 1979 is the result of domestic learning processes and the studies of international examples and experiences and underlying therefore frequent modifications.

Chinese formal EIA-procedures are according to the international standards but however their implementation is often doubtful. In some cases the EIA got published and approved only after the hydropower project was constructed. Additionally public disclosure of the entire EIA report is not necessary and only summaries of EIAs have to get published. For strengthening the EIA-procedure many hydropower projects have now to get reviewed and approved by either provincial or even central authorities. Despite the often still weak EIA-implementation a few cases spurred the debate on hydropower projects in Yunnan:

1. In June 2009 China's Ministry of Environment Protection halted two large hydropower projects along the Jinsha river (Upper Yangtze). The move was considered as the severest punishment in the country's environmental appraisal history as it involved two large state-owned conglomerates Huadian (with the 2,200 MW Ludila project) and Huaneng (with the 1,700 MW Longkaikou). But in November 2010 the mighty National Development and Reform Commission (NDRC) gave official clearance. In that context the 2,400 MW Jin'AnQiao project on the same river, which is the first and largest private funded LHP in China, got in 2010 the formal EIA-approval only just before finishing the construction.
2. The development of the Nu river cascade is halted since 2004 and subjected to environmental investigation. The Nu or Salween is an international river and its development is therefore a central government responsibility under the charge of the Yangtze Water resource commission. Although first ideas about the Nu-projects came up in the 1970s, only in 1999 the NDRC adopted a plan of a cascade based on 13 dams with a combined capacity of 23 GW. The proposed developer, state owned Huadian Corp., tried together with the provincial government of Yunnan to rush China's State Council into approving the projects before the new EIA-law could come into effect in 2003. China's former SEPA (now Ministry of Environmental Protection) and a Beijing-based NGO began to organize national and even international opposition to the project. This resulted in an unique situation for China, establishing a link between researchers,

NGO activists and also politicians. Huadian was forced to conduct an EIA and submit it to the SEPA. Finally in 2004 then Prime Minister Wen Jiabao put the project on hold due to an insufficient EIA. In between the EIA report had been completed and approved by the SEPA for a reduced number of dams, also their number is contradictional. The present situation is still diffuse, despite many hints of an upcoming official approval of the stop are existing. But on the other side, preliminary work is under progress and beside Huadian is now also Guodian active in the region, as seen in a field trip by the author in early 2011.

3. Beside the Nu-cascade the most controversial hydropower project in China is the 2,800 MW Upper Hutiaoxiao project in Yunnan. The proposed dam, with a height of 276m would officially relocate more than 100,000 people, mostly minorities; unofficial the number is much higher. The location is on the Tiger Leaping gorge on the Jinsha (Upper Yangtze), one of the worldwide most spectacular canyons. The proposed dam was also aimed at diverting water from the Jinsha to the provincial capital Kunming which lacks drinking water. The plan opposed by Chinese NGOs caused an unexpected public outcry, which was supported by Chinese Media and it received also international attention. Therefore the plan has been shelved since 2004 and in late 2007 the project was cancelled due to a report by The South China Morning Post. But so far no official statement was given.

These examples about environmental and socio-economic issues present a mixed picture about China's hydropower decision making process. On the one side the process is still secretive, top-down oriented and authoritarian, but on the other side democratic procedures get introduced and in certain cases individuals and civil society organisations can organize effective protests and even stop controversial projects. How is the comparable situation in India?

3.3.2 Environmental and resettlement policy in India (northeast India)

India ranks fourth in the world in terms of the number of its large dams, but after China it ranks second in terms of displaced people. The main objective for dam construction is irrigation, but in the last years there is also clear trend towards hydropower. Although the government is the owner of almost all those dams, there exists no official data concerning resettlements. It is estimated that in India between 32 and 56 million people were displaced up to the new millennia, most of them in tribal areas (Choudhury, 2010).

Compared to China's top-down authoritarian decision making on dams, in India it is more akin of a polycentric process which involves at various stages multiple actors.

In its young history India faced many serious tensions about the utilization of water resources, most of them were interstate conflicts. But with the controversy over the Sardar Sarovar Multipurpose project in the late 1980s began in India to emerge an era of social and environmental movements and civil-society-driven consciousness-building. The civil-society became the major actor working towards progressive and gradual changes in dam related policies (Choudhury, 2010).

The role of public participation is historically limited, similar to China's situation. But what changed in India is the relevance of the environmental sphere, namely the compulsory public hearing, relevant for the EIA documents. Those public hearing became one of the most contested and controversial arenas, strongly affected by civil society organisations and environmental and social activists as well as the local communities.

These civil societal organisations and activists criticize the quality of many EIA-reports, which are poor in many social and environmental aspects, their inadequate baseline information and additionally act often only as a post-clearance study. Contrary to China, India's 'Right to Information Act' binds the government to make all the relevant documents public accessible (e.g. Feasibility study, Enviromental Impact Assessment, etc.).

In case of NHPC's 3,000 MW Dibang-Multipurpose project in Arunachal, which will have India's highest gravity dam (288m), it came also to serious opposition and public awareness campaigns. For the first time civil society organisations from an upstream state (Arunachal) jointly engaged in agitation with organisations from a downstream state (Assam).

Also as a reaction to the strong criticism of the civil societal organisations towards the public hearing and the EIA-documents the then Union Minister of State for Power, Jairam Ramesh, now Environment Minister, raised concern about the 'MoU-virus' of Northeastern States like Sikkim and Arunachal Pradesh. He referred to the large number of projects handed out by these states to private hydropower developers, whereas most of these agreements have been accompanied by huge monetary advances. Often they were signed before compulsory public consultations or mandatory clearances (Vagholikar & Das, 2010).

As a result to those critiques many hydropower projects in Northeast India got seriously delayed. None of the many large (> 500 MW) hydro projects by private players are finished so far. Also the state-owned NHPC, which is on the forefront of hydropower development, commissioned till now only the 510 MW Tista-V project; and in 2012 the 2,000 MW Lower Subansiri is following.

Also resulting from those critiques with the well-known environmental and socio-economic impacts of large scaled hydropower projects, a few Indian state governments encouraged the development of run-off-river (RoR) type projects. In that context quite ~~a few~~ LHPs had being converted and split from proposed storage schemes in the Pre Feasibility Reports to one and often even more than one RoR-schemes, which have been allotted to private developers for project implementations. On the other side there is strong opposition of conversion of more storage schemes into RoR-schemes, because the multipurpose-function of reservoirs (flood control, irrigation, drinking water supply) gets reduced which results in higher power tariffs.

4. Hydropower development in transboundary watersheds of SE-Asia

Finally a short overview should be given about international hydropower ambitions of both countries as well as about implications for transboundary watersheds.

4.1 China and India overseas hydropower activities

Over the last decades Chinese dam builders have accumulated a vast expertise and knowledge base in hydropower construction. Getting economically strong and based on growing financial reserves, Chinese state owned dam builders and financiers appeared on the global hydropower market with a bang in the early years of the new century. They started to take on large and often destructive projects in countries like Myanmar (former Burma) or Sudan, which had previously been shunned by the international community. Their emergence threatened to roll back progress regarding human rights and the environment which civil society had achieved over many years (Bosshard, 2010; Mc Donald et al., 2009).

However, the situation is changing and Chinese dam builders and financiers are trying to become good corporate citizens rather than rogue players on the global market. Presently about 40 foreign hydropower stations are completed and more than 200 are under construction or planned. Chinese hydropower activities are global, but the majority of the projects are in Southeast Asia and Africa where China has fostered strategic regional and bilateral ties. The major actors and drivers of Chinese overseas hydropower ambitions are the six major state owned power suppliers, but also companies like Sinohydro, China Southern Power Grid, various Chinese Banks (mainly China Exim Bank) and Yunnan based companies.

For Chinese hydropower activities abroad Yunnan plays a crucial role due to its geopolitical position close to Southeast Asia. A number of projects in Myanmar are of direct relevance to Yunnan, like the cascade of six dams along the Upper Irrawaddy (11,160 MW); two dams on the Salween (Upper Thanlwin - 2,400 MW and Tasang - 7,100 MW); as well as four other dams on the Shweli and Dayin/Tarpein rivers. Some of the projects caused political tensions between the countries, because a number of hydel projects are in areas which are controlled by various rebel groups (Shan, Kachin, etc.). These groups are partly blaming Chinese companies for doing other activities than hydropower construction and in 2009 even attacked a hydropower construction side. Further Myanmar surprisingly suspended in late 2011 a key project, the $3.6 billion Myitsone dam, causing diplomatic irritations.

Most of Chinese hydropower projects in Myanmar are on a 40 years BOT basis and therefore they get directly connected to Yunnan's power grid for transmission which transfers Myanmar's electricity to the load centers along the Chinese coast. Beside the strong activities in Myanmar, Yunnan sends further electricity to energy hungry North-Vietnam and in future also to Thailand.

Contrary to Chinese global hydropower activities, India's dam building abroad is quite moderate and only in the beginning. Except a relative active global hydropower consultancy business (mainly by the state-owned WAPCOS), India's foreign dam projects concentrate only in the neighborhood (e.g. Afghanistan, Nepal, Bhutan and Myanmar).

In particular India and Bhutan have agreed to realize 10 LHPs with a combined capacity of 11.6 GW which should feed in future into the Indian Grid. So far only three projects are under construction and one is finished, despite India has committed to draw already 10 GW by 2020 from Bhutan. Even more difficult is the situation with Nepal, which has with 42 GW by far the highest economic exploitable potential in the Himalayan region, but has still only an installed capacity of 0.7 GW and faces therefore an ongoing power crisis (cp. Fig. 3). Although India has been assisting Nepal for long in the development of its SHP sector, but the prestigious projects which are already for many years under discussion at various mutual interest levels have not been realized so far. These large projects should also straight feed into the Indian grid.

Compared to China's very active hydropower activities in Myanmar, India pursues presently only two large projects of a strategic venture in Myanmar, the Tamanthi 1,200 MW and Shwesayay 600 MW. Although the MoUs were already signed in 2004, the projects are still not realized.

Comparing the global hydropower activities, it is striking that in both countries the drivers are the state owned power generation, transmission and design as well as consultancy companies. Hydropower is seen as a strategic venture. China became here over the last decade very successful and has been realizing many projects, both in its neighborhood as well as globally. Contrary India's engagement in its neighborhood is characterized more by discussion instead of realization. Additionally is China entering into the hydropower markets of India's Himalayan neighbors, mainly Nepal.

4.2 Transboundary watershed policy

Further a short overview about implications for transboundary watersheds should be given. In particular for the Lancang/Mekong river exist quite a few studies e.g. Grumbine & Xu, 2011; Osborne, 2009. China is with Yunnan and Guangxi part of the GMS-initiative but it is not part of Mekong River Commission. In particular the serious drought which affected in 2010 the region, produced controversial discussions and argumentations between upstream Yunnan (or China) and downstream states about the causes and effects of damming up Yunnan's Lancang/Mekong River. Beside these discussions also the downstream Mekong states plan the construction of another 13 major dams in the lower part of the Mekong. The developers of those projects are mainly from China and from Thailand, but also from Vietnam and France.

For the binational watershed of the Nu/Salween- river it is striking that China stopped so far damming up its own part of the river but is again, together with Thailand, very active in the hydropower development along the Salween in Myanmar.

The hydropower-development of the transboundary and geopolitically conflictual West-Himalayan regions is based on the Indus-Water treaty from 1960. Contrary to all the tensions between India and Pakistan there are so far no serious geopolitical tensions about the realization of all the projects along the Indus and its tributaries, which are shared between the two nations. The situation may change, because recently tensions arose about the proposed 330 MW Kishanganga hydroproject along the Neelum River where India and Pakistan disagree on the application of the provisions of the Treaty. In 2011 Pakistan had approached the international court of arbitration constituted by the United Nations against the Indian move to construct the hydel project by diverting the Neelum River.

Pakistan's ongoing power shortage should be faced by a massive hydropower development of its economic exploitable potential of more than 55 GW. Contrary to India is China here involved in a few hydropower projects and the number may soon increase. However in 2011 caused the 1,100 MW Kohala-project on the Jhelum-river political tensions between the two countries.

More serious is the oncoming conflict on the Yarlung Tsangpo where China and India are directly involved. The River which is called Brahmaputra in India and Siang in Arunachal is so far hardly dammed up, but huge and very prestigious projects are in the pipeline. In early 2011 China officially announced the first dam along the Yarlung Tsangpo which caused a political outcry in India. In direct response to China's plan of constructing a proposed 40 GW project (double of the Three Gorges Dam) on the famous bend of the Yarlung Tsangpo, India revived its idea of constructing a 10 GW project on the Upper Siang. This would be by far India's largest hydel project. New Delhi is pushing state-owned power companies (namely NTPC and NHPC) and the Arunachal government to speed up with their own projects on the river. Background is the doctrine of prior appropriation, which indicates that the priority right for the river falls to its first user.

5. Conclusion

China as well as India are characterized by the globally fastest growing hydropower development. Over the past decades China gained an engineering and technical expertise, which made it by far the worldwide most advanced hydropower market. Additionally China is economically so strong that it can realize a big number of large and expensive hydel projects. Since the new millennia China's expertise and financial strength is also globally seen, because it is constructing and funding a fast growing number of hydro power projects worldwide.

Contrary to China is India still depending on foreign technology. This is true enough in particular for very large projects or projects in difficult terrains. Additional India is, despite various political initiatives, not able to finance a larger number of big hydel projects parallel. The path of India's hydro capacity generation, both in India itself but also due to its strategic partnerships in neighboring states, has been very slow. In particular for Northeast-India with its huge hydropower potential, projected implementation has been dismal so far.

In both countries hydropower used to be a part of classical multipurpose projects, whereas power generation was often the means for large scaled irrigation projects. Today (hydro-) power generation is essential for sustaining the economic growth and therefore large hydro projects are part of regional development initiatives. This results also from the fact that the highest hydropower potential is far away from the load centers and their economic and urban hubs. While China uses this opportunity for developing economically peripheral regions and developing and fostering the globally most advanced grid and transmission systems India is not able to do similar.

The driver for China's as well as in India's hydropower development is the government, both on the central as well on the provincial /local level. In particular the NE-states of Sikkim and Arunachal Pradesh signed plenty of Memoranda of Understandings/Agreements with private players, but still many of that signed projects have not been realized so far.

The other major drivers are the state-owned power generation, transmission and design companies. Private companies play in both countries for large projects only a minor role so far. However in India's context the government tries eagerly to involve also the private sector, mostly on BOT basis. But due to the high initial investments, private companies are still hesitating.

Compared to the LHP-sector the small hydropower development is in both countries quite successful pushed by the governments. Most of the investments came here from the local private sector, and many of the recent SHP-projects are applying for CDM.

Both countries reject officially the WCD guidelines but their decision making process differentiates. China is following in its hydropower policy a classical authoritarian top-down approach. The decision making on dam development is based on domestic laws, which adopt gradual global norms. Despite this progress the environmental implications and to some extend also the social consequences of hydropower decision making are still not transparent at all. Despite this critique, get some projects surprisingly halted up for a while or some even stopped.

Contrary to China, is India's hydropower decision making more akin of a polycentric process which involves at various stages multiple actors. Despite the critique of civil societal organisations and environmental as well as social activists over poor EIA-reports is India's decision making quite transparent, in particular that all the relevant documents are publically accessible. However due to the quite active civil society structures, the implementation of many hydropower projects has been slow and causes often a restructuring of many projects.

The main critique by the author is that in both states hydropower development is seen as the only tool for watershed development. There is no serious assessment for developing a concept of the whole watershed which goes far beyond hydropower. This concept should create an innovative management model, which considers water, environment, livelihood, food security, etc. and therefore enhancing public participation and transparency as well as financial benefits for the people in the watershed.

This is underpinned by the fact that most of the rivers in Asia's watertower (from the Karakorum over the Himalayas to the Hengduan Mountains) are transboundary rivers. Therefore should be a shift from a sovereign state-focused watershed development to an integrated transboundary watershed approach which also recognizes political sensitivities of the whole region.

6. References

Agrawal, D.K.; Lodhi, M.S. & Panwar, S. (2010). Are EIA studies sufficient for projected hydropower development in the Indian Himalaya region? *Current Science*, Vol. 98, No. 2 (Jan 2010), pp. 154-161, Available from www.ias.ac.in/currsci/25jan2010/154.pdf

Bosshard, P. (2010). *China's dam builders clean up overseas. In: Asia Times (12 May 2010)*

Chang, X.L.; Liu, X. & Zhou, W. (2010). Hydropower in China at present and its further development. *Energy*, Vol. 35 (2010), pp. 4400-4406, Available from doi:10.1016/j.energy.2009.06.051

Choudhury, N. (2010). Sustainable dam development in India. Between global norms and local practices. *Discussion Paper / Deutsches Institut fuer Entwicklungspolitik*, (10/2010), ISSN 1860-0441, Bonn, Germany

CEA (2008). Hydro Development Plan for 12th Five Year Plan (2012-2017), Central Electricity Authority of India, (Sept. 2008), New Delhi, India

Dore, J. & Xiaogang, Y. (2004). Yunnan Hydropower Expansion. Working Paper, 38 pages, Available from: http://www.sea-user.org/download_pubdoc.php?doc=2586

Grumbine, E.R. & Xu, J. (2011). Mekong Hydropower Development, *Science*, Vol. 332 (April 2011), pp. 178-179, Available from 10.1126/science.1200990

Heggelund, G. (2006). Resettlement Programmes and Environmental Capacity in the Three Gorges Dam Project. *Development and Change*, Vol. 37, No. 1, 2006, pp. 179-199, ISSN: 1467-7660

Hennig, Th. (2007). Energie für Chinas Wachstum. Eine kritische Betrachtung zur Nutzung der Hydroenergie in China. *Geographische Rundschau*, Vol. 59, No. 11, (Nov. 2007), pp. 44-53, ISSN 0016-7460

Hennig, Th. (2009). Entwicklungspotentiale in Südwestchina. Hintergründe und Auswirkungen zu Yunnans ambitioniertem Hydroenergie- und Verkehrsinfrastrukturausbau, *Asien,* No. 112-113, (Oct. 2009), pp. 103-122, ISSN 0721-5231

Hennig, Th. & Linde, L. (2008). Bewegung an der Peripherie. Hydroenergieentwicklung, Landnutzungsdynamik und Infrastrukturentwicklung in den südostasiatischen Grenzregionen Chinas und Indiens. *Geographische Rundschau,* Vol. 60, No. 4, (April 2008), pp. 42-51, ISSN 0016-7460

Hensengerth, O. (2010). Sustainable dam development in China between global norms and local practices. *Discussion Paper / Deutsches Institut fuer Entwicklungspolitik,* (4/2010), ISSN: 1860-0441, Bonn, Germany

Huang, H. & Yan, Zh. (2009). Present situation and future prospect of hydropower in China. *Renewable and Sustainable Energy Reviews,* Vol. 13 (2009), pp. 1652-1656, Available from doi:10.1016/rser.2008.08.013

IUCN (2002). Three Parallel River of Yunnan Protected Areas, Available from http://unesdoc.unesco.org/images/0013/ 001322/132231e.pdf

Jianchu, X. (2007): The Highlands. A Shared Water Tower in a changing climate and changing Asia. *Working Paper / World Agroforestry Centre,* ISBN 978-99946-853-6-3

Kwai Wong, K. (2005). Diverse Botanical Communities in Yunnan and the Yangtze River Shelter Forest System. *Geography,* Vol. 90, No. 3, pp. 288-293, ISSN 0016-7487

Li, F. (2002): Hydropower in China. *Energy Policy,* Vol. 30 (2002), Available from PII:S0301-4215(2)00085-X

Magee, D. (2006): Powershed Politics. Yunnan hydropower under Great Western development. *China Quarterly,* 185 (March 2006), pp. 24-41, Available from doi:10.1017/S0305741006000038

Mc Donald, K; Bosshard, P. & Brewer N. (2009). Exporting Dams. China's hydropower industry goes global. *Journal of Environmental Management,* Vol. 90 (2009), pp. S294-S302, Available from doi:10.1016/j.jenvman.2008.07.023

Nautiyal, H.; Singal, S.K. & Sharma, A. (2011). Small hydropower for sustainable energy development in India. *Renewable and Sustainable Energy Reviews,* Vol. 15 (2011), pp. 2021-2027, Available from doi:10.1016/rser.2011.01.006

Osborne, M. (2009). The Mekong. River under Thread. *Lowy Institute Paper 27 / Lowy Institute for International Policy,* ISBN: 978-19210-043-8-4, Australia

Ramunathan, K. & Abeygunawardena (2007). Hydropower Development in India. A sector assesment. Asian Develop-ment Bank, Manila, Philippines, Publication Stock No. 031607

Rao, V.V.K. (2006). Hydropower in the Northeast. Potential and Harnessing Analysis. Background Paper No. 6, Available from http://mdoner.gov.in/writereaddata/sublink3images/69706470684.pdf

Tullos, D.; Brown, Ph.H.; Kibler, K.; Magee, D.; Tilt, Br. & Wolf, A.T. (2010): Perspectives on the Salience and Magnitude of Dam Impacts for Hydro Development Scenarios in China. *Water Alternatives,* Vol. 3, Issue 2, pp. 71-90

Vagholikar, N. & Das, P.J. (2010). Damming the Northeast. Kalpavriksh, Aaranyak and ActionAid India, Pune/Guwa-hati/New Delhi, India

WWF (2001). Ecoregions-based conservation in the eastern Himalaya. Identifying important areas for biodiversity conservation. WWF, Washington D.C.